EDUCATION IN PROVINCIAL FRANCE
1800–1914

EDUCATION IN PROVINCIAL FRANCE, 1800–1914

═══

A Study of Three Departments

by
ROBERT GILDEA

Clarendon Press · Oxford
1983

Oxford University Press, Walton Street, Oxford OX2 6DP
London Glasgow New York Toronto
Delhi Bombay Calcutta Madras Karachi
Kuala Lumpur Singapore Hong Kong Tokyo
Nairobi Dar es Salaam Cape Town
Melbourne Auckland
and associated companies
Beirut Berlin Ibadan Mexico City Nicosia

OXFORD is a trade mark of Oxford University Press

Published in the United States
by Oxford University Press, New York

British Library Cataloguing in Publication Data
Gildea, Robert
Education in provincial France 1800-1914.
1. Education—France—History—19th Century
I. Title
370'.944   LA691.7
ISBN 0-19-821941-5

Library of Congress Cataloging in Publication Data
Gildea, Robert.
Education in provincial France, 1800-1814.

Bibliography: p.
1. Education, Rural—France—History—19th century.
2. Catholic Church—Education—France—History—19th century.
I. Title.
LC5148.F8G54 1983 370'.944 83-2100
ISBN 0-19-821941-5

Printed in Great Britain
at the University Press, Oxford
by Eric Buckley
Printer to the University

For Anne

Acknowledgements

The financing of research was made possible by the Social Science Research Council, London University, St. Antony's College, St. John's College, and Merton College, Oxford. For advice and support in the writing of the book I am indebted to so many people that I fear I shall not be able to name them all. My first thanks go to Charles Lecomte, *bleu de Bretagne*, former history master at the Lycée Émile Zola, Rennes, who was my companion and guide during the year I spent in Rennes in 1975-6. I am grateful for guidance and criticism of my work on the West to Michel Denis of the Université de Haute-Bretagne, Henri Goallou, and Michel Lagrée of the Lycée Jolit-Curie, Rennes; for advice on the Midi to Gérard Cholvy and Raymond Huard of the Université Paul Valéry at Montpellier and to Guy Dupré of the Lycée Alphonse Daudet at Nîmes; and for enlightenment on the Nord to Yves-Marie Hilaire of the Université of Lille III. Françoise Mayeur, of the same university, who patiently read the manuscript of the book, I am proud to count a fellow-traveller on the byways of French education and my sternest critic. In this country, I must thank Richard Cobb, Douglas Johnson, John Roberts, Liam Smith, and Theodore Zeldin for their constant encouragement. Lastly, I must express my gratitude to those archivists whose unseen work is too often insufficiently appreciated by historians, and yet whose welcome and assistance I have continuously exploited. Among these are MM Charpy, Debas, and Robinet, archivists respectively of Ille-et-Vilaine, the Gard, and the Nord; Père Delahaye of the Eudist order at Rennes and Père Hugues of the Jesuit congregation at Lille; Abbé Desreumaux, archivist of the diocese of Lille and Abbé Marchand, vicar-general of the diocese of Nîmes; and those who owe nothing to Combes but a residence on the Celestial City: Brother Bertrand of the Petits Frères de Marie, Brother Hazell of the Frères des Écoles Chrétiennes, and Brother Libert of the Frères de Ploërmel. For having failed to put to good use the advice of all these, I am alone responsible.

The author and publishers would also like to thank the British Academy for providing a subsidy towards the publication of this volume.

Contents

List of Figures

List of Maps

List of Abbreviations

A.A.	Archives archiépiscopales.
A.A.A.	Archives des Augustins de l'Assomption.
A.C.	Archives consistoriales.
A.D.	Archives départementales.
A.E.	Archives des Eudistes.
A.Ep.	Archives épiscopales.
A.F.E.C.	Archives des Frères des Écoles Chrétiennes (also, Frères de la Doctrine Chrétienne).
A.F.M.	Archives des Frères Maristes.
A.F.P.	Archives des Frères de Ploërmel (also, Frères de l'Instruction Chrétienne).
A.H.R.F.	*Annales historiques de la Révolution française.*
A.M.	Archives municipales.
Am. H.R.	*American Historical Review.*
A.N.	Archives nationales.
Annales H.E.S.	*Annales d'histoire économique et sociale.*
Annales E.S.C.	*Annales, économie, société, civilisation.*
A.P.I.C.	Archives des Pères de l'Immaculée Conception.
B.M.	Bibliothèque municipale.
B.N.	Bibliothèque Nationale.
D.E.S.	Diplôme d'études supérieures.
E.H.R.	*English Historical Review.*
E.P.S.	École Primaire Supérieure.
F.H.S.	*French Historical Studies.*
I-et-V.	Ille-et-Vilaine.
J.E.H.	*Journal of Economic History.*
J.M.H.	*Journal of Modern History.*
M-et-L.	Maine-et-Loire.
P. & P.	*Past and Present.*
R.H.	*Revue historique.*
R.H.E.F.	*Revue d'histoire de l'Église de France.*
R.H.E.S.	*Revue d'histoire économique et sociale.*
R.H.M.C.	*Revue d'histoire moderne et contemporaine.*
S.I.E.	Société pour l'instruction élémentaire.

Introduction

Some time during the Second Empire, it is said, the Minister of Public Instruction pulled out his watch and announced to the Emperor that he could state what, at that precise moment, all the children in France were studying. This story, encapsulating as it does the notion of the complete centralization of the French educational system, has become rooted in popular mythology. Yet it is a prejudice that cannot be sustained and it is the object of this book to show why.

Until now, works on education in nineteenth-century France have been of two sorts. The first are national in scope, illustrating regional variations by means of bald statistics that give no impression of the complex realities of provincial life, or by adducing picturesque evidence to support a particular point that gives no guarantee of being typical. The second kind of study are local monographs which are highly sensitive to developments in one chosen area but offer no possibility of generalizing about France as a whole.

My thesis on education in a single department of Brittany, Ille-et-Vilaine, clearly fell into the second category. It seemed to me important to move away from Paris and to study various problems within a defined area, a department, where the intractable nature of the raw material might force a revision of received ideas. And yet I became increasingly aware that the department that I had chosen was unrepresentative and that any conclusions reached would have to be weighed against conclusions drawn from work undertaken in other areas. The fashion of departmental studies has to some extent given way to the fashion of provincial studies—Aquitaine, Limousin, Franche-Comté—in accordance with the view that the historian should start not with the artificial administrative unit of the department but natural, historical units corresponding to the ancient provinces of France. A similar study of Brittany or the Grand Ouest would not, however, have served my purpose. It was necessary not to plough ever wider furrows in a given region, a task that tends to be subject to the law of diminishing returns, but to scratch at the soil in what were, for me, virgin territories, and those territories should be as different as possible, in terms of economic development and political and religious mentality, from the Upper Brittany that had provided my starting-point. There is no such thing as a 'typical' or ordinary department in the mosaic of French provincial life, and therefore no point in trying to find it and place it under the microscope. Moreover, the conclusions that I hoped to reach by comparing three very different departments would arise not from eliciting their lowest common denominator but by allowing each highly individual example to mirror itself in the others, so that light would be thrown on both similarities and contrasts.

The provincial perspective also makes possible an attempt to gauge the importance and function of education within the pattern of everyday life. This is not another general treatment of the ideology of anticlericalism or of the conflict between Church and state over education, but a study of how conflicts over the school arose out of the framework of the political and religious options of given communities and helped in turn to define them. It does not deal with pedagogic theory nor involve a scholastic analysis of changing curricula which may be limited to the four walls of the classroom but tries to discover the social values on which education was based and to compare them with the attitudes to school of families and the eventual careers of those who passed through the various levels of the school system. Lastly, it tries to examine the connections between education and work by focusing on the much neglected question of technical or professional education, on the initiatives that were proposed by communities or imposed by the state.

It becomes clear that the evolution and application of education policies was less a question of blueprints devised in an abstract manner in Paris and proclaimed Delphic-fashion for the provinces to follow, than a continual dialogue between Paris and the provinces. Initiatives were taken sometimes at ministerial level, sometimes by prefects, sometimes by municipalities, the churches, or private speculators. Sooner or later, governments tried to systematize education policies but their enforcement was still a matter of negotiation. The provinces had their own political and religious attitudes, their own patterns of ambition, their own rhythm of economic development, and these would have to be taken into account, or they would assert themselves in spite of government policy.

The first part of the book examines the dialogue in the political and religious context. Clearly the loyalty of the population to the regime was far more important for governments than the level of their productive competence and here government policies were asserted from the start. Education may be seen in the first instance as an instrument of political colonization whereby the provinces were brought into line with successive regimes in Paris. The century after the French Revolution saw the elaboration of a public system of education, firstly secondary education, designed to train the civil and military élite, then, especially after the introduction of universal suffrage in 1848, primary education which concerned itself with the mass of the population. It would be wrong to imagine that public education was identified with the state and private education with the Church, for the Church managed to gain control of the public system of education in the early nineteenth century, both through the secular clergy and religious congregations, and exercised massive influence in support of royalist or reactionary regimes. A precondition of the success of liberal or democratic forms of government would be secularization, or the removal of the Church from influence in public education.

The provincial perspective allows a reasoned assessment of political colonization and its limits. Our choice of departments may be considered somewhat idiosyncratic: all of them are situated on the periphery of France and all manifested a vigorous Catholicism, albeit a Catholicism of very different kinds. The Catholicism of Ille-et-Vilaine was deeply influenced by a counter-revolutionary, *frondeur* mentality and by Breton superstition; that of the Gard was cast in the mould of the Counter-Reformation, its mission to eliminate the Protestant presence from the Cévennes; that of the Nord to preserve the independence of the Flemish population. In all three areas, indeed, Catholicism and provincial separatism were very closely identified, each serving to reinforce the other.

In 1800, the department as an administrative unit had only ten years' prescription. Very rarely did the department correspond to the realities of provincial life, historical, geographical, or economic. These contours invariably cut across departmental boundaries. The tension between French- and Flemish-speaking areas in the Nord extended beyond France and was common to the Low Countries as a whole. The conflict between Catholics and Protestants in the Gard was situated on an axis of heresy that stretched from the Spanish border to the Alps and into Central Europe. The Catholic royalism contested by a 'thin blue line' of revolutionary strongholds in Ille-et-Vilaine was the general pattern throughout the Grand Ouest of Brittany, Anjou, and Maine, studied by André Siegfried. Yet the departmental approach can be defended, for it affords a lens through which these political, religious, and linguistic frontiers can be studied in great detail, together with the impact on those frontiers of the school question. That question was invariably coloured by historical tensions that already existed; indeed the protagonists in the conflict over the school tended to latch on to existing rivalries in order to further their cause, and this in turn helped to modify that existing geography of attitudes. Paradoxically it is the detailed analysis that best demonstrates the general thesis, for after a century of conflict between clerical and anticlerical communities over the politics of education, it can be shown that very little had been done to alter the frontiers between those communities. By the time the revolutionary decade was over in 1800, the contours forged by years of recurrent conflict were deeply engrained, and would be very difficult to modify. In the slow-moving provinces of France, the past weighed heavily on the minds of the living.

The purpose of education was not only to ensure loyalty to successive regimes by inculcating religious obedience or civic virtue. It was also to order society into some kind of hierarchy, on the one hand to train a governing élite, on the other to bring the mass of the labouring population within the pale of civilization, to guarantee against social dislocation.

Again, the provincial perspective throws significant light on the problem. For while the school system was organized into two tiers, secondary and primary, distinguished not by the age but by the social class of their clienteles, passage between the two remaining fairly difficult, French society fell broadly into three social categories. The élite included a nobility and bourgeoisie united by wealth, office-holding, and a classical education. The lower class was composed of wage-earning labourers, whether agricultural or industrial. In between them were the *classes moyennes*, a petty bourgeoisie of rich farmers, artisans, tradesmen, *employés*, and *petits fonctionnaires*, and it was the thickness of this stratum that distinguished France from other far more stratified countries such as Spain or Italy. These *classes moyennes* fitted easily neither into the system of secondary education, which was often too exalted, nor into the system of primary education, which was usually inadequate for their needs.

In order to obtain hard evidence on school attendance it is possible to use a number of surveys carried out at the national level, which may include breakdowns for individual departments or towns. But the categories used in such surveys are often pre-established, the data arbitrarily fixed, and the tally extant for a particular department or town not very large. So alongside these surveys use must be made of the records of individual schools, notably the enrolment registers, some of which have been transferred to departmental archives while others—such as those of *écoles normales* or Catholic colleges—remain in the school establishments, though many have been destroyed. These registers record the profession of the father in about two-thirds, sometimes three-quarters of cases, offer the researcher a choice of periods to sample in order to compare the intake of several schools, and even, where the records of *associations d'anciens élèves* can be used in conjunction, to trace the careers of a certain proportion of graduates into later life.

To accept the neat division of the education system into primary and secondary is to fall into the trap of writing the history of institutions, to gloss over the complexities of the relationship between various social strata and the schools. If however we start not with schools but with social classes,

we place ourselves in a position to examine the choices open
to those groups, and to measure the success with which different schools adapted themselves to attract a clientele.

The bourgeoisie, for example, taken broadly to include
rentiers, the liberal professions, public officials, bankers,
négociants, and industrialists, tended to favour a long, costly,
classical education that separated them off from the petty
bourgeoise and led to the *baccalauréat*, that passport to the
faculties, liberal professions, and office. The *classes moyennes*
had a choice, opting either for a Latin education (the quicker
and cheaper the better) or for an effective, French-based
commercial education that would prepare them for technical
schools and the world of agriculture, trade, transport, and
industry. A close study of a number of departments at different levels of economic development and with different
occupational structures illustrates the way in which *lycées*
and municipal colleges, private *pensionnats* and commercial
schools, the many-headed education system of the Catholic
Church and that extension of elementary education, the
higher primary schools, competed for these markets. And
success at the beginning or middle of the nineteenth century
did not necessarily mean success at the end.

The role of the primary-school system itself can best be
examined by pinning it down in the local context. To bring
the population within the pale of enlightenment was only
one task and perhaps not the most important. The idea of
'moralization', an inner mission to civilize the lower orders,
was at the centre of elementary education. A certain tension
existed between those who believed that there could be no
social stability without the influence and sanction of religion,
and those who looked forward to a social harmony achieved
by training rational economic men. But what each side had in
common was a belief that fundamental social problems such
as crime, violence, unemployment, begging, slums, and
disease were not environmental but moral, and could be solved
not by improving the world in which men lived but by reforming man himself. Another constant, however, was the
gulf between the official view of education as a civilizing and
integrating force and the popular view that schooling should
provide knowledge in order to further a career or gain material

benefit. Some Catholic populations sent children to school only to learn the catechism in order to undergo that important rite of passage, the First Communion. But for most of the lower classes education was a form of moral discipline that seemed tangential to both the world of work and to that of leisure.

In the end, the impact of education has to be measured against the economic and social structure. The campaign to civilize the lower orders was dependent in the end not on the number of schools set up or the extent of free education, but on levels of wealth and poverty and the demands on child labour in the local economy, which regulated rates of school attendance and literacy. And though bourgeois notables feared that legions of middling sorts would erupt into the secondary-school system, challenge their ascendancy, or fail at the hurdle and turn into dangerous malcontents, social mobility depended finally not on schooling but on local conditions of employment. Time and again we are led to shift our gaze away from the schoolroom in order to consider the realities of provincial life, realities which varied from one end of the country to the other.

This is nowhere more explicit than in the case of technical education, the use of education as an instrument of economic progress. At the beginning of the nineteenth century, the idea was popularized by a few brilliant minds that technical education was of key importance as a means of dragging France out of her economic backwardness, to take a place of honour among the European economies and to ensure that she became an imperial nation rather than suffered colonization itself. By the end of the century a heated debate was taking place between partisans of technical education and those who believed that France could be saved from economic distress only by the erection of high tariff walls around the country. New sectors did develop in response to the economic depression at the end of the nineteenth century, but the relative importance of education, protection, or entrepreneurial initiative must be assessed. Economic progress, both in industry and agriculture, was far more advanced in the north of the country than in the west or Mediterranean region, and though professional education might propagate

new techniques to effect in backward areas it is more doubt-
ful that it acted as a driving force in regions that were
economically advanced. Again, technical schools might raise
expertise at critical sectors of the economy, but there was
also the danger that they might be preoccupied by technology and methods that were already being consigned to the
museum of industry. In economic organization, as in the
rest of society, there were three elements in the hierarchy,
the employers, the workers, and between them an expanding
sector of foremen, draughtsmen, accountants, engineers, and
managers. It was important that professional education
should concentrate its resources at the most important
levels. Was it in a position to train entrepreneurs, did it have
the means to improve the mass of unskilled workers, or
would returns be greatest if it concentrated on the managers
and foremen? Lastly, the very sense of professional education
should be closely examined, for while it may seem obvious
that it was on the side of progress, there was also the possibility that it served not to initiate rapid economic change
but to clear up the mess afterwards, not to prepare the way
for a brave new world but to shore up an old world that at
the end of the nineteenth century was fast disintegrating.

But what of the conditions that existed in the three
departments that have been chosen? In the Nord, the demo-
graphic explosion had certainly made itself felt. In 1804
it was the most densely populated department in France,
with 145 inhabitants per square kilomotre, against a national
average of 55.[1] Pauperism, unemployment, overcrowding,
and disease were therefore harsh realities in the Nord, reaching horrendous proportions in poor, urban parishes like
those of Saint-Maurice and Saint-Saveur in Lille. Alban
de Villeneuve-Bargemon, a prefect of the Nord during the
Restoration, calculated in 1834 that one inhabitant of the
Nord in six was on poor relief, rising to one in three in
Lille.[2] And yet in order to sustain such a dense population at

---

[1] R. Blanchard, *La Densité de la population du départment du Nord* (Lille, 1906), p. 11.
[2] Alban de Villeneuve-Bargemon, *Économie politique chrétienne, ou re-cherches sur la nature et les causes du paupérisme en Europe* (Paris, 1834), ii, 50–6.

all, the economy, both agricultural and industrial, must have been robust and expanding. The north of France was indeed exposed to all the advantages of a market economy. Canals criss-crossed the *plat pays* of the north, carrying coal, while the railway-building boom of 1848-59 linked by rail to Paris and to the sea-coast at Dunkerque both the textile centre of Lille-Roubaix-Tourcoing and the coal-basin of Douai-Valenciennes, the development of which railway mania in turn helped to stimulate.[3] Demand provoked an early and thorough agricultural revolution in the Nord. Wasteland was largely enclosed and improved. Fallow land was eliminated by crop-rotation with grasses, sugar-beet, and colza, by animal–arable husbandry, especially in the pasturelands of Flanders, or by the injection of large doses of fertilizer, notably in the wheat-growing chalk-lands of the Cambrésis. Agricultural prosperity reduced the frequency of runs of bad harvests that caused bread prices to soar and ate away at the resources of the mass market, upon which industry relied. Demand kept up and industry replied by mobilizing its resources. A labour force could be found in the congested countryside, especially the armies of hand-loom weavers required to make up the machine-spun yarn. When these were scythed down after 1840 by the introduction of the power-loom, many—including migrants from over the Belgian frontier—moved into the mill-towns or towards the expanding coal and iron industry in the south of the department. Entrepreneurs were recruited from the rural bourgeoisie, from cloth merchants like Jean-François Motte-Clarisse (1750-1822), who became involved in the 'put out' system of textile manufacture, and even from foremen with technical expertise.[4] Capital was transferred from one sector to another, as entrepreneurs moved from farming, grain-trading, or

---

[3] F. Caron, *Histoire de l'exploitation d'un grand réseau: la Compagnie du Chemin de Fer du Nord, 1846-1937* (Paris–The Hague, 1973), pp. 157-62.

[4] J. Lambert-Dansette, *Quelques Familles du patronat textile de Lille-Armentières, 1789-1914* (Lille, 1954), pp. 295-332; J. A. Roy and J. Lambert-Dansette, 'Origine et évolution d'une bourgeoisie: le patronat textile du bassin lillois, 1789-1914', *Revue du Nord*, xl (1958), 49-69, xli (1959), 23-38; Gaston Motte, *Les Motte, étude de la descendance Motte-Clarisse* (Roubaix, 1952); C. Fohlen, *L'Industrie textile au temps du Second Empire* (Paris, 1956), pp. 84-7.

distilling into textiles, and from textiles into coal.[5] Most were family firms that financed themselves by exploiting the work-force and ploughing back profits. Only large metal-lurgical concerns like the Société Denain–Anzin of the 1840s required the mobilization of capital by means of the limited-liability company.

The accumulation of capital dramatically affected the balance of social power in the Nord. The landed nobility was weaker than in other parts of France and though the magis-tracy of Douai and the liberal professions were resilient, influence was shifting, especially after 1830, to the *négociants* and industrialists. This *patronat*, furthering its business interests by dynastic intermarriage and Catholic solidarity, formed a veritable aristocracy of money. At the other end of the social hierarchy were the farm-hands and agricultural labourers, the hand-loom weavers working on linen in the valley of the Lys, producing woollen cloth in the satellites of Roubix at Croix, Wattrelos, and Wasquehal, and finer lawns and cambrics in the case of the *mulquiniers* of the Cambrésis. The small peasant proprietors of this area not only spent the winter weaving in their cellars but were also forced to find extra work in the summer by travelling in gangs of harvesters to the cornfields of central France. In the south-east of the department, stone was quarried near Solre-le-Château and iron worked in traditional Catalan forges at Trélon, using the charcoal-fuel and water-power of the Ardennes. But the work-force was becoming increas-ingly divorced from the land and crowded into towns, whether we consider the thread-makers in the attics and cellars of Lille, the textile workers herded into large-scale factories like the 'filature monstre' built by Louis Motte-Bossut at Roubaix in 1843, or the metallurgical workers of the new iron towns of Denain and Maubeuge, their furnaces and mills fuelled by the coalfield. And yet social stratifica-tion did not exclude the presence of a thick wedge of mid-dling sorts. In the countryside there was a rural bourgeoisie

---

[5] F. Caron, *Economic History*, p. 81; M. Gillet, *Les Charbonnages du Nord de la France au XIXe siècle* (Paris–The Hague, 1973), pp. 99–101; Y.-M. Hilaire, 'Révolution industrielle et libéralisme (1814-1851)', in L. Trénard, *Histoire des Pays-Bas français* (Toulouse, 1972), p. 397.

of rich farmers, owning plough-teams and hiring labour, together with millers, brewers, and innkeepers. And in Roubaix, a report to the municipal council in 1852 stated that 'our population, as everywhere else, is composed of three classes; the rich and well-to-do, the *classe moyenne*, and the working class, the poor and the indigent'. The *classe moyenne* was defined as 'tradesmen, small manufacturers, foremen, grocers, bakers and in general all those with a trade and paying a patent'.[6] Contemporary research tends to confirm these impressions. A study at Lille based on probate archives has calculated that in 1856-8 the élite of the city, composed of landowners, industrialists, merchants, officials, and the liberal professions, made up 8 per cent of the population but owned 90 per cent of its wealth. The *classes moyennes* —artisans, shopkeepers, *employés*, and *petits fonctionnaires*—accounted for 32 per cent of the population and 9.5 per cent of its wealth. The rest, the popular classes, made up 60 per cent of the mass but owned a mere 0.5 per cent of the collective wealth.[7]

The pattern of political and religious life in the Nord, embedded deep in past experience, was also original. Flanders and Franche-Comté had been part of the Burgundian patrimony, inherited by the Habsburg monarchy after the death of Charles the Bold and passing to Spain in 1556. Conquered by Louis XIV between 1667 and 1678, Flanders nevertheless looked back to a 'golden age' under Burgundian or Imperial rule, and the Austrian occupation of July 1793 was not resented. Local notables collaborated with the enemy in juntas set up by the Austrians and suspended revolutionary legislation. Henri Dubois-Fournier, a cambric and lace merchant who almost lost his head for collaboration, became a vigorous patron of Catholic schools in the Nord.[8] Separatism was reinforced by the language question, for the part of

[6] A. D. Nord, IT 77/2, report to municipal council of Roubaix, 5 Nov. 1852. On this question, see L. Moulin and L. Aerts, 'Les classes moyennes', R.H.E.S., xxxii (1954), 168-81, 293-309.

[7] F.-P. Codaccioni, *De l'inégalité sociale dans une grande ville industrielle. Le drame de Lille de 1850 à 1914* (Lille, 1976), p. 93.

[8] P. Dubois, *Un patriarche: vie de M. Dubois-Fournier, 1768-1844* (Lille, 1899), pp. 46-7; J. Peter and C. Poulet, *Histoire religieuse du départment du Nord pendant la Révolution, 1789-1802* (2 vols., Lille, 1930-3), pp. 305-9.

Flanders that lay north of the Lys was Flemish-speaking,
and the tension between Flemish and Walloon districts was
as sharp in the Nord as it was in Belgium (Map 1). For the
French revolutionaries, other languages and patois had to be
driven from French soil, for they favoured the ascendancy
of priest and noble over the ignorant peasant and therefore
represented aristocracy, fanaticism, and counter-revolution.
But for the Flemish-speakers, French was the language of
scepticism and materialism, of military and bureaucratic
power, while Flemish was the language of the community,
of traditional custom and the Catholic religion. In Flanders,
separatism, a national language and Catholicism supported
and reinforced each other, and the intractable Catholicism of
the Flemish population was an obstacle with which many
secularizing authorities would have to contend. Because it
was not annexed to France until 1678, Flanders had been
subjected to a 'Spanish' Counter-Reformation, undertaken
under the auspices of the Archdukes Albert and Isabella
(1598-1633). This movement involved the rule of zealous
bishops, the expansion of religious orders such as the Jesuits,
who established colleges at Cambrai, Valenciennes, Lille,
Bergues, Bailleul, Cassel, Maubeuge, Armentières, and Le
Cateau, and the training and disciplining of the secular clergy
by seminaries, visitations, synods, and deanery assemblies.[9]
There was very little Protestantism in the Nord, and what
existed in the Cambrésis was effectively stamped out by
Louis XIV's *dragonnades* and forced conversions, leaving a
general indifference to religion. Missionaries from Geneva
made some impact on the countryside around Valenciennes
after 1818, but the pattern that emerges is one of an irreligious and anticlerical south of the department and a Flemish
north that was both pious and willing to accept the leadership of the clergy. The fact that the Church owned up to
40 per cent of the land in the Cambrésis before the Revolution, but very little in Flanders, helped to create a popular
anticlericalism in the south.[10] On the other hand the peasantry
of the Cambrésis and Avesnois, forming associations in order

[9] A. Pasture, *La Restauration religieuse aux Pays-Bas catholiques sous les Archiducs Albert et Isabelle, 1596–1633* (Louvain, 1925).

[10] A. Lajusan, 'La Carte des opinions françaises', *Annales E.S.C.*, 4, 1949, 408-9.

Map 1. The Nord, indicating Flemish-speaking region

Region habitually speaking Flemish, 1856–7

Gravelines
Dunkerque
Bourbourg
Bergues
Quaëdypre
Hondschoote
Wormhoudt
Killem
Cassel
Steenvoorde
Hazebrouck
Caestre
St Jean Cappel
Morbecque
Méteren
Bailleul
Merville
Nieppe
Estaires
Lys
Armentières
Bousbecques
Comines
Halluin
Quesnoy-sur-Deûle
Mouscron
Tourcoing
Wattrelos
Marcq
Beaucamps
Roubaix
Fournes
Haubourdin
Lille
Lannoy
Hellemmes
La Bassée
Wattignies
Seclin
Froyennes
Pont à Marcq
Cysoing
Wannehain
Moncheaux
Templeuve
Orchies
Douai
Péruwelz
Marchiennes
St Amand
Vieux Condé
Pommereul
Aniche
Fresnes
Condé
Arleux
Anzin
Denain
Petite Forêt
Escaut
Valenciennes
Bouchain
Bantigny
Iwuy
Cambrai
Saulzoir
Frasnoy
Bavai
Le Quesnoy
Carnières
Solesmes
Mercoing
Sambre
Maubeuge
Clary
Berlaimont
Hautmont
Landrecies
Le Cateau
Taisnières en Thiérache
Solre le Château
Avesnes
AISNE
Trélon
Fourmies

PAS-DE-CALAIS

BELGIUM

N

to make collective purchases of Church lands, managed to corner two-thirds of the lands sold off during the Revolution, and were therefore committed to the principles of 1789.[11] Reactions to the revolutionary attempt to reform the Church, the Civil Constitution of the Clergy, confirm this pattern. Only 15 per cent of clergy in the Nord took the oath to the Civil Constitution that was required under the decree of 27 December 1790, and these were concentrated in the southern extremity of the department, the Avesnois, furthest away from the fanatically clerical *Flandre flamingante*.[12] In future, Flanders would fight valiantly to protect Catholicism against any offensive by the state.

In the Gard, the main source of conflict was that between Catholics and the Protestants, who formed about a third of the population in the mid-nineteenth century (Map 2). In a sense the Wars of Religion were still being fought out. Persecution of Protestants intensified when France was at war with Protestant powers like England and Prussia, notably during the War of the Spanish Succession, but resistance was ferocious. Between 1703 and 1713 the Protestants of the Cévennes joined in revolt with those of the Vaunage outside Nîmes and in Nîmes itself, and these Camisards wreaked vengeance on the Catholic populations in a series of guerrilla campaigns.[13] Left without adequate protection from royal troups, the Catholics formed their own groups of partisans called Cadets de la Croix or Florentins, the latter because so many came from Saint-Florent in the canton of Saint-Ambroix. The geography of confrontation was becoming defined with increasing boldness. And just as the Camisard tradition lived on among the Gard Protestants of the nineteenth century, so did the Catholic crusade among the Cévenol populations bordering on the Ardèche and, beyond Le Vigan, in the Causses. Emmanuel d'Alzon, founder of the Assumptionist order (1810–80), did not forget his ancestors who died fighting the Hugenots in 1580 and 1704.[14]

---

[11] G. Lefebvre, *Les Paysans du Nord pendant la Révolution française* (Paris-Lille 1924), pp. 10–26, 443, 501–3.

[12] J. Peter and C. Poulet, op. cit., p. 130.

[13] P. Joutard, *La Légende des Camisards* (Paris, 1977), p. 29.

[14] S. Vailhé, *Vie du P. Emmanuel d'Alzon* (Paris, 1926), i, 5–8.

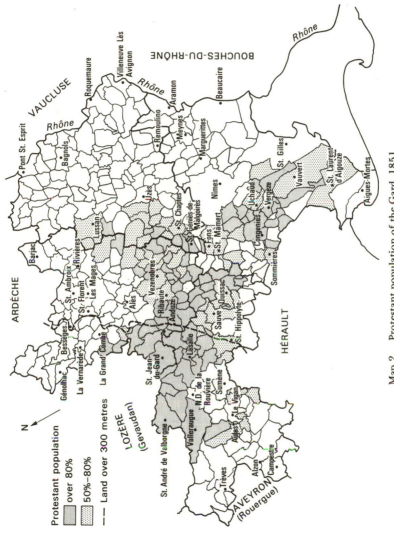

Map 2. Protestant population of the Gard, 1851

The Revolution completely altered the balance of power in the Gard. The Protestant élite, granted civil liberties and profiting from the propertied franchise, established its supremacy in the department in the years 1790-3, while the Catholic opposition, fanatical nobles leading bands of depressed textile workers and peasants, suffered a severe setback in the 'Bagarre de Nîmes' of 13-15 June 1790, during which three hundred of them were killed. Catholic counter-revolutionaries encamped at Jalès, in the southern Ardèche, and won support in traditionally anti-Protestant centres nearby, such as Saint-Amboix and Barjac.[15] After 1793, radical Montagnards took over both in Paris and Nîmes. They threw the Protestants into the federalist movement to protect their interests (executing the father of the future Minister of Public Instruction, Guizot) and persecuted Catholics in campaigns of 'deChristianization' that provoked fanatical resistance around Saint-Amboix in the autumn and winter of 1793. The Thermidorean reaction allowed Catholic extremists to have their revenge on the sansculottes. Marauding *égorgeurs du Midi*, organized into bands like the Compagnons du Soleil, were often co-ordinated by royalist agents and received tacit support from the constituted bodies of the Directory.[16] Under Napoleon, the Protestant aristocracy of wealth returned to public prominence, and made its hatred of royalist restoration explicit during the Hundred Days of 1815. But the White Terror that accompanied the Second Restoration, a Catholic retaliation for the Bagarre de Nîmes, drove the Protestants into liberal opposition until 1830. The Gard certainly deserved a reputation as the Ulster of France.

The struggle for power between Catholic and Protestant to a large extent reflected the pattern of social relationships in the Gard. The landed nobility was in general of modest means and Catholic, but exercised a monopoly of political power under the Ancien Régime and during the Restoration.

---

[15] G. Lewis, *The Second Vendée: The Continuity of Counter-revolution in the Department of the Gard, 1789-1815* (Oxford, 1978), pp. 16-32. See also the articles by James N. Hood in *J.M.H.*, 43, 1971, 245-75, and 48, 1976, 259-93, and *P. & P.*, 82, 1979, 82-115.

[16] Lewis, op. cit., pp. 81-3.

The Protestants, excluded from public office before 1789, built up a position of strength in trade and silk-manufacture, making use of their connections with the Calvinist banking houses of Geneva. The Protestant élite of Nîmes seemed to prove the Weber thesis as much as the Catholic *patronat* of Lille disproved it. Beneath them, the commerce passing through Nîmes, Beaucaire, and Alès threw up a stout *classe moyenne*. The hosiers of Nîmes, élite of the silk workers, claimed to have enjoyed the right to carry swords before 1789.[17] The rural Protestant population included the rich, easy-going farmers of the Vaunage district outside Nîmes and the austere smallholders trying to scratch a living out of the Cévennes. Small property ownership tended to predominate in the countryside, the economy buoyed up by the silk-weaving of a chain of small towns in the foothills of the Cévennes: Le Vigan, Saint-Hippolyte, Anduze, and Saint-Jean-du-Gard, with woollen cloth at Sommières. From higher up, and from the mountains of the Rouergue and Gévaudan, peasants descended into the plains of Lower Languedoc for the cereal harvest and *vendanges*.[18] Alternatively, they took advantage of the expanding mining and metallurgical industries at Alès, Bessèges, and La Grand' Combe, at the gateway to the upper Cévennes, sometimes to find winter employment, sometimes to settle. The population of Bessèges and its agglomeration grew from 4,700 in 1846 to 13,600 in 1866.[19] In Nîmes itself, crowded into 'Les Bourgades' on the northern fringes and the western quarter of La Placette, 15,000 silk workers were employed in 1832, shortly before the decline set in around 1840.

The revolution in communications had a massive influence on the economy of the Gard. The river-port of Beaucaire acted as a lung for the silk industry of Nîmes and the Cévennes in the early part of the century, but the Paris-Lyon-Marseille railway made Lyon the silk capital of France at the expense of Beaucaire, while the pebrine disease which

---

[17] A. Audiganne, *Les Populations ouvrières et les industries de la France* (Paris, 1860), ii, 158.

[18] R. Dugrand, *Villes et Campagnes en Bas-Languedoc* (Paris, 1963), pp. 432-5. E. Le Roy Ladurie, *Paysans de Languedoc* (Paris, 1966), p. 102.

[19] R. Lamorisse, *La Population de la Cévenne languedocien* (Montpellier, 1975), p. 129.

affected the silk-worm larvae after 1855 undermined local silk-raising in the Gard. On the other hand it was the railway that made possible the expansion of the heavy industry of the Gard: Alès was reached by rail from the coast in 1842, Bessèges in 1857. And again it was the railway that transformed the agricultural economy. There was of course no comparison with the agriculture of the Nord. That of the Gard was distinctly Mediterranean. It produced only a third of the grain necessary for its inhabitants; the rest was either imported from Sicily or Burgundy or replaced by the chestnut, 'l'arbre à pain', grown on the slopes of the Cévennes. Here was a polyculture based on chestnuts, mulberries to provide leaves for the insatiable silk-worms, and sheep-pasture in the mountains, grain and wine on the plain.[20] But the growing national demand for wine after 1848 and easy communications stimulated the cultivation of cheaper wines along the Mediterranean coast of Languedoc and brought about a shift from polyculture to a wine-growing monoculture, from small proprietors to an agricultural proletariat.[21]

The economy of Ille-et-Vilaine received nothing like the same stimulus. The population of Brittany was scanty and poor, the great trading days of Saint-Malo were over, while on the southern coast Redon had decayed to a sluggish river-port. In the interior, the habitat was dispersed, communications along the sunken roads which wound through the *bocage* were difficult, and Rennes had to wait until the Second Empire before the railway from Paris reached it. Agriculture remained backward and often of a subsistence nature. *Terre froide* or wasteland, moorland, and heath, frequently covered 35 per cent of the surface of cantons in the Redonnais, an area described by Taine as 'the Scotland of the north of France, without mountains'.[22] It is true that wasteland was of immense use to marginal populations for

[20] S. Grangent, *Description abrégée du département du Gard* (Nîmes, An VIII), pp. 12-15. H. Rivoire, *Statistique du départment du Gard* (Nîmes, 1842), ii, 226-81.

[21] R. Laurent, 'Les Quatres âges du vignoble' in *Économie et Société en Languedoc-Roussillon de 1789 à nos jours* (Montpellier, 1978), pp. 11-18; Léo Loubère, *The Red and the White* (Albany, 1978), pp. 140-7.

[22] H. Taine, *Carnets de voyage: notes sur la province, 1863-65* (Paris, 1895), p. 251.

rough pasture, fuel, building materials, and fertilizer made of decomposing reeds, ferns, and heather, but it kept agricultural productivity at very low levels.[23] To compensate for low yields, cultivated land was cropped to exhaustion under cereals, resulting in what Arthur Young called 'the thraldom of regular fallows'.[24] The Redonnais was still confined to a two-field system, and scientific crop-rotation and the raising of livestock that could produce manure for arable farming were sorely in need of development. Industry was similarly backward. At Châteaubourg and Châteaugiron, hemp was woven for rough clothing and the sails ordered by the navy, while in the forest of Paimport, there was some traditional iron-smelting. But these old industries declined, unable to compete with the modern textile mills and metallurgical works of northern and eastern France, while the shoe industry of Fougères did not take off until the end of the century. Labour was in short supply, for the tiny agricultural surplus did not permit the support of a large urban population. The few entrepreneurs came from outside the department: Jean-Jules Bodin, who manufactured agricultural machinery beside the Farm-School of Trois-Croix, was a Sarthois; François-Charles Oberthür, who set up a printing-works in Rennes in 1842, was a lithographer from Strasbourg. The tanner, Edgar Le Bastard, mayor of Rennes between 1881 and 1892, was a Norman. Capital was hard to come by and where it existed would be invested in land or building, not in industry.

The backwardness of the economy to a large extent determined the social structure. In Ille-et-Vilaine it was the landed nobility both military and *parlementaire* in its origins, that declared its supremacy. A city like Rennes was dominated by the Church, the army, the Palais de Justice, the administration, and the university, and lived on government stipends and rents drawn from the countryside.[25] The middle class had made some impact during the revolutionary period, but after 1815 the social gulf was reinforced by political antagonism. The commercial and industrial classes remained outsiders

[23] J. Chombart de Lauwe, *Bretagne et pays de la Garonne* (Paris, 1946), pp. 43–6.

[24] A. Young, *Travels in France* (Dublin, 1793), ii, 135.

[25] M. Denis, 'Rennes au XIXe siècle, ville parasitaire?', *Annales de Bretagne*, lxxx (June 1973), 403–39.

and had no corporate identity: Oberthür tried to find respectability by espousing royalism; Le Bastard became a radical and sought to challenge the aristocratic and clerical establishment. There was of course a class of shopkeepers, artisans engaged in luxury crafts for the consumption of the élite, and substantial farmers who leased the estates of absentee landowners in the basin of Rennes, but in the west of France society was rigidly stratified. Those who could struggled to obtain a Latin education that would take them in the direction of teaching, the professions, or, most likely, the priesthood. Otherwise they would be condemned to the *immobilisme* of the lower classes, the condition of whom few would envy. In the towns, poor weavers, casual labourers, and paupers were crowded into low-lying suburbs like that of Toussaints in Rennes. In the countryside, rural industries that once afforded an ancillary income were declining. Along the Channel coast, fishing as far as the cod-banks of Iceland offered some prospects for the rugged population, while in the summer the labourers of the north-east of Ille-et-Vilaine travelled to bring in the harvests in the wheatfields of Normandy, Beauce, and the Île-de-France. But poor harvests were all too regular, and often only the staple diet of porridge or pancakes made from buckwheat stood between the Breton peasant and starvation. When agricultural prices collapsed at the end of the century, the countryside emptied as the rural populations moved in search of employment towards Paris.

The border between Brittany and the rest of France was of the utmost importance from the political point of view. Brittany formed a duchy independent of France until it was annexed by a pact of union in 1532, and the marchland fiefs established by the dukes of Brittany, especially the baronies of Vitré and Fougères, left traces that were still evident in 1913 when André Siegfried wrote, 'hierarchical and feudal, it is the atmosphere of Maine which continues to reign in this royalist and Catholic milieu'.[26] Even after annexation, Brittany remained a *pays d 'état* with its provincial estates and tax privileges. When these were violated by a monarchy pressed

---

[26] A. Siegfried, *Tableau politique de la France de l'Ouest sous la Troisième République* (Paris, 1913, 2nd ed., 1964), p. 97.

for money, revolts like that of the *Bonnets rouges* in 1675 broke out.[27] Breton separatism in Ille-et-Vilaine was not reinforced by the Breton language, the limits of which had shrunk back to the mid-point of the Morbihan and the Côtes-du-Nord. But the defence of the Catholic religion was an integral part of the maintenance of Breton integrity. When Calvinists seized the fortress of Vitré for the Protestant king, Henri of Navarre, in 1589, the Catholic Ligue laid siege to it, drawing on support in an area that corresponded exactly to a future bastion of legitimism in the countryside around Vitré. And parishes like Saint-Marc-le-Blanc, Coësmes, and Châteaugiron, against which the Ligueurs launched punitive expeditions, would subsequently embrace the revolutionary cause.[28] The Catholicism of the western flank of the department was as vital, but quite different in tone. As late as 1954, Canon Boulard observed 'zones of religious solidarity, stalwart islands in the midst of indifferent regions . . . which correspond exactly to the limits of former dioceses',[29] that is, dioceses which were abolished in 1790 when these were reorganized to correspond to the new departments. One of these—in fact briefly resurrected by the ill-fated Concordat of 1817—was that of Saint-Malo, its intense Catholicism worked up since the Counter-Reformation by a line of 'good bishops', pastors rather than political prelates, preaching and teaching orders, seminaries, and regular visitations. Its ghost, both in the eastern Côtes-du-Nord and Morbihan and the west of Ille-et-Vilaine, would be revealed repeatedly in the nineteenth century as an area more pious, more fertile in vocations to the priesthood and congregations, more attached to Church education than surrounding areas (Map 3).[30]

The French Revolution both illustrated and reinforced this

---

[27] Y. Garlan and C. Nières, *Les Révoltes bretonnes de 1675, papier timbré et bonnets rouge* (Paris, 1975).

[28] M. Lagrée, 'La Structure pérenne, événement et histoire en Bretagne orientale, XVIe–XXe siècles', *R.H.M.C.*, xxiii (July–Sept. 1976), 394–407.

[29] F. Boulard, *Premiers Itinéraires en sociologie religieuse* (Paris, 1954), p. 49.

[30] M. Lagrée, *Mentalités, religion et histoire en Haute Bretagne au XIXe siècle: le diocèse de Rennes, 1815–1848 (Paris, 1977)*, pp. 119–23. C. Langlois, *Un diocèse breton au début du dix-neuvième siècle: Vannes 1800–1830* (Paris, 1974), pp. 583–93.

Map 3.   Ille-et-Vilaine in the Year VII

emerging geography of attitudes. The oath to the Civil Constitution of the Clergy was refused by all but 17 per cent of
the clergy in Ille-et-Vilaine, those who believed that religion
and the principles of 1789 could be reconciled coming from

parishes around Saint-Aubin-du-Cormier in the north-east of the department, and around Janzé and Retiers in the south-east.[31] While these parishes were left untroubled by the revolutionary authorities, those whose clergy refused the oath were not only deprived of their faithful non-jurors, driven out by National Guard contingents if necessary, but also had to suffer the imposition of a *curé intrus*, an apostate, on their religious life. It has been pointed out that 'all the parishes which had been the victims of National Guard attacks in 1792 were classed as *chouan* in 1795 in Ille-et-Vilaine'.[32] The Chouans, or Catholic and counter-revolutionary guerrillas, also had much to complain about concerning the scale of Church lands in the department, most of which was acquired by bourgeois intruders and speculators, confirming the impression that the Revolution had been 'imported by the townsmen'. It was nevertheless the attempt of the Convention in March 1793 to raise an army of 300,000 men to defend the frontiers of the Republic that really triggered off the Chouan wars. By the end of the 1790s the pattern of 'white' resistance to the Revolution and the small pockets of 'blue' partisans was clearly defined, and that pattern of hostilities would emerge time and again in the course of the following century (Map 3).[33]

---

[31] J. Bricaud, *L'Administration du départment d'Ille-et-Vilaine au début de la Révolution, 1790-91* (Rennes, 1965), p. 371.

[32] T. S. A. Le Goff and D. M. G. Sutherland, 'The Revolution and the rural community in eighteenth-century Brittany', *P. & P.*, 62 (1974), 116.

[33] R. Dupuy, *La Garde Nationale et les débuts de la Révolution en Ille-et-Vilaine, 1789-mars 1793* (Paris, 1972), p. 274.

PART I
POLITICS AND EDUCATION

The Catholic Renaissance, 1800–1860

The nineteenth century in France saw the gradual substitution of the state for the Church as the principal medium by which the youth of the country were educated. Under the Ancien Régime, education was virtually a monopoly of the Church, albeit under the supervision of the state, but the situation was transformed by the Revolution of 1789. The constitutional monarchy, let alone the Republic of 1792, in no way resembled the absolutist and aristocratic polity that had preceded them, and so that the regime might survive, the people had to conform to the constitution, their customs to its laws. It went without saying that the education monopoly of the Church, encrusted in Ancien Régime society as it was, had to be dismantled and the school system recast to train a new generation of citizens.[1] But despite the largely destructive role of the revolutionary assemblies and the positive achievement of Napoleon, the supremacy of the state was far from assured in 1814, at which point the Church sought to recover its hegemony.

The struggle for supremacy fought out over the nineteenth century was evenly balanced, although it fell clearly into two halves. In the first part of the century, until 1860, the state laid the foundations of a public system of education, first at the secondary level, then at the primary. But this public education did not remain in lay hands: the Catholic Church came to appropriate the public system of education to its own advantage, while maintaining a flourishing private sector in its own right. In the second half of the century, the state took on an anticlerical aspect, and launched repeated offensives in order to destroy the ascendancy of the Church in public education. More and more the Church was obliged to fall back on private education, until this citadel was itself besieged at the beginning of the twentieth century.

[1] In general, see H. C. Barnard, *Education and the French Revolution* (Cambridge, 1969).

In the period 1800–60, three attempts were made to break the hold of the clergy on the education system. Firstly, in 1808, Napoleon founded the University, a lay corporation that was to control and staff the faculties, *lycées*, and colleges. But having restored the Catholic Church under state supervision by the Concordat of 1802, which made the clergy into an arm of the bureaucracy, Napoleon was obliged to leave the direction of seminaries to the episcopacy. The seminaries came to afford Catholic secondary education to a clientele much broader than that training for the priesthood and at the Restoration the University, though it was not dismantled, was fully colonized by the Catholic clergy. Secondly, after the Revolution of 1830 an attempt was made to purge the University of its clerical masters and to establish a nation-wide system of elementary education. But the teaching congregations managed to secure a grip on the elementary schools, while Catholic pressure-groups inaugurated a campaign for *La Liberté de l'enseignement*, demanding the right to set up their own schools and colleges outside the state monopoly. Thirdly, an attempt was made during the brief period of elation in the spring of 1848 to liberate the primary-school teachers from the influence of the parish clergy and to turn them into missionaries of republicanism. But fear of social revolution encouraged the ruling classes to look with new favour on the Catholic Church as a force for order, and in 1850 the state monopoly was abrogated and private education at both secondary and primary levels consecrated. The latter part of the Second Republic and the period of the authoritarian Empire (1852–60) witnessed the apogee of Catholic education in the nineteenth century, largely private at secondary level, both public and private in the primary schools.

The watershed of the nineteenth century as far as the politics of education is concerned was not 1880 or 1870, but, in the middle of the Second Empire, in 1860. In that year, Napoleon III lent his support to the unification of Italy by Piedmont at the expense of the Papal States, and ruptured the old alliance between throne and altar. Henceforth French governments, predominantly anticlerical in complexion, would seek to drive the clergy out of the schools. Again, three phases may be defined. In the first, during the 1860s, a bid was made to eliminate those clergy who remained in the colleges of the University, to build up lay education in the primary schools, and to launch state secondary education for girls, hitherto dominated by the convents. These policies might have continued but for the war of 1870 and the reactionary period of Moral Order government in 1873-7. The landslide victories of the republicans at the polls in 1876 and 1877 opened the way to a new anticlerical offensive, at both secondary and primary levels, in order to underpin the new regime with appropriate civic education. But the forces of conservatism rallied round a policy of 'religious defence' to gain ground in the elections of 1885, and faced by the rising tide of socialism in the 1890s, the ruling groups shelved the divisive schools question in order to combine in defence of what they had in common, private property. In the third phase, following the Dreyfus Affair, in which the Catholic clergy played a significant counter-revolutionary role along-side the army—the alliance of the sabre and holy-water sprinkler—the Republic, now in the hands of the radical-socialist *Bloc des Gauches*, devoted itself not only to eliminating the Church from primary education, public or private, once and for all, but to divesting itself of the obligation to maintain the Catholic Church financially, which it had done for the preceding century.

The basic argument of this section is that, whatever the policies of the central government, those policies had to contend with the ideological aspirations of the provinces, which had firm historical roots and could not be altered by the whim of a law or decree. In subsequent chapters, the conflicts and compromises that took place in each of our three departments will be considered. To begin with, how-ever, it is worth looking in greater detail at the successive phases of government policy and analysing, in a general sense, the pattern of obstacles that interfered with their application.

Before the Revolution, very little distinction was made between the education of sacred and secular élites: both attended colleges that were in the hands of the teaching orders. Seminaries were scarcely more than centres of retreat for ordinands or boarding-houses for poor clerks who would also attend the local college, and though founded by the episcopate they might also be entrusted to religious orders. Elementary education too was very much under the control of the Church. The appointment of *maîtres d'école* was the responsibility of those who paid them: in the countryside, the *curé*, seigneur, or village community; in the towns, the municipalities in the case of *petites écoles* and the founders in that of charitable schools, the most effective of which were run by the Frères des Écoles Chrétiennes. That said, the bishop always had the right to approve the nomination of schoolteachers, and these clerks were very much part of the fabric of the Church, under the thumb of the *curé*, not only teaching (especially the catechism), but also ringing the bells, serving at mass, and assisting at funerals.

The work of the Revolution was in the first place only destructive. The oath to the Civil Constitution of the Clergy was imposed not only on college masters but also on *maîtres d'école*; many refused it and were dismissed. Then the colleges had to face the hammer-blows of the abolition of the religious orders, attacks on the episcopacy and Parlements which broke up their governing bodies, the eradication of the seigneurial revenues with which they were endowed, and the sale of properties that belonged to them, designated *biens nationaux*. It was not until 15 April 1791 that a law was passed entrusting colleges to municipal authorities, paving the way for the revival of some of them.

In the domain of elementary education, the gulf between the ambitions of the revolutionary assemblies and the sorry reality of the application of reforms was even more pronounced. Free, compulsory, lay education, commonly associated with the Ligue de l'Enseignement after 1866, was in fact prescribed by the Convention in the law of 19 December 1793. Political resistance apart, it soon became clear that major problems dogged the establishment of a nation-wide system of elementary education. Enlightened notables had to be mobilized in the localities to enforce legislation, and *jurys d'instruction* to undertake this task were tried first at district, then at departmental level, but with little success. Schoolteachers had to be provided with adequate maintenance, but solutions varied between the

Convention's policy of a fixed salary, paid in the event in depreciated assignats or not at all, and the Directory's view that teachers should cobble together an income from *écolage* or school pence, boarding fees and pluralism, doubling with other jobs. Lastly, some form of pedagogic training was required for the *instituteurs* of the Republic, but the only step in this direction was the foundation in 1794 of an *école normale* in Paris, which could cope only with a tiny minority of candidates.

The only major educational achievement of the Directory was the establishment of the *écoles centrales*. These schools were of significance for three reasons. Firstly, they represented an attempt to train a new republican élite, enlightened and virtuous defenders of the Constitution of 1795, talent replacing birth and favour in the battle for public office, genius permitted to flower in a world free of despotism and fanaticism. Secondly, whereas the colleges of the Ancien Régime had insulated themselves from the Enlightenment, weighed down by ancient languages, with logic, the 'sciences' (physics, metaphysics, mathematics), and Thomist theology crammed into the last two years of study, the *écoles centrales* attempted to marry education and Enlightenment. Courses were offered in three cycles: drawing, natural history, and ancient languages in the first two years (between twelve and fourteen); mathematics and physics and chemistry in the second; general grammar, *belles-lettres* (rhetoric and poetry), history, and legislation in the third. Finally, in their organization the *écoles centrales* presented the aspect of faculties rather than colleges. They were based on lecture courses for which pupils enrolled according to inclination. They had no central direction but the *conseil général* of professors elected a directorate of two or three for a fixed period. They provided no boarding facilities, which may have been in the Jesuit tradition but certainly did not correspond to the 'vogue des pensionnats' which established itself in college life in the pre-revolutionary decades.[2] The weaknesses of the *écoles centrales* were thus patent: not only were they identified with the Directorial regime but they failed to take account

[2] R. Chartier, M. M. Compère, and D. Julia, *L'Éducation en France du XVIe au XVIIIe siècles* (Paris, 1976), p. 215.

of the 'immense gap' which existed between their own high standards and the quite rudimentary primary-school system, and they failed to provide the moral supervision, religious education, and *répétition* of lessons which *pensionnats* could provide. One sign of this failure was the continuing growth of *pensionnats* at the turn of the century, set up by ex-religious, masters of the *écoles centrales*, or by speculators pure and simple.

Where there was a need for education, schools tended to spring up to meet it. One of the tasks facing successive governments, therefore, was not so much the provision of schools as the imposition of some kind of order and uniformity on the school equivalent of anarchy. It was this that Bonaparte, as First Consul, tried to do by the law of 1 May 1802 (11 *floréal* An X). Under this system the *écoles centrales* were transformed into *lycées*, the latter to be fed by a combination of *écoles secondaires communales*, maintained by municipalities and run by *bureaux d'administration* dominated by municipal officials, and of *écoles secondaires particulières*, private schools which achieved a standard that permitted them to prepare pupils for free places at *lycées*, and which came ultimately under the surveillance of prefects and sub-prefects. Order, authority, and hierarchy were to be restored in the secondary-school system as they were in the country at large after the 18 *Brumaire*. While the law of 1 May 1802 had provided for the teaching of 'la morale' or ethics independent of religious dogma, the Concordat that declared Catholicism to be the religion of the majority of Frenchmen was reflected in the *arrêté* of 10 December 1802 which said nothing about 'morale' but made the appointment of a Catholic chaplain, the saying of common prayers, and religious instruction compulsory in every *lycée* and college. The influence of the Enlightenment waned as the study of ancient languages was restored to its position of pre-eminence in the syllabus. And the military tenor of the regime was echoed in the autocratic administration of the schools, the discipline of the *internat*, the wearing of school uniforms, and the rhythm of the school day beaten out on the drum.

Yet Bonaparte did not conjure up a secondary-school system *ex nihilo*. As in so many things, he innovated less

than he imposed a pattern on materials to hand. Continuity
with the past was of the first importance. The *lycées* were
only *écoles centrales* in a new guise, and took a large proportion of their masters from them. *Écoles secondaires
communales* were colleges either of the Ancien Régime
or of the revolutionary period decorated with a new name.
And *écoles secondaires particulières* were only a creaming of
the mass of *pensions* and private institutions already in
existence.

The corner-stone of the Napoleonic system of education
was the decree of 17 March 1808, founding the University.
According to this measure, no school could be established
without the authorization of the Grand Maître nor anyone
teach in higher or secondary education without a degree from
one of the Faculties of Arts, Sciences, Law, or Medicine.
Given the prevailing misconceptions about the centralized,
monolithic structure of education in France, it must be
underlined from the beginning that the University never
succeeded in fulfilling these vast ambitions. Indeed, the
decree involved a built-in compromise with the Catholic
Church which was now an integral part of the imperial
edifice. At the top, the anticlerical chemist and *convention-
nel*, Fourcroy, who had been Directeur général de l'instruction publique since 1802, was passed over for the post of
Grand-Maître by the royalist *émigré* Fontanes, a figure
agreeable to the Catholic clergy. Among the Rectors, as
among the bishops under the Concordat, there were not only
men with a revolutionary past but non-jurors, who had
refused the oath to the Civil Constitution of the Clergy. In
addition, the Napoleonic *corps universitaire*, like the clergy,
was heterogeneous: it seems that the masters of *écoles centrales* who streamed into the *lycées* in 1802 were increasingly
diluted by masters from Ancien Régime colleges and universities, from private *pensionnats* and *écoles secondaires*.

The clearest sign of Napoleon's compromise with the
Catholic Church was the privilege accorded to seminaries.
The Catholic clergy, massively depleted by the revolutionary
episode, needed no prompting to set about renewing the parish
priesthood as soon as the risks of persecution diminished,
while for Napoleon a restructured parish ministry was as

important a part of the state apparatus as the civil administration and police. Under the decree of 17 March 1808 seminaries were therefore entrusted to the episcopacy rather than to the University, while a decree of 30 September 1807 not only provided scholarships for seminaries but granted exemption from the University tax (one-twentieth of the boarding fee, imposed on 'secondary' schools, where Latin was taught) to poor seminarists and to those attending the vicarage schools, where devoted *curés* instructed handfuls of Latinists with a view to sending them to the seminary. But since these seminaries were the only loophole in the University monopoly, they swelled out of all proportion as those seeking a true Catholic education, and virtually free Latin instruction, though with no intention of taking holy orders, flooded in through their gates. Next door to them, the *lycées* and colleges of the University vegetated. But, as in 1860, it was an Italian campaign and the absorption of the Papal States into a kingdom of Italy that snapped the bond between a Bonaparte and the Catholic Church. A war of attrition against seminaries began. Under the decree of 9 April 1809, 'seminaries' was taken to mean *grands séminaires*, and a *baccalauréat* delivered by one of the Faculties of Letters was required as a passport to them. The draconian decree of 11 November 1811 did recognize *petits séminaires* but limited them to one in each department, and that to be in the same town as a *lycée* or municipal college. In addition, pupils in seminaries, *pensions*, or private institutions over the age of nine were required to attend the classes of the *lycée* or municipal college in their town and to wear its uniform; seminaries were reduced to little more than boarding-houses.

It might be supposed that the Restoration of the Bourbon Monarchy, reviving the alliance of throne and altar, would have provoked the abolition of the University. An ordinance of 5 October 1814 did release the seminaries from all restrictive legislation while that of 17 February 1815 replaced the Grand Maître by a Conseil royal de l'Instruction publique. But Napoleon, returning for a hundred days, restored the University to the status quo of 1808, and the Second Restoration did not put the clock back to 1789. For while the ultra-royalists, dominating the 'Chambre Introuvable' between

August 1815 and September 1816, ranted against the modern state, state power remained in the hands of 'ministerialists' who would make but few concessions to clerical and feudal reaction. The University survived, run by a Commission de l'Instruction publique that was in the hands of moderates like the *doctrinaire* Royer-Collard and the Protestant Baron Cuvier; a die-hard Catholic, the abbé Frayssinous, felt obliged to resign in May 1816. In the provinces, the worst revolutionaries among the Rectors were purged and *lycées* were transformed into *collèges royaux*, under the direction of Catholic clergy. But if the University was Catholicized, it would in no way relinquish its monopoly for the benefit of the seminaries. Hence the stringent ordinance of 15 August 1815 which sought to ensure that seminaries confined themselves to the training of priests and did not become covert ecclesiastical colleges: seminarists were to wear ecclesiastical habit and seminaries should not accept day-boys—the colleges of the University existed for their purpose.

Though the 'ministerialists' claimed that the University was as Catholic and royalist as might be desired, it was a watery compromise that agreed neither with the ultra-royalists nor with republican youth. The *collégiens* were, after all, the *enfants du siècle* who had grown up under the Empire, eager for glory and adventure, and were now forced to sheath their swords for lack of argument. Stifled by the monastic atmosphere of the Catholicized colleges, yet touched by the wave of liberal unrest which rocked Europe after the Restoration, they expressed their anger in conspiracy and protest in 1819. It was in the light of the emergent liberal-republican menace that the *ultras* were able to seize the initiative again. The Duc de Berry was murdered on 14 February 1820, and the moderate Decazes ministry tumbled. In the elections of November 1820, the Right wing triumphed and the reactionary Villèle-Corbière ministry was installed. An ordinance of 27 February 1821 gave bishops a right of surveillance over colleges, in part restoring to them the authority they had wielded before the Revolution. On 1 June 1822, Mgr Frayssinous, bishop of Hermopolis, became Grand Maître of the University, and developments in secondary education became increasingly clerical and conservative. In the first place, existing seminaries came into their own. Secondly, small towns which had been campaigning to transform their colleges into seminaries, either to shift the burden of maintenance on to the diocese and obtain exemption from University tax, or to obtain a Catholic education more attuned to the sentiments of the region, had their wishes fulfilled. And lastly, many municipal colleges in lay hands were handed over to priests, and duly flourished.

The connection between political conflict and educational enterprise was equally clear in the field of elementary education. It is significant that Napoleon took no initiative to school the people until the Hundred Days, after the ordeal of the First Restoration had demonstrated the fragile basis of his rule. In the period 1800–15, popular education was abandoned to private initiative or to the communes, who would pay an *instituteur* or not, according to their inclination or resources. Glimmers of light on the horizon were few and far between. Initially, improvement originated in the Church and the teaching congregations. Perpetual vows and religious orders bound by them were prohibited under a law of February 1790, and Napoleon maintained this prohibition, but a decree of 22 June 1804 did permit the reconstitution of religious congregations that presented their constitutions for examination, under imperial decree. The first congregations to re-establish themselves were those which concerned themselves with the care of the sick and the relief of poverty: running hospitals and orphanages shaded off naturally into the foundation of charity schools for pauper children in the work of the Ursulines (authorized by a decree of 9 April 1806), the Sisters of Saint-Thomas-de-Villeneuve, and the 'sœurs grises' of Saint-Vincent de Paul. In some Catholic parts of the countryside, the *tiers ordres* of 'bonnes sœurs des campagnes', 'du foyer', 'en plein vent', or 'trottines' as they were variously called, laced into the rural community instead of being enclosed, distinguished from it only by their chastity, provided a spiritual leaven, visiting the sick, teaching catechism, propagating religious devotions.[3] It was very often the nexus between a group of pious country girls, some of whom were members of a *tiers ordre*, and a non-juring priest, that prepared the way for the foundation of a new religious congregation.

Anxious to secure his rule on more popular foundations after his return from Elba, Napoleon not only held a plebiscite on a constitution reformed under the Acte Additionnel, but realized the urgent need to enlighten the mass of the population, and as soon as possible. Mutual education was imported from England by his Minister of the Interior, Lazare Carnot, following a visit to the training-school of Joseph Lancaster's British and Foreign School Society at Borough Road, Southwark, by a number of Carnot's emissaries, including the abbé Gaultier, the geographer Jomard, and the economist J.-B. Say.[4] It represented an attempt to apply the principles of political economy to education: the division of labour by the monitorial system and the encouragement of competition would permit the education of the greatest number of children in the shortest possible time with the minimum of resources. Carnot produced a report to the Emperor on 10 April 1815, and under the decree of 27 April 1815 an École normale was set up in Paris to train mutual *instituteurs* and a Société pour l'instruction élémentaire, modelled on the British and Foreign School Society, was organized under the *idéologue*, Baron de Gérando, to supervise the campaign.

At the Second Restoration, mutual education was naturally abandoned as an official project, and the ministerial royalists concentrated on the thorny problem of rallying the notables to the work of organizing a system of popular education to train law-abiding, God-fearing Frenchmen. An ordinance of 29 February 1816 required the establishment of a web of *comités cantonaux* to propagate schools in hitherto deprived areas and to see that order, morality, and religion were

---

[3] C. Langlois, *Structures religieuses et célibat féminin au XIXe siècle: les tiers ordres dans le diocèse de Vannes* (Lyon, 1972); G. Cholvy, 'Une école pauvre au début du XIXe siècle: "pieuses filles", béates ou sœurs des campagnes' in D. Baker and P. Harrigan, eds., *The Making of Frenchmen* (Waterloo, Ontario), 1980, pp. 137–41.

[4] R. Tronchot, *L'Enseignement mutuel en France de 1815 à 1833* (thesis, Lille, 1973), i, 105.

properly inculcated. As in the University, the clergy were put in positions of command: cantonal *curés* (rural deans) were to convene and preside over the committees, which were generally composed of landowners, mayors, and *juges de paix*, with a smattering of notaries and tax-collectors, entirely representative of slow-moving, traditional, rural life.

Entrenched within the official elementary-school system, the Catholic clergy launched a crusade against the mutual-education enterprise that threatened their monopoly. For though no longer government-inspired, mutual schools were now being propagated by the Société pour l'instruction élementaire and its missionaries, together with an élite of moderate royalist prefects who were convinced by the validity of the new method. Confronted by the progress of mutual education in the large towns, the clergy of cathedral and populous urban parishes, rather than the episcopate, together with ultra-royalist municipalities, campaigned to bring back male teaching congregations like the Frères des Écoles Chrétiennes, in order to strengthen the Catholic, hierarchical structure of local society. New congregations of teaching nuns were also founded, but progress was slow until the law of 24 May 1825 which permitted the proper authorization of religious communities of women already in existence, and endowed them with 'civic personality' which enabled them, as communities, to acquire and receive property. The fortunes of Catholic education had been boosted by the elections of February and March 1824, which returned the 'Chambre Retrouvée' in which liberal opposition was negligible. The main outcome was the ordinance of 8 April 1824 which transferred from Rectors to bishops both the right to authorize *instituteurs*, and that on submission of a certificate of religious instruction, and the surveillance of elementary schools, either directly or, in the case of the largest ones, by committees of their nominees. The *comités cantonaux* of 1816, where they survived, were scrapped.

The progress of the liberal opposition, as opposed to clericalism as it was to absolutism, would nevertheless clip the wings of the Church in matters of education. In the elections of November 1827 the liberals gained a resounding success, and the Martignac ministry was installed. As a result, the domination of primary education by the episcopate was shattered. Under the ordinance of 21 April 1828, the right to authorize *instituteurs* was restored to Rectors, and new supervisory bodies, the *comités gratuits*, were set up, composed of the cantonal *curé*, mayor, and *juge de paix*, ex officio, together with two nominees each of the Rector and prefect, as well as two of the bishop's. In addition, the ordinances of 16 June 1828 hit out at the *petits séminaires*, prohibiting unauthorized congregations (as usual a synonym for the Jesuits) from staffing them, and confining them to their original purpose of training priests. Lastly, some concessions were made to Protestants, permitting Protestant as well as Catholic chaplains in *collèges royaux* that had a large Protestant population.

The Revolution of July 1830 snapped the bond between throne and altar and inaugurated an official campaign against the Church in education, in order to underpin the constitutional monarchy. At the same time the displaced legitimist notables, going into internal exile in their country residences, were resolved to use the provincial influence to defend the positions gained by Catholic education. On the government side, the first task was to laicize the University, to exclude priests from the headship and staffs of *collèges royaux* and *collèges communaux*. Not that this would prevent the clergy from continuing the war from the citadels of private education. In elementary education, the clergy were all but eliminated from the new supervisory committees, the *comités d'instruction primaire* set up under the ordinance of 16 October 1830. A decree of 18 April 1831 required teaching brothers to pass an examination for the *brevet* rather than obtain it simply on the presentation of their letters of obedience, as nuns still did. And in many places newly elected liberal municipalities cut off public funds to the Frères des Écoles Chrétiennes. But again, notables and clergy in Catholic areas rallied round to finance them out of private resources.

The central achievement of the July Monarchy in matters of education was of course the Guizot law of 28 June 1833. Though in some respects anticlerical, at a deeper level it can be seen to be reaching some sort of accommodation with the Catholic Church. The law was concerned with three principal areas of activity: organs of supervision and inspection, adequate maintenance for *instituteurs*, and teacher-training. In the first place, the law fought its way out of the under-growth of supervisory committees by dividing them into two levels. Abandoning the cantons, it prescribed *comités d'arrondissement* which, based on sizeable towns, would be able to draw on an educated urban élite and were packed with a large proportion of officials, including the prefect or sub-prefect, *procureur du Roi, juge de paix*, the cantonal *curé*, and, for the first time, members of the teaching body. The *comités d'arrondissement* were more powerful than preceding committees: their function was not only to spread and improve elementary education, but also to appoint *instituteurs*, the nominations being ratified by the Rector, and to discipline them in cases of infraction. Members of the committees were required to inspect schools locally, but leisure and zeal were at a premium, and from 1835 official salaried *sous-inspecteurs des écoles primaires* were attached, first to departments, then to *arrondissements* or groups of *arrondissements*. At the lower level of supervisory com-mittees, on the other hand, the clergy retained a strong influence. The *comités locaux*, committees of the municipal council together with the *curé*, were often swayed by the *curé*, for whom the schoolteacher was little more than the magister of the Ancien Régime, bell-ringer, choir-master, and sacristan, as well as *maître d'école*.

The second major stipulation of the Guizot law was to require every commune in France to maintain an elementary school, provide minimum *traitement* or stipend of 200 francs for the schoolteacher, fix the monthly *rétribution scolaire* or school pence to be paid by pupils, and provide that 'indigents' who could not pay the fee be educated as charges on the parish. There was often a considerable gap between legis-lation and application, not least because many communes simply did not have the resources to support a school. The meagre stipend could nevertheless make all the difference between surviving and succumbing for a poor rural teacher, so that the post of *instituteur communal* was to be coveted. In the event, Guizot envisaged no riot of new appointments. *Instituteurs* who were equipped with the *brevet de capacité* and were in receipt of a *traitement* or housing allowance, however small, when the Guizot law was passed, were deemed under a circular of 9 December 1833 to be *communaux*. But teaching congregations might also be put in charge of the *école communale*, so that often the battle for communal status took on the trappings of a religious war. It should be noted that a law of 23 June 1836 extended the provision of the Guizot law to the education of girls, but without the obligation to maintain a second school, which the vast majority of rural communes could ill afford. In towns, girls had the choice between convents or lay *pensionnats* if they were rich, charitable schools run by nuns or *petites écoles* of widows or spinsters if they were poor; in the countryside, colonized only slowly by the religious congregations, girls made do with dame schools or went to the local *école communale*, which thus became 'mixed'.

Since there was no major purge of schoolteachers already in place, who included social misfits thrown up by revolutionary turbulence and economic depression, unfrocked priests, demobilized NCOs, unemployed artisans, and even ex-convicts, intellectual standards were not going to be high. The Guizot law sought to raise standards, abolishing the third-degree *brevet*, which assessed proficiency only in reading, writing, and counting and was seen to be an invitation to ignorance, and requiring an *école normale* to train teachers to be set up in every department. Some departments already had *écoles normales*, and the noviciates of the male teaching congregations also played their part, but it was clearly necessary to reach the mass of teachers already exercising their profession in the villages. For these, conferences of *instituteurs* were organized at the cantonal level, meeting monthly not only to exchange views about pedagogic method but also to raise solidarity and morale amongst 'humble apostles' dedicated essentially to ploughing a solitary furrow. But these meetings tended to be little but a pooling of ignorance or to degenerate into political debate or drinking, and were replaced by 'summer schools' at the *école normale*, supported by publicly subsidized pedagogic reviews. Ignorance among *institutrices* was even more pronounced, but the desire to raise standards was counterbalanced by fears of over-education, especially among women.

In the field of elementary education, Catholics made up some of the ground they had lost since the anticlerical Revolution of 1830 in dogged and unspectacular fashion. Their performance in secondary education, the crusade against the University monopoly to establish *La Liberté de l'enseignement*, was more dramatic. Under the July Monarchy, the authorities could and did keep Catholic education in chains; in the last instance, its claims were blocked at the highest level. Nevertheless, Catholic leaders were able in this period to undermine the foundations of the University monopoly and build the basis of a school system of their own which would come into its own under the Falloux law. In general, four alternatives were open to them. They could take the road to exile—a road that was already familiar to some of them. They could colonize the colleges of the University with priests in the manner of the Restoration. They could defend the seminaries—once the only breach in the wall of the monopoly—against attempts to remove them from the control of bishops and subordinate them to the University. Finally, the *pensions* and institutions that were kept in tutelage to the University by the requirements that they should be staffed by directors and masters with the appropriate University degrees and that pupils should attend the rhetoric and philosophy classes of University colleges in order to obtain the *baccalauréat* could nevertheless be developed behind a front of legality ready for the moment when they could emerge as the great Catholic colleges of the mid-nineteenth century.

That moment came with the Revolution of February 1848. The originality of this revolution was that, unlike that of 1830, it was not in the first instance anticlerical. The overthrow of the Voltairean and Protestant oligarchy might augur well not only for republican intellectuals and artisans, but also for those Catholics campaigning for liberty from the University monopoly. In the event, they would have to wait, because the provisional republican government was somewhat extreme. *Collèges royaux* were renamed *lycées*. The Minister of Public Instruction, Hippolyte Carnot, son of Napoleon's minister during the Hundred Days, sought to emancipate *instituteurs* from the influence of the clergy, making them apostles of republicanism in the communes, intermediaries between the republican intellectual élite and the popular classes. But the elections to the Constituent Assembly on Easter Day 1848 took place before republican sentiment had really penetrated the countryside, so that universal suffrage played into the hands of the conservative notables. In Paris Carnot pressed on regardless, introducing into the Assembly a bill on free, compulsory civic education and requiring nuns who opened schools to have the *brevet de capacité* instead of merely showing letters of obedience. But after the June Days the propertied classes and rural masses rallied to the side of order: Louis-Napoléon was elected president of the Republic in December 1848, and in May 1849, of 750 deputies elected to the Legislative Assembly, 550 were conservatives.

The Second Republic of 1849–51, like the Moral Order regime of 1873–7, was a period of the 'Republic without republicans' during which conservative notables crusaded to exorcise the demon of revolution. Universal suffrage was restricted by the law of 31 May 1850 and attacks were launched on lay education and the University. The first victims of this White Terror were the *instituteurs*, now pilloried as village demagogues, a poisonous leaven in the Catholic and conservative countryside. The most intriguing aspect of this reaction of the notables was that it translated a fear not only of over-education, but indeed of any popular education at all, when it was not compensated by the deeply moralizing force of religion. The *comités d'arrondissement* had been vigilant in their pursuit of 'red' *instituteurs*; now there were attacks on the *écoles normales*, portrayed as training-camps of revolutionary agitators.

Despite profound disillusionment, the debate in the mid-century was not so much between enlightenment and ignorance as to what extent education should be under the patronage and control of the Catholic Church. After 1848 the Church's role as a defender of the existing social order against democracy and anarchism was looked on with new favour by conservative notables, and the Falloux law of

15 March 1850 was the product of this movement of 'défense religieuse et sociale'. This law eroded the monopoly of the University in secondary education. Any man armed with the *baccalauréat* over twenty-five years of age and with five years' experience in a public or private school was entitled to open a secondary school. *La Liberté de l'enseignement* had been achieved, and the Catholic clergy lost no time exploiting its position. Colleges that had flourished in exile closed, but only because of the new prosperity of Catholic education at home. *Petits séminaires* were released from many of the constraints of 1828. Catholic *pensions* and institutions that had campaigned for *plein exercice*, the right to educate their pupils up to the *baccalauréat*, now achieved their goal, and began to undermine the position of *lycées* and municipal colleges. Lastly, municipalities were empowered to turn their colleges over to private persons or associations, while still subsidizing them to the tune of 10 per cent of the college's expenditure. In this way municipalities could be relieved of mounting debts, Catholic populations could obtain Catholic colleges, and Catholic education could to some extent be financed out of public funds.

The Falloux law concerned elementary as much as secondary education. In many ways it represented the repentance of the notables for the over-education of the people that the Guizot law had brought about. The penalty had been paid in 1848, and the restoration of a society in which the popular classes obeyed the clergy and deferred to their natural superiors could be achieved only if education became again what it had once been: moral, religious, and under the control of the Church and the traditional ruling groups. The Guizot law was therefore partly dismantled in 1850, an assertion that may be substantiated if we examine the three areas of its provisions: supervision and surveillance, the responsibility of communes for maintaining schools, and the *écoles normales*.

The shattering of the University's monopoly in education involved a severe reduction of the influence of *universitaires* on the bodies charged with the supervising of education. Whereas the *conseils académiques* of the Guizot law, presiding over several departments, were dominated by an official élite of school inspectors, *universitaires*, and magistrates, who could offer professional advice to the Rector, the *conseils départementaux de l'instruction publique* set up in each department by a law of 14 June 1854 were heavily weighted with representatives of the Churches and the *conseils généraux*. It was these bodies which appointed the *commissions de surveillance* of the *écoles normales* and the examining commissions for the delivery of *brevets de capacité*. In addition, the *conseils départementaux* appointed the local organs of supervision. The *comités d'arrondissement*, those urban and enlightened instruments of the Guizot law, were abolished along with the *comités locaux*, and replaced by *délegations cantonales*. This, as their name suggests, was very much a step backwards to 1816. Representatives of the liberal professions, public service, and the world of trade and industry were reduced to a minimum, and *universitaires* were purged completely. Landowners, clergy, *juges de paix*, and local mayors accounted for about two-thirds of their composition. The functions of the delegations were not vast—mainly to ensure that teaching in schools was not contrary to morality, the constitution, and the laws—but amateurism replaced professionalism and *querelles de clocher* between mayors and *curés* were institutionalized. Moreover, reactionary forces were given an influence they had not enjoyed since the Restoration. One of the only restrictions on their influence was the retention by the Rector of the right to appoint *instituteurs*. Some municipalities, notably those dominated by conservative forces, wrongly claimed that the right of appointment had now devolved upon them. Their initiative was limited to instances where a vacancy was occasioned by the death, resignation, or dismissal of the incumbent; then they might opt in principle between a lay teacher and a religious congregation.

The obligation of communes to maintain an elementary school did not change, but the principle of *La Liberté de l'enseignement* was to apply there as well: in future, any Frenchman (or French woman) over twenty-one with a *brevet de capacité* might open a school, after making a *déclaration d'ouverture* to the local mayor, the sub-prefect, the Rector and the *procureur de la République*. Given the

absence of resources, especially in the countryside, to maintain schoolteachers from school pence, this liberty was a
mixed blessing. Those confined to private teaching were, for
the most part, nomads, irregulars, and social failures, ex-
seminarists, ageing corporals of the Grande Armée, Dickensian
figures who flitted between desks in mail-coach offices and
trading-houses to desks in schoolrooms. The goal was still
to obtain the post of an *instituteur communal*, not least in
the case of the religious congregations, which attained the
apogee of their influence in the 1850s. It was not unusual
to see a private school run by *frères* or nuns establish itself
in a commune where a lay *école communale* already existed
and force the resignation of the lay teacher by offering
seductively low rates, by threatening to fail his pupils in
catechism class which would disqualify them from First
Communion, or by discrediting the *instituteur* for irreligion
or his pupils for offences such as poaching. Then they could
be sure that a Catholic municipality would hand over the
école communale.

One modification introduced by the Falloux law was the
obligation on communes of over 800 inhabitants to maintain
a school for girls. This provision filled a major gap in the
legacy of Guizot, but it included the fatal reservation that
lack of resources was an acceptable excuse. The result was
that the education of girls remained overwhelmingly in
private hands. For lay *institutrices* this was a recipe for
desolation, but the teaching congregations of nuns pro-
liferated, and in some comfort. The Catholic clergy took a
leading role, realizing that the education of infants was in
the hands of mothers, who must be pious and devout. Moreover, women represented stability and continuity in French
society, and never was there more need to reinforce these
elements than in the wake of the 1848 Revolution. On the
financial side, the parish clergy organized voluntary subscriptions, made untold sacrifices of their own—helped by
the fact that as celibates they had no heirs—and encouraged
'pious foundations' by rich and charitable persons, whether
legitimist nobles or mining companies. From a popular point
of view, moreover, while the unmarried, isolated school-
mistress from outside was looked on with suspicion and

distrust, the nuns fitted into the fabric of rural life. They were teachers of the poor, but also minders of crèches, house-keepers to the *curé* who would also clean the church, rural nurses who would visit the sick and distribute medicaments, and a source of alms for the indigent. It would be fair to say that the *deux jeunesses* in mid-nineteenth-century France were not so much the clienteles of lay schools and those of the teaching congregations but, which amounted to the same thing, boys and girls.

The third major achievement of the Guizot law was to systematize the establishment of *écoles normales*. A powerful movement of opinion in favour of their abolition in 1849 has been mentioned. They remained, but their regime was severely modified under a decree of 24 March 1851. A rustic simplicity was preferred to cleverness, and every effort was made to discourage urban youths and college rejects from applying. Within the schools, the syllabus was pruned back to a bare minimum of requirements. Lastly, the *écoles normales* were cut off from the evil influences of the outside world by suppressing monthly outings and reducing holidays to a fortnight in summer, while the religious atmosphere of the *internat* was intensified. The 'red' *instituteur* would become no more than a phantom of the past; once again, the school-master would be the secular arm of the *curé*.

By 1860 there had been three attempts to establish a secular, public system of education, and all had failed. The University monopoly lay in ruins and the Catholic Church had a firm grip of the education of the élite, its war machine consisting of colleges not unlike English public schools, seminaries, and the municipal colleges that had been sur-rendered to it. At the elementary level religious congregations were invading the *écoles communales*, whether they were for boys, mixed or for girls. The education of girls remained by and large in private hands, but here too the nuns had a commanding lead over the poor lay spinsters who vegetated in village schools or the more stylish lay *pensionnats* of the towns. From the bourgeoisie to the very poor, the Catholic Church controlled the education of girls. Once the Second Empire had broken with Rome, this monopoly posed a threat to the security of the regime.

The Response in the Provinces (1)

The ideological contours, the relief map of political and religious attitudes that had been established in our three departments by 1800 not only formed a backdrop but constituted the raw material with which education reforms had to contend. For one of the main tensions in nineteenth-century France was that which existed between Paris and the provinces, between official policies which could change with kaleidoscopic rapidity, sensitive to every shift in political power, and the slower, more measured rhythm of the departments, which could accept, transform, or openly resist such initiatives from the centre. This was not merely a question of inertia: initiatives from the centre flourished or died according to the mentality of the region in question. Reactions depended on whether the area was French-speaking or non-French-speaking, Protestant or Catholic, pious or indifferent, republican or *chouan*. There was, moreover, another side to the picture. It would be wrong to imagine that the various departments only re-sponded passively to developments in Paris. Each had its own internal dynamics, each bloc of cantons or communes searched for a pattern of education which reflected its own political and religious character. Lay education suited some areas, education in the hands of religious congregations others. Unfortunately, the realization of one scheme in preference to another depended in the last instance on Paris, and on the web of patronage and politics that linked Paris and the provinces. Hence the dual origin of conflict over education policies: offensives from the government hierarchy which found no response in the localities; campaigns in the locality which found no response at the centre. In 1914, Frenchmen would be scarcely nearer a solution than they had been in 1800.

(a) The Nord: French and Flemish

The Nord stands out in the nineteenth century as the department of textile mills and coal-mines, of the first industrial revolution. Yet despite heavy industrialization, it should not be seen as a department in which the tone was set by a de-Christianized working class. The most 'deChristianized' part of the department was the agricultural south, the Cambrésis, in which the eradication of Protestantism had left the scars of religious indifference, and the Avesnois, with its high proportion of *curés constitutionnels*. The Catholic strongholds were in the north, in Flanders. But Flanders was divided into rural *Flandre flamingante* and industrial, French-speaking Flanders. In both the Catholic influence of the Austrian Netherlands to which they had once belonged was still present: it was as if the weight of the cultural past could not be cast off, even by momentous economic and social change (Map 1).

A continuity of historical geography from the Ancien Régime was clear in the pattern of schooling. The network of colleges was dense in what was to become the department of the Nord, the Jesuit presence in Flanders particularly marked.[1] On the other hand, the study of Fontaine de Resbecq, intended to demonstrate how complete the fabric of primary schools was even before the Revolution, shows that the least traditional south of the department, especially the Avesnois, was the best provided for. More recent studies have confirmed that the literacy rate of the Avesnois was far higher than that of other districts in the Nord.[2]

Even in the École Centrale of Lille, that revolutionary innovation of the Directory, there was continuity with the past. This was clear not only in the case of 'classical' subjects but also in the new disciplines. The development of free public instruction in drawing, mathematics, and in some cases architecture and botany by the large towns after 1750 provides the key. Public drawing courses were designed to

[1] J. Peter, *L'Enseignement secondaire dans le départment du Nord pendant la Révolution, 1789–1802* (Lille, 1912, pp. 15-36); P. Marchand, 'Le Réseau des collèges dans le nord de la France en 1789', *Revue du Nord*, lviii, no. 229, 225-34.

[2] F. Furet and J. Ozouf, *Lire et écrire* (Paris, 1977), i, 233-5. This is based on E. Fontaine de Resbecq, *Histoire de l'enseignement primaire avant 1789 dans les communes qui ont formé le département du Nord* (Lille–Paris, 1878).

supplement the poverty of apprenticeship in the workshop, since drawing was deemed to be the basis of all mechanical arts and indispensable to the artisan élite. Louis Watteau, nephew of the famous painter, was assistant at the École gratuite de dessin at Lille in 1756 and drawing-master of the École Centrale, succeeded there on his death in 1798 by his son.[3] The work of Linnaeus had opened up botany as a science and encouraged municipalities to organize botanical gardens and found chairs of botany. François-Joseph Lestiboudois (d. 1815), author of *Botonographie belge*, who had succeeded his father as professor of botany at Lille, became professor of natural history at the École Centrale.[4] This was a dynasty of botanists that would stand for a more enlightened, less classical, approach to secondary education in the early part of the nineteenth century.

The reorganization of the secondary-school system by Bonaparte was carried out with existing materials. Of twenty-six colleges which studded the future department of the Nord in 1789, only four were swept away for good. But while in Flanders it was continuity with the Ancien Régime that was most pronounced, in the south it was continuity with the revolutionary period. At Tourcoing, the burgeoning wool town of Flanders, the Recollets who had run the college between 1666 and 1790 returned in the person of the director and another master, padded out with former Carmelites and Benedictines, after 1802.[5] But at Cambrai, where the Jesuits had been replaced in the 1760s by secular priests, the struggle to reconstitute the college in 1802 was led by Pety and Marchant, masters in the turbulent years 1790-3.[6] It is interesting that to break the links with the École Centrale, the Lycée of the Nord was founded not at Lille but at Douai. In this way, it slumbered in the atmosphere of the old *parlementaire* city and its masters

    [3] B. M. Lille, Fonds Lemaire II, 53; J. Peter, p. 137.

    [4] J. Peter, pp. 137, 140-1, 151-2. His father, Jean-Baptiste (1715-1804), was the founder of the botanical gardens of Lille and the town's first lecturer in botany in 1770.

    [5] H. Leblanc, *Histoire du Collège de Tourcoing, principalement sous l'administration de M. l'abbé Lecomte* (Tourcoing, 1870), pp. 51-4.

    [6] A. Durieux, *Le Collège de Cambrai, 1270-1882* (Cambrai, 1882), pp. 107-9, 135-7, 142.

could be drawn from former professors of the University of Douai.

The University regime of 1808 gave rise to an original situation in the Nord. On the one hand, the Rector of the Academy of Douai was Taranget, sometime professor at its University and a non-juror; on the other, the bishop of Cambrai was Louis Belmas, a *curé constitutionnel*, hailing from the Catalan border and constitutional bishop of the Aude in 1800.[7] But the battle-lines were already drawn: Taranget was to fortify the University against Catholic education, Belmas was to insist on the training of priests outside the *lycées* and colleges. The diocese of Cambrai was not in a privileged position as far as the recruitment of clergy was concerned: it was not until 1807 that Belmas was able to open a seminary, and the decree of 9 April 1809 obliged him to vow that no *petit séminaire* existed, while concealing its existence from the authorities, euphemistically calling it an 'école préparatoire' to the Grand Séminaire.[8] From 1811 the seminarists had to attend the municipal college of Cambrai, and it was not until after the triumph of the *ultras*, in 1823, that Mgr Belmas was able to withdraw his seminarists from the college.

With the Restoration came the crusade by liberal elements to propagate mutual education. The historian of mutual education has noted that 'the department of the Nord seems to be the only one where there was no strife'.[9] The missionary of the Société pour l'instruction élémentaire in the Nord was Benjamin Appert, a convinced Bonapartist who had been deprived of his post as assistant master at the École impériale de dessin by the royalist regime. Lack of strife may be explained by the fact that his invitation came from the Compagnie des mines of Anzin and that though mutual schools were set up in the towns dominated by conservative notables they made their impact on the coast at Dunkerque and Gravelines, in the coal-basin of the Escaut at Anzin,

---

[7] See L. Mahieu, *Mgr Louis Belmas, ancien évêque constitutionnel de l'Aude, évêque de Cambrai, 1757–1841* (2 vols., Paris, 1934).

[8] Mahieu, ii, 22–31; Boussemart, *Histoire du Petit Séminaire de Cambrai* (Cambrai, 1902), 78, 110–20.

[9] Tronchot, ii, 351.

Valenciennes, Fresnes, Condé, and Vieux-Condé, and spread into the black country around Douai and the new industrial region of Maubeuge.[10] But in the old towns the royalist die-hards were quick to react, bringing in the Frères des Écoles Chrétiennes to cut off the progress of mutual education. The crusade was led by the Comte de Muyssart, mayor of Lille, followed by Édouard Deforest de Lewarde, of an old *échevin* and *parlementaire* family, who brought the Frères to Douai at the end of 1816,[11] and Henri-Joseph Dubois-Fournier, an ultra-royalist and ultramontane merchant of Valenciennes, who was at daggers drawn with the constitutional, Gallican Mgr Belmas.[12]

It was in the towns, similarly, that the regular education of girls first got under way. The prefect of the Nord, the Comte de Rémusat, set up committees in 1819 to examine *institutrices* for the *brevet*, but he remarked that he was concerned 'only with the *institutrices* of the towns and the very small number of those who might establish themselves in other communes'.[13] As likely as not, those *institutrices* would be the nuns of newly founded teaching congregations, the crucial link being that between pious girls and parish priests. Natalie Doignies, a peasant girl of Moncheaux-en-Pévèle who became a servant in an aristocratic household of Lille, took as spiritual confessor the chaplain of the central prison, Louis Detrez, taught in an *école dominicale* founded by the Bureau de bienfaisance, and with Detrez founded the Filles de l'Enfant-Jésus in 1827.[14] Other female congrega-tions authorized under the law of 1825 in the Nord included the Sœurs de Sainte Thérèse of Avesnes, who had particularly

---

[10] B. Appert, *Dix Ans à la Cour du Roi Louis-Philippe et souvenirs du temps de l'Empire et de la Restauration* (Berlin–Paris, 1846), i, 91-8; P. Pierrard, 'L'Enseignement primaire à Lille sous la Restauration', *Revue du Nord*, lv, no. 217 (April–June 1973), 123-33; Tronchot, ii, 377.

[11] A.F.E.C. NC 504, Deforest de Lewarde to superior-general, 31 Jan. 1816 and prospectus of school, Oct. 1816; F. Capelle, *Éloge historique de M. de Forest de Lewarde* (Douai, 1852).

[12] P. Pierrard, 'L'Établissement des Frères des Écoles Chrétiennes à Valen-ciennes, 1824-48' (*Mém. d'études supérieures*, Lille, 1949) and P. Dubois, *Un patriarche: vie de M. Dubois-Fournier* (Lille, 1899).

[13] A. D. Nord 1T 85/16, prefect of Nord to Min. Int., 12 July 1819.

[14] L. Detrez, *Mère Natalie, Fondatrice de la Congrégation des Filles de l'Enfant-Jésus* (Paris–Lille, 1930).

impressed the examining commission of the Comte de Rémusat in 1819; the rural *arrondissement* of Avesnes was second only to the urban one of Lille for the proportion of qualified *institutrices*. [15]

In the Nord, where the old landed nobility was weak, the Revolution of 1830 was well received: the commercial and industrial bourgeoisie rallied as a general rule to Orleanism. What resistance there was came from a section of the *patronat* which adhered resolutely to legitimism and clericalism, headed at Lille by the sugar-refiner Charles Kolb-Bernard and his brother-in-law the abbé Charles Bernard, from old *parlementaire* families of Douai, and from the Flemish clergy in the *arrondissements* of Dunkerque and Hazebrouck. [16] As a result, the campaign to laicize the University and the municipal schools met with some concerted opposition. In the colleges the old priest-principals had enjoyed much prestige, while their lay successors had too little distinction. The new principal at Lille was Édouard Gachet, a former workshop master who had opened a *pension* in 1824 as a result of losing an arm, though he knew no Latin. [17] At Tourcoing, the abbé Louis Flajolet, who since 1823 had built the success of the municipal college on a Catholicism that appealed to Flemish populations, was dismissed, but he took all but one of his boarding pupils to a new college that he had set up over the Belgian frontier, at Mouscron. The task of the young, married principal brought in from Condé to a region where 'a sort of sacerdotal fascination' reigned, was almost impossible. [18]

The laicization of the University was accompanied by a bid by the new liberal municipalities to laicize the municipal schools, cutting off the subsidies with which their legitimist predecessors had endowed the Frères des Écoles Chrétiennes,

[15] A. D. Nord 1T 85/16, 'État numérique des institutrices de filles qui ont satisfait à l'examen du jury', (?)1819; Mahieu, ii, 130-2.

[16] A.-J. Tudesq, *Les Grand Notables en France, 1840-49* (Paris, 1964), pp. 162-5; see also P. Pierrard, 'Un grand bourgeois de Lille: Charles Kolb-Bernard, 1798-1888', *Revue du Nord*, xlviii, no. 190, 1966, 381-5.

[17] A. D. Nord 2T 404, reports of inspector Caillat, 1828, and of Insp. Acad. Denfert to Rector of Douai, 4 July 1831.

[18] H.-J. Leblanc, op. cit., p. 108; A. D. Nord 2T 404, Insp. Acad. Denfert to Rector of Douai, 10 July 1831.

and building up the mutual-school system. Cambrai was the only town of importance that kept up payments to the Frères: between 1830 and 1832, Frères were evicted at Lille, Roubaix, Douai, and Valenciennes. There was, of course, all the difference between an administrative act and the effective elimination of clerical education. Notables and clergy who supported the Frères at once launched subscriptions and found new premises for them. At Valenciennes, Dubois-Fournier took charge of operations,[19] while at Lille, a Commission de soutien des écoles chrétiennes, which included the *curés-doyens* and 'clerical *haute bourgeoisie*'— Kolb-Bernard, Henri Bernard, Anatole de Melun, the publisher Lefort—not only rehoused the Frères but ensured that their pupils multiplied from 1,462 in 1831 to 2,066 in 1838.[20]

Of our three departments, it is possible to argue that the Guizot law made least impact in the Nord, because of what had previously been achieved. For the prefect, looking back over ten years in 1844, 'the law of 1833 was useful in this department only to regularize what existed already'. Twenty-six *comités d'instruction primaire* were functioning under the law of 16 October 1830, including not only landowners and officials but also merchants and industrialists making up 11 per cent and the liberal professions accounting for 33 per cent of the membership. Moreover, it took until 1839 to limit the number of committees to one per *arrondissement*, and even then 37 per cent of the membership was carried over from 1830.[21] The *comités locaux* were much less impressive, indeed the inspector reported in 1838 that in the three Flemish *arrondissments* of Lille, Hazebrouck, and Dunkerque the clergy who were supposed to animate them 'are restricting elementary education as much as they can to habits of narrow devotion'.[22]

The wealth of the Nord, based on a combination of

[19] P. Pierrard, 'L'Établissement des Frères des Écoles Chrétiennes à Valenciennes, 1824-1848', pp. 31-7.

[20] P. Pierrard, 'L'Enseignement primaire à Lille sous la Monarchie de Juillet', *Revue du Nord*, lvi, no. 220, Jan.-Mar. 1974, pp. 7-8.

[21] A. D. Nord 1T 111/3, *Comités gratuits*, 1830; 1T 99/1, *Comités de'arrondissement*, 1834.

[22] A. D. Nord 1T 107/1, Inspector Carlier to prefect of Nord, 1838.

agriculture and industry, made the provision of a school in each commune relatively painless: only 12 per cent of communes had to be taxed ex officio in 1833 to complete the provision. Again, no overnight renewal of the teaching body was intended, even had it been possible. The Rector of Douai warned Guizot that despite the large numbers of 'old *maîtres d'école'* still teaching, barely qualified at all, and 'faithful to the defective method of individual education', there were nevertheless 'long years of service to consider, even to reward; infirmities and distress to support and relieve; ancient, laborious and useful lives to respect and conserve'.[23] Guizot's circular of 9 December 1833 confirming serving *instituteurs* in their posts was a stalwart defence of prescriptive rights that tended to fly in the face of municipal councils that were anxious to take advantage of the new legislation in order to get rid of dead wood. In the *arrondissement* of Cambrai, fifty-three *instituteurs* petitioned the Minister in desperation in 1835, fearing that the mayors and the Comité d'arrondissement were conspiring to remove them.[24] This they could not do, but at the other end of the department a new threat was making itself felt in the shape of a congregation of teaching brothers, the Petits Frères de Marie. In 1842 these were brought into the Nord by the Comtesse de la Grandville, the unhappily married daughter of a former president of the Estates of Artois, and established on her estate at Beaucamps, in the countryside to the west of Lille. By 1845, they had founded a school at Quesnoy-sur-Deûle and looked set for considerable expansion in the industrial region of French Flanders. Neither could lay *instituteurs* view with indifference the percolation into the countryside of female congregations, since many of them educated girls in their 'mixed' classes. The arrival of the Sœurs de la Sainte Famille of Amiens at Premesques, near Armentières, in 1841, was greeted by the protest of the *instituteur* 'that by depriving him of the little girls who paid, they were going to take away his means of existence'.[25]

[23] A. D. Nord 2T 32/3, Rector of Douai to Min. Instr. Pub., 24 Dec. 1832.

[24] A. D. Nord 1T 84/2, petition of *instituteurs* of *arrondissement* of Cambrai to prefect of Nord and Min. Instr. Pub., 28 Feb. 1835.

[25] A. D. Nord, 1T 124/11, *Gazette de Flandre et d'Artois*, 26 Nov. 1841.

In the Nord, the little group of missionaries who had pioneered mutual education had also done much to promote teacher-training. But the prefect was reluctant to provide public funds for the S.I.E. and it was May 1835 before an *école normale* opened at Douai for the Nord and Pas-de-Calais, under a director who had attended not only the École normale of the S.I.E. but that established in Paris at the end of 1794.[26] The problem of the mass of schoolteachers in the villages, ageing and rusty but secure in their jobs, had yet to be conquered. Cantonal conferences sprang up in French Flanders, the Cambrésis, and the Avesnois (three alone in the canton of Avesnes-Nord in 1839), but made no impact on *Flandre flamingante* or the coal-basin of Douai–Valenciennes. There was never enough money, whether from the subscriptions of participants or meagre Conseil général grants, to keep up their small libraries. The school-inspector reflected that the conferences were well served by an élite of *instituteurs*, but that those teachers who genuinely needed training never attended, for fear of publicizing their ignorance.[27] In time, they were replaced by summer courses at the École normale of Douai and a review, the *Instituteur du Nord et du Pas-de-Calais*, produced between 1841 and 1848.

As far as the training of *institutrices* was concerned, the attitude of the notables was ambivalent. Standards, especially in the countryside, were appallingly low, and *brevets*—when they had been acquired—had often been delivered as a matter of grace. On the other hand, conservatives felt that educating girls beyond the bare minimum was unnecessary and unnatural. The Conseil général of the Nord told the prefect in 1842 that 'rural *institutrices* are not supposed to train *femmes savantes*. The important thing for a village girl is that she knows how to sew, read, write and reckon, that she has a firm grasp of religious education and has a perfect understanding of all those manual skills which might be useful to her.'[28] Compromise was reached by entrusting the training-

---

[26] A. D. Nord 1T 69/5, J.-P. Boulanger, *maître de pension* at Saint-André, Lille, to prefect of Nord, Mar. 1833.

[27] A. D. Nord 1T 107/1, Inspector Carlier to prefect, 1839.

[28] A. D. Nord 2T 1075, Conseil général of Nord, 20 Sept. 1842.

course for girls set up at Douai in 1845 to nuns, the Bernardine Dames de Flines.

It was nevertheless in the domain of secondary education that the Catholics made their first breakthrough. All options were exploited, whether the road to exile, restoring the colleges of the University to priests, defending existing seminaries, or building up private institutions. As a frontier department, exile was for the Catholics of the Nord a viable solution. The story started on 29 July 1830 when Saint-Acheul, the Jesuit noviciate and study-centre at Amiens, was attacked and pillaged by crowds shouting 'Vive la Charte!' 'A bas Charles X!' 'A bas la Calotte!' 'Mort aux Jésuites!'[29] Saint-Acheul, established as a *petit séminaire* and *externat* under the Jesuits, 'nightmare of the liberals', had been closed by the government under the ordinances of 16 June 1828, and was now closed again by popular violence. Legitimist families who wanted a Jesuit education for their sons now had to send them to Le Passage, near San Sebastián in Spain, or to Fribourg in Switzerland. After July 1830, two men started to campaign for 'a resurrected Saint-Acheul' over the border in Belgium, itself now liberated from the Orange yoke by a Catholic–liberal revolution. One was Achille Guidée (1792-1866), who succeeded Loriquet as rector of Saint-Acheul in 1828. The other was Dubois-Fournier, several of whose twenty-one children had been educated at Saint-Acheul. Both entered negotiations with the Jesuit provincials of France and Belgium and with the bishop of Tournai, and Dubois-Fournier purchased an old convent at Brugelette, between Ath and Mons, where a college opened in October 1835.[30] The college, attended by Carlists not only of the north of France, but also from Lorraine, the Paris region, Brittany, Gascony, and Provence, started with fifty pupils but had 330 by 1845.[31] The co-director was Adolphe Pillon (1804-85), born near Amiens and educated

[29] F. Grandidier, *Vie du R. P. Achille Guidée de la Compagnie de Jésus* (Amiens–Paris, 1867), pp. 102-10.

[30] On this, see Gradidier, pp. 131-56; P. Dubois, op. cit., pp. 151-8; P. Delattre, *Les Établissements des Jésuites en France depuis quatre siècles*, vol. i, (Engheim, 1948), pp. 944-8; H. Beylard, 'Les Jésuites à Lille au XIXe siècle', *Revue du Nord*, liii, no. 208, 1971, p. 65.

[31] P. Dubois, op. cit., p. 158; P. Delattre, op. cit., p. 949.

at Saint-Acheul, whose peripatetic teaching career had taken him to Dôle, Aix, Le Passage, and Mélan in Savoy, before his appointment to Brugelette.[32]

Clearly only a minority of 'integral' Catholics, well enough heeled to afford the boarding fees of 790 francs a year, made the pilgrimage to Brugelette. More important therefore was the campaign of the municipalities and Catholic clergy of Flanders to restore municipal colleges to the priests. In this field, the outstanding achievement was that of two Flemish priests who, growing up under the Restoration, had suffered the Revolution of 1830 as a traumatic experience, trained together at the Grand Séminaire of Cambrai shortly afterwards, and served as curates in adjoining parishes of Douai between 1834 and 1838. The abbé Dehaene (1809-82), a native of Quaëdypre near Bergues in Flemish-speaking Flanders, was the son of a peasant-farmer who had fled to Belgium during the Terror, had been educated at the municipal college of Hazebrouck when it was in the hands of the priests, and returned in January 1838 at the invitation of the municipality to save the college from the decline it was suffering in lay hands. The abbé Lecomte (1810-69), the son of rich peasants of Bousbecques (Tourcoing-Nord), was requested to return to Tourcoing by Auguste Cordonnier, *négociant* and mayor, and Philippe Motte, spinner and member of the College's Bureau d'administration, and he took over as principal in December 1838.[33] The College of Tourcoing, situated in a Catholic bastion of French Flanders which would put out Belgian flags rather than *tricolores* in 1843 to greet the archbishop of Cambrai, came into its own under an ecclesiastical principal. Lecomte secured first-class status (with classes up to *baccalauréat* level) for the college and by 1844 an inspector placed it next after Lille in the pecking order of municipal colleges, pointing out that it captured numerous youths who would otherwise continue to Brugelette.[34] In 1845, Lecomte brought off the first of

---

[32] Le R.-P. Orhand, *Le R. P. Pillon et les Collèges de Brugelette, Vannes, Sainte-Geneviève, Amiens et Lille* (Lille, 1888), pp. 10-49.

[33] See the fundamental biographics by J. Lemire, *L'Abbé Dehaene et la Flandre* (Lille, 1891) and H.-J. Leblanc, *Histoire du Collège de Tourcoing* (Tourcoing, 1870).

[34] A. D. Nord 2T 405, report of Insp. Acad. Le Bailly, 9 June 1844.

his colonizing enterprises, the foundation of Notre-Dame des Victoires at Roubaix, headed by one of his own staff. One important measure of the success of priest-principals like Lecomte was the decision by the academic authorities to raise the municipal college of Lille to the status of *collège royal*. Only in this way could it meet the clerical challenge effectively. This steeling of the University also involved moves towards a monolithic *corps universitaire*. The Inspector of the Academy of Douai recognized in 1841 that the appointment of Édouard Gachet as principal of the College of Lille had been 'a mistake', not only because of his past as a thread-maker but because of his administrative in-effectualness and his close relations with the Catholic and legitimist élite at Lille.[35] In August 1842 Villemain dis-missed him and replaced him by the principal of the College of Cambrai. Gachet, offered the post of director of the École normale at Douai, turned it down, went back to being a *maître de pension*, and threw himself into the Catholic camp, campaigning for the renewal of the municipal subsidy to the Frères des Écoles Chrétiennes.[36]

The initiative in the foundation of private institutions on French soil, as in the case of Brugelette, came above all from the powerful Catholic laity, this time of Lille, led as usual by Kolb-Bernard, Henri Bernard, and Édouard Lefort. In the interests of the textile *patronat* of Lille, Roubaix, Tourcoing, and Armentières, who found no solace in the municipal college of Lille, they purchased the château of Marcq, in the countryside outside the town, from the Comte de Muyssart, who had been mayor of Lille between 1816 and 1830, and established the Institution libre of Marcq-en-Barœul in 1840. It was entrusted to the Société Saint-Bertin, an association of secular clergy who took no vows and had no noviciate, but which also owned ecclesiastical institu-tions at Saint-Omer and Dohem in the diocese of Arras, and César Wicart, a master at Saint-Omer who had the necessary qualifications was put up as *directeur légal*. Negotiations to acquire *plein exercice* were opened but successive ministers

---

[35] A. D. Nord 2T 404, report of Insp. Acad. Vincent, 9 May 1841.
[36] Édouard Gachet, *Œuvres diverses* (ed. H. Lefebvre, Lille, 1946), pp. 323-7.

stonewalled the project, and the appropriate decree was not obtained until 6 February 1848 after the intervention at Paris of Mgr Giraud (1791-1858), archbishop of Cambrai since 1842.[37]

It was of course only after the 1848 Revolution that Catholic education really asserted itself. In the spring of 1848, it seemed that Catholicism would flower in the sunlight of liberty. The abbé Dehaene in Flanders amassed considerable popular support when he stood for the Constituent Assembly in the name of religious and political liberties and for an 'improvement of the condition of working men'.[38] But the marriage between Catholic and republican was uneasy, even in the early days. The *commissaires du governement* appointed by the new Minister of the Interior, Ledru-Rollin, had little patience with clericalism. The barrister, Désiré Pilette, deputy *commissaire* to the republican journalist and former conspirator Charles Delescluze in the Nord, purged the Comité local of Lille and dissolved that of neighbouring La Madeleine, because of their opposition to lay schools.[39] At the École normale of Douai, the republican director, Casimir Giroud, made the civic education of the pupils one of his main tasks.[40] The tide was nevertheless moving swiftly to the right. In the Constituent Assembly elections, the radical list which included Delescluze, Pilette, the doctor Achille Testelin, and the journalist Adolphe Bianchi was well beaten. That December, Ledru-Rollin secured only 7 per cent of the popular vote in the presidential elections, though the proportion rose to 28 per cent at Lille. In May 1849 the republican bloc secured only six seats

[37] See L. Marchant, *L'Institution libre de Marcq-en-Barœul* (Lille, 1948) and F. Capelle, *Vie du Cardinal P. Giraud* (Lille, 1852). On the Société Saint-Bertin, see Y.-M. Hilaire, *Une chrétienne au XIXe siècle? La vie religieuse des populations du diocèse d'Arras, 1840-1914* (Lille, 1977), i, 362-7.

[38] J. Lemire, op. cit., pp. 196-205. On this question, see also A. Cobban, 'The influence of the clergy and the *instituteurs primaires* on the elections of the French Constituent Assembly, April 1848', *E.H.R.*, lvii, 1942, 334-44.

[39] A. D. Nord 1T 99/11, *arrêtés* of *commissaire adjoint*, 25 Mar. and 3 April 1848. See also, in general, A.-M. Gossez, *Le Département du Nord sous la Deuxième République, 1848-1852* (Lille, 1904); M. Dessal, *Un révolutionnaire jacobin: Charles Delescluze* (Paris, 1952), pp. 51-68.

[40] A. D. Nord 2T 1043, director of École normale of Douai to Commission de surveillance, 15 Sept. 1848.

against eighteen that went to Bonapartists, Orleanists, and to Catholic monarchists like Kolb-Bernard and Behagel.[41] The repression of republican intellectuals and, worse, *demi-savants*, came swiftly. Giroud was relieved of his post at the École normale and under the law of 11 January 1850 fifteen *instituteurs* were dismissed and eleven transferred, the *arrondissements* of Lille, Douai, and Cambrai suffering most.[42]

The reverse of the medal was the triumph of *La Liberté de l'enseignement*. In the Nord, this took place under the auspices of the new archbishop of Cambrai, René-François Regnier (1794-1881), an Angevin who had been *proviseur* of the College of Angers in the clerical heyday of the 1820s, then bishop of Angoulême.[43] The *émigré* College of Brugelette in Belgium, a product of the University monopoly, could not survive in the new atmosphere of liberty. By the time it closed its doors in 1854, Achille Guidée had returned to the revived Saint-Acheul and Père Pillon to open a Jesuit College at Vannes.[44] But while the Jesuits had managed to open a residence in Lille under the protection of Mgr Giraud in 1843, Regnier was adamant that the diocese of Cambrai should remain as free of Jesuits as it had been under the Gallican Mgr Belmas: they were allowed neither *internat* nor *externat*.[45] By contrast, Mgr Regnier himself set about founding ecclesiastical colleges, notably Saint-Joseph at Lille, Saint-Charles at Cambrai, and, rivalling the Lycée, Saint-Jean at Douai.

As in the 1840s, Flanders was a hive of clerical activity. Having appropriated two colleges of the University, the two priest-principals Lecomte and Dehaene now exploited their position of strength to establish a string of ecclesiastical colleges outside the University, a crusade that would not endear them to the Academy. From Hazebrouck, the abbé

---

[41] Gossez, op. cit., pp. 141-2, 328-32, 338.

[42] A. D. Nord 1T 84/4, prefect of Nord to Min. Instr. Pub., 15 April 1850.

[43] See C.-J. Destombes, *Vie de Son Eminence le Cardinal Regnier, archevêque de Cambrai* (2 vols., Lille–Paris, 1885) and P.-M. de la Gorce, 'Un grand évêque et un grand diocèse, le Cardinal Regnier, 1850-1881', *Revue des questions historiques*, no. 119 (Mar. 1934), 548-59.

[44] P. Delattre, op. cit., pp. 978-80.

[45] Destombes, i, 311-13; H. Beylard, art. cit., p. 76.

Dehaene set up the Institution Notre-Dame des Dunes at Dunkerque, challenging the municipal college, placing the only member of his staff with a *baccalauréat* and five years' teaching experience, the abbé Delelis, at the head of it, and opening it in December 1850. In November 1857, he moved Delelis to Gravelines for the purpose of opening the Pensionnat Saint-Joseph there, seeking to undermine the lay Pensionnat Sainte-Barbe. From Tourcoing, the abbé Lecomte had been active well before the Falloux law, founding the Institution Notre-Dame des Victoires at Roubaix in 1845. Now he ventured into the deChristianized 'bad lands' of the south of the department, delegating priest-*professeurs* on the staff of the College of Tourcoing to head ecclesiastical institutions that he founded at Solesmes in 1849, Valenciennes in 1850, Notre-Dame des Anges at Saint-Amand in 1851, and Bavai in 1852.[46] But not only was Lecomte, the principal of a University college, founding private ecclesiastical schools to undermine the University, he also tried to bully the University into appointing a full contingent of clergy to staff the municipal college at Tourcoing. The contradiction could not endure for ever, and in January 1856 Lecomte was squeezed out of Tourcoing, retreating to the Alps to become a Chartreux.

There was always a risk that such a rash of foundations would place an immense strain on available resources, whether of money or staff. In 1853 Mgr Regnier presided over the foundation of an association of diocesan priests, like that of Saint-Bertin which ran Marcq and Bergues, this time under the patronage of Saint-Charles. Designed to stabilize personnel, collect funds, own school property, and negotiate from strength with municipalities for subsidies, it came to include the staff of all ecclesiastical colleges and institutions in the diocese outside the Société de Saint-Bertin, except Hazebrouck.[47] Even then, the network of colleges was difficult to run. In 1858, the Institution of Roubaix was saved from disaster only by a parents' association, the Société du Collège de Roubaix.[48] In the following year,

[46] Leblanc, pp. 259–73.   [47] Ibid., pp. 357–62; Destombes, i, 366–7.
[48] T. Leuridan, *Histoire de l'Institution Notre-Dame des Victoires de Roubaix* (Roubaix, 1891), pp. 67–8.

Mgr Regnier considered transforming the vegetating ecclesiastical institutions of Solesmes, Bavai, and Saint-Amand into
petits séminaires, partly to make possible diocesan direction
and the injection of diocesan funds, partly to increase the
intake into what would be 'mixed' institutions, half seminaries
and half day-schools. But this recourse to the fourth
ecclesiastical strategem, combining the prestige of seminaries
with the scope of day-schools charging minimum fees, in the
manner of the Restoration, was fiercely opposed by the
Academy, which now feared for the survival of the little
colleges in the south of the department—Maubeuge,
Avesnes, Le Quesnoy, Saint-Amand, Le Cateau, and even
Cambrai.[49]

The Falloux law also gave a great boost to Catholic influence in elementary education. On the *délégations cantonales* of the Nord in 1851, the representation of 'enlightened'
interests among the liberal professions was cut to 11 per cent,
with merchants and industrialists standing at 8 per cent,
while clergy, landowners, and peasants made up 48 per cent
of the membership.[50] Moreover, the sub-prefect of Hazebrouck protested that his suggestions had been ignored and
that the list for the *arrondissements* of Dunkerque and Hazebrouck had been drawn up 'with the exception of two or
three names, entirely in a clerico-legitimist interest', including
clergy who were fanatical defenders of teaching in Flemish.[51]

Liberty to open schools was of course subject to official
scrutiny, and anticlerical candidates had to be put firmly
aside, especially in the south of the department. The school-
inspector of Cambrai warned the Rector against the can-
didacy of one Antoine Châtelain, who 'has no religious
principles . . . never attends any church services and has
often, in the cabarets of Neuvilly, railed against religion and
its ministers'.[52] Such anticlericalism was especially worrying
in the 1850s, because Protestant propagandists were seeking
to fan the ashes of Calvinism that existed in pockets around

---

[49] A.D. Nord 2T 1094[1], Insp. Acad. to Rector of Douai, June 1859.

[50] A.D. Nord 1T 66/10, *Délégués cantonaux*, 1851.

[51] A.D. Nord 1T 110/2, sub-prefect of Hazebrouck to prefect of Nord,
29 Dec. 1850.

[52] A.D. Nord 1T 124/3, primary-school-inspector of Cambrai to Insp. Acad.
Douai, 4 Oct. 1861.

Saint-Amand and in the cantons of Clary and Le Cateau, and were said to be making progress especially 'in those communes where the population is in conflict with the *curé* or in those stirred most strongly by the spirit of '48 '.[53] At Vieux-Condé, the *curé* appealed to the superior-general of the Marists to send brothers to found a school, in order to save the souls of 'a large number of Catholic children . . . exposed to the danger of becoming heretics'.[54]

On the other hand, freedom of private education did not prevent the teaching congregations from directing their energies towards the conquest of *écoles communales*. In Flemish communes such as Caestre, *curé*, landowners, and shopocracy mobilized their forces and browbeat the subprefect into appointing a Marist to the public school.[55] In Flanders, where there were clearly not enough Marists to go round, the parish clergy satisfied themselves by demanding teachers who were in the mould of the pre-revolutionary *maîtres d'école*. The *curé-doyen* of Bailleul north-east, who was also *délégué-cantonal*, advised the prefect to send a teacher who could sing plainchant, play the organ, and speak Flemish, otherwise the post of lay clerk at Saint-Jean Cappel, with its income of 800 francs a year, would not be his.[56] It may even be argued that in the 1850s the Flemish clergy managed to conquer the École normale of Douai. The reactionary director of the school, Juge (1850-5), barred Voltaire from the school library, attached a chaplain to the school, and promoted the study of plainchant and organ-playing, partly because 'its touching sounds predispose so nicely to religious feelings and to prayer', but more specifically to train ideal *clercs-instituteurs* for the parish clergy of Flanders.[57] His successor in 1855, a priest,

---

[53] A.N. F17 2649, Rector of Douai to Min. Instr. Pub., 12 Nov. 1858.

[54] A.F.M. BEA 660, Vieux-Condé, *curé* to superior-general, 25 Oct. 1852.

[55] A.D. Nord 17 89/19, mayor of Caestre to prefect of Nord, 26 Feb. 1860; B. Ménager, *La Laïcisation des écoles communales du Nord 1879-1899* (Lille, 1971), p. 14.

[56] A.D. Nord 1T 110/4, abbé Bacquert to prefect, 17 June 1857. On *clercs laïques*, see also Y.-M. Hilaire, *Une chrétienté au XIXe siècle?*, i, 353-8.

[57] A.D. Nord 1T 69/7, Juge, director of École normale of Douai to Commission de surveillance, 15 July 1855; R. Hemeryk, 'La Congréganisation des Écoles normales du département du Nord au milieu du XIXe siècle, 1845-1883', *Revue du Nord*, lvi, no. 220, 1974, 17-19.

the abbé Loizellier, achieved the *coup* of securing the confidence of the clergy of the diocese in the École normale. They sent him their protégés, together with letters congratulating him for fortifying *normaliens* with a morality which would save them from the temptations of the drinking-houses and *ducasses* of the Nord.[58]

The Nord had certainly proved by 1860 that it was a difficult nut for the forces of secularism to crack. Deeply Catholic, even in industrial areas, its Catholicism was reinforced by a separatism that reflected its previous incarnation in the Spanish Netherlands and by the Flemish language that was still spoken north of the Lys. The north of the department in fact appeared as a veritable bastion of Catholicism, prepared not only for defence but also for offensive forays.

The French Revolution swept little away in the Nord and the Napoleonic regime found itself building on a network of Ancien Régime colleges, such as that of Tourcoing in French Flanders, to which the Recollets returned between 1802 and 1830. When the University was set up in 1808, the Academy of Douai was placed in the hands of a non-juror who defended state secondary education, sufficiently Catholicized as it was, against a 'constitutional' bishop who might have looked far more plausible as a supporter of the regime. The attempt to laicize the University in 1830 was a sorry manœuvre. The Catholics hammered away at it from Belgium, now liberated from the Orange yoke of the Netherlands, from the great municipal colleges of Flanders dominated by the secular clergy, and from the stylish Marcq-en-Barœul, against which the College of Lille stood little chance. Once the monopoly had been broken in 1850, Catholic foundations sprang up along the coast and in the less religious lands of the south, colonized from the episcopal capital of Cambrai but most effectively from Flanders.

The Catholic grip on elementary education was even tighter. Mutual education succeeded in only a few centres in the south, and provoked a counter-crusade by the religious

---

[58] A.D. Nord 1T 69/7, letter of Ploquart, *curé* of Frasnoy, transcribed in report of abbé Loizellier, director of École normale of Douai, 31 July 1858; Hemeryck, art. cit., p. 20.

congregations. The municipalities cut their subsidies to the *frères* in 1830 to no effect; they returned with renewed strength, funded by a Catholic laity. The public system of primary schools was gradually taken over by the Catholic Church, especially in Flanders, but also in the south, where anticlericalism and traces of Protestantism had to be stamped out. This was achieved both with the help of religious congregations and with that of lay teachers trained in 'clerical' *écoles normales*, for *institutrices* were educated by the Bernardine order after 1845 while the *instituteurs* of the 1850s graduated from a training college that scarcely differed from a seminary. The task of declericalizing public education after 1860 would certainly not be easy.

(b) The Gard: Protestant and Catholic

Abutted against the chain of the Cévennes, with one toe touching the Mediterranean, the Gard was unequivocally a southern department, but in no way typical of the 'Midi rouge'. Its peculiarity was explained by the strong Protestant presence, both in the mountains and in the rich Vaunage to the south-west of Nîmes. The Catholics reigned supreme in the plain bordering the Rhône, but religious conflict was rife in those areas where the different populations overlapped: in the foothills of the Cévennes, from Le Vigan to Sommières, and in that part of the Cévennes bordering the Vivarais, later the Ardèche, from Génolhac to Saint-Ambroix. Religious antagonisms reinforced political conflicts: in 1789 the Protestants adopted the principles of the Revolution that emancipated them, while the Catholics fought for monarchy and the Ancien Régime, although the Catholic camp also clearly included revolutionaries opposed to aristocracy and the hierarchy of the Church. The early nineteenth century witnessed rapid industrialization in the coal-basin of Alès, 'capital of the Cévennes', but, as in French Flanders, industrialization did not bring deChristianization automatically in its wake: the miners were usually peasants who had descended from the Massif Central and the ascendancy of Catholic *patronat* was difficult to contest. Anticlerical feeling was stronger in urban centres like Nîmes or, as the century advanced, among the wine-growers of the plain.

In education, then, the conflict between clerical and anti-clerical views was complicated by rivalry between Catholics and Protestants. Even before the Revolution, it was clear that the Catholic Church was developing its elementary-school system above all in those areas where the *nouveaux convertis*, as Protestants were officially called, were numerous: in the diocese of Alès, specifically set up as a missionary station, in the west of that of Uzès (Saint-Ambroix, Lussan, Saint-Chaptes), and around Sommières and Saint-Hippolyte, in the diocese of Nîmes. Where the Catholics were more secure, fewer schools were founded.[59] But the tension within the Catholic Church was just as significant. When in 1762 the Jesuits were expelled from the College of Nîmes, their place was taken by the Doctrinaires, a congregation influenced by Jansenism, who were already teaching at the busy river-port of Beaucaire. The advanced views of the Doctrinaires made them willing to take the oath to the Civil Constitution of the Clergy and, amenable to the revolutionary municipality, they remained in charge of the College of Nîmes until their principal abjured the priesthood in March 1794.[60] Former Doctrinaires continued to be found in prominent positions in the schools. The professor of ancient languages at the École Centrale of Nîmes was an ex-Doctrinaire, Joseph Guérin, who set up a *pension* when it closed. Another ex-Doctrinaire, Jean Roman, a Provençal like Guérin, opened his *pensionnat* in 1798, had it recognized as an *école second-aire particulière* in 1803, and ran it until 1825.[61]

As in the Nord, continuity with the past was very apparent under Bonaparte's reorganization of the secondary-school system in 1802. At Bagnols, a little town embedded deep in the Catholic plain, where a college founded in 1784 by the seigneur, the Prince de Conti, had been entrusted to the Missionnaires de Saint-Joseph of Lyon, another Josephist,

---

[59] M. Laget, 'Petites écoles en Languedoc au XVIIIe siècle', *Annales E.S.C.*, 26 (1971), 1416-17.

[60] E. Goiffon, *L'Instruction publique à Nîmes: le Collège des Arts, les Jésuites, les Doctrinaires* (Nîmes, 1876), pp. 122-3.

[61] A.D. Gard 2T 83, *Institutions et pensions, dossiers personnels, N-Z.*

Dumas, took over as director in 1802, along with three other former members of the order.[62] At Alès, the college had been transformed in 1786 into an École de la Marine, under the direction of secular clergy, since the Marquis of Castries, Minister of the Marine, was Comte d'Alès and coseigneur of the town. The college was resurrected and between 1804 and 1807 the principal was the abbé Fleury, who had been a master there before the Revolution. But the revolutionary tradition soon made itself felt. In 1809 direction of the college passed to Louis Portier, an ex-Doctrinaire, now married to a Protestant, who had been *maître de pension* from 1794 at Anduze and Saint-Hippolyte. He brought a team of Protestant regents like François Mourgues with him and encouraged the recruitment of Protestant pupils. There was no lack of conflict between this 'new guard' and the old guard of regents who had taught in the Ancien Régime college, the latter led by the irascible abbé Albert and determined to fight the Camisard menace.[63] A similar clash of traditions took place at the Lycée of Nîmes. While the *proviseur* was ex-religious and the climate one of restored Catholicism, seven of the eight masters appointed in May 1804 came from *écoles centrales* (five from that of Nîmes) and one from Louis-le-Grand,[64] and the pupils had been drawn from the Prytanée Militaire, the Écoles Centrales of the Gard and the Ardèche, and sons of army officers, officials, *ingénieurs*, and notables with a revolutionary-patriotic record. 'Weaned on revolutionary principles' but fearing that they would be turned into capucins, the *lycéens* rioted in April 1805, smashed windows and chanted the Carmagnole.[65] As a result, the direction of the Lycée was entrusted to Tédenat, a tough former mathematics master at the École Centrale of Rodez, who had links with d'Alembert's pupils.

When the University was set up, the man who was chosen to be Rector of Academy of Nîmes was Tédenat, a man with

[62] A.D. Gard 2T 14, municipal council of Bagnols, 9 July 1802.

[63] A.D. Gard 2T 69, report of Inspector Felix, 1809; A.N. F17 8212, Albert to Grand-Maître, 8 Dec. 1814.

[64] A.D. Gard 2T 4, imperial *arrêté* of 31 May 1804.

[65] A.D. Gard 2T 4, Reydelles, *censeur* of Lycée, to prefect, 29 April 1805; G. Maurin, 'Instruction publique sous le Premier Empire', *Revue du Midi*, 39, 1906, 345.

a revolutionary past, unlike Taranget at Douai. Moreover, in contrast to the Academy of Douai, the Catholic threat was at present minimal, partly because the Gard was not fertile soil for the recruitment of priests, more specifically because as yet there was no bishop of Nîmes. Accordingly, the return of Napoleon in 1815 was greeted with enthusiasm. Rector Tédenat reported that 'the tricolour standard flies from sunrise to sunset on the Lycée buildings and all the masters and pupils are sporting national cockades'.[66]

But the days of revolutionary ascendancy were numbered. After Waterloo, violence reached new heights in the Midi as royalist volunteers organized into bands of *miquelets* took their revenge on Bonapartist Protestant notables and officials. This White Terror was connived at and even encouraged by th 'occult government' of the Duc d'Angoulême in Toulouse, ruling a 'southern kingdom' dominated by royalist ultras, more or less independent of Paris. A massacre of Protestants in Nîmes on 1–2 August 1815 was followed by a new wave of violence on 19–20 August, leading up to the elections to the 'Chambre Introuvable'. The Catholic royalists wrought vengeance for 1790, but the Protestants were able to retaliate, organizing for self-defence in the Vaunage and Gardonnenque. In January 1816, a Catholic schoolmaster of the college of Alès, Perrin, was found murdered and the Protestant regent of mathematics and former *fédéré*, Mourgues, was arrested. He was subsequently released for lack of evidence, but as Gwynne Lewis has said, 'the death of Perrin fanned the flames of religious hatred in the Gard to white heat'.[67]

Tension between the Catholic royalist atmosphere of the *collèges royaux* and the continuing revolutionary sentiments of many of their inmates was also clear. In 1819 pupils of the Collège royal of Nîmes clashed with pupils from the *pension* of the ex-Doctrinaire, Guérin. Two years later, the College was in the grip of a mutiny against its clerical staff, and in 1823 there was trouble when one pupil was discovered in possession of 'republican writings preaching regicide',

[66] A.D. Gard 2T 8, Rector Tédenat to prefect, 13 May 1815.
[67] G. Lewis, 'The White Terror in the department of the Gard, 1789–1820' (Oxford D.Phil. thesis, 1966), p. 261 and *The Second Vendée*, chaps. 5, 6.

including Voltaire's *Contes*, Rousseau's *Émile* and *Contrat Social*, and Antoine Court's *Histoire des Camisards*.[68] But matters were being taken firmly in hand by the bishop appointed in 1821 to the diocese of Nîmes, newly constituted and incorporating the small Ancien Régime sees of Alès and Uzès. Mgr Petit-Benoît de Chaffoy (1752-1837), the son of a *conseiller* of the Parlement of Besançon and educated at the College of Pontarlier in the most Catholic part of the Franche-Comté, was the first in a series of devout bishops that Nîmes would receive from that province and one who, having suffered the rule of the 'constitutional' bishop Le Coz at Besançon, did not tolerate schismatics lightly.[69] Chaffoy's first concern was the recruitment of priests, and in 1822 he came to an agreement with the town of Beaucaire for the establishment of a *petit séminaire*: the municipality purchased a château and gave the bishop usufruct of it, on condition that he admit day-boys from the town as well as aspirants to holy orders.[70] But in this period of the Restoration there was a steady trickle into the priesthood from the colleges of the University,[71] if not from the Collège of Uzès, deserted by its staff of secular priests in 1825, then from Bagnols and Alès, the latter flourishing since 1813 under the abbé François Reynaud, nicknamed the 'Rollin des Collèges communaux'.[72]

It was during the Restoration that conflict over elementary education first became acute. Abandoned until then to private initiative, the task of *instituteur* provided a bare or supplementary income for a marginal population thrown up by revolutionary turbulence, war, and economic depression. A report of 1809 on the previous (or continuing) occupations of *instituteurs* in the three *arrondissements* of Alès, Uzès, and Le Vigan bears witness to the way in which poor peasants

---

[68] A.D. Gard 2T 61, Rector of Nîmes to abbé Raynal, *proviseur* of Nîmes, 16 Jan. 1819; Rector to prefect of Gard, 9 Nov. 1821; M. Bruyère, 'Le Collège royal de Nîmes sous la Restauration', *Mémoires de l'Académie de Nîmes*, li, 1936-8, 28-9.

[69] See E.-A. Couderc de Latour-Lisside, *Vie de Mgr de Chaffoy, ancien évêque de Nîmes* (2 vols., Nîmes, 1856-7).

[70] A.D. Gard V 89, municipal council of Beaucaire, 29-30 Mar. and 16 July 1822.

[71] A.D. Gard 10T 30, Rector Felix to Conseil académique, 13 June 1820.

[72] M. Bruyère, *Alès, capitale des Cévennes* (Nîmes, 1948), p. 651.

under-employed artisans, ex-priests, and seigneurial officials, disbanded soldiers, and even former convicts drifted into teaching.

Fig. 1 Previous Occupations of *Instituteurs* in the Gard, 1809[73]

bourgeois	0.4
officiers de santé, pharmacists, druggists, barber-surgeons	2.2
ecclesiastics: parish priests, cantors, lay brothers	7.1
instituteurs or *écrivains*	12.4
small officials, *greffiers*, clerks	14.2
tradesmen, shopkeepers	3.6
textile workers	19.1
other artisans	13.3
cultivateurs	12.9
soldiers, sailors	12.0
wage-labourers, servants	2.7

Under the ordinance of 29 February 1816 primary education was to be set on a new footing, but now Protestants dazed after the fall of the Empire and Catholics resurgent after the White Terror struggled to obtain control of the schools in order to reinforce their own position. In mixed Catholic and Protestant cantons, two *comités cantonaux* were supposed to be set up, one for each religion, but the *curés* were in a position to insist that the *instituteurs* appointed be Catholic. On the other hand, mutual education was propagated in the Gard essentially as a Protestant enterprise. The missionary of the S.I.E. was Pastor Martin of the British and Foreign School Society, who had been made director of the mutual École normale in Paris in June 1815 and, sent to the Midi in 1817, founded a model school at Saint-Hippolyte-du-Fort, in the foothills of the Cévennes, which was to train *instituteurs* for the Gard and the Hérault.[74] The prefect appointed to the Gard in February 1817, the Baron d'Argout, representative of the 'ministerial' tendency in royalism, wanted to promote mutual education as a way

[73] A.D. Gard 1T4, statistics provided by prefect of Gard for Grand Maître Fontanes, 1809. The sample is of 225.
[74] A.D. Gard 1T 8, mayor of Saint-Hippolyte to sub-prefect of Le Vigan, 27 May 1817; A.N. F17 11772, prefect of Gard to Min. Int., 3 Nov. 1817; Tronchot, ii, 566–71.

of mixing the children of both religions, in order to exorcise the horrors of the White Terror.[75] But the legacy of conflict made this impossible. The Catholic population, led by the Marquis of Vallongue, ultra-royalist mayor of Nîmes, entered into negotiations for the return of the Frères des Écoles Chrétiennes.[76] The Protestants, realizing that they would have no place in Frères' schools maintained by Catholic municipalities, seized on mutual education as an efficient and neutral system that could be adopted for their purposes. Most mutual schools in the Gard, organized by Protestant consistories and notables, were situated in the Protestant cantons of the Vaunage and the Cévennes. Against them were ranged, in addition to the lay Catholic *instituteurs*, the Frères des Écoles Chrétiennes brought in by the Catholic clergy and municipalities of Beaucaire and Uzès, Nîmes, and Alès, the patron of the Frères in Alès being the abbé Taisson, director of the École de la Marine before the Revolution and now *curé* of the cathedral.[77]

It was in the 1820s, of course, that the influence of Catholic interests reached their zenith. The Petit Séminaire of Beaucaire swelled with its contingent of day-boys while at the Collège royal of Nîmes the proportion of Protestant scholars, deprived of a chaplain of their own religion, fell. The bishop of Nîmes, charged with the surveillance of religious education in the colleges under the ordinance of 27 February 1821, acquired supreme powers to supervise primary education under that of 8 April 1824. The *comités cantonaux* were disbanded and representations by Protestant consistories in matters of education were ignored. Catholic *instituteurs* were favoured in mixed communes, unauthorized private Protestant schools were closed down, and Protestant teachers who did continue their functions were frequently harassed in the same way as 'immoral'

---

[75] A.D. Gard 1T8, prefect to Min. Int., 16 May 1818; Tronchot, ii, 586. On d'Argout, see Lewis, 'White Terror', pp. 317–70 and B. Fitzpatrick, 'Catholic Royalists in the department of the Gard, 1814–51' (Warwick Ph.D. thesis, 1977), pp. 107–40.

[76] A.P.E.C. NC 716/1, mayor of Nîmes to Gerbaud, director of Frères des Écoles Chrétiennes, 27 June 1816.

[77] M. Bruyère, *Alès*, p. 649. Copy of act between abbé Taissan, ceding property for school, and municipality of Alès in A.D. Gard 1T 513.

teachers against whom witch-hunts were ruthlessly organized.

The July Revolution of 1830 entirely reversed this tendency. The legitimist nobility was overthrown, despite an attempt led by the Comte de Narbonne to exploit the landing of the Duchesse de Berry at Marseille to launch an insurrection,[78] and the Protestant bourgeoisie was returned to power. The July Monarchy, with its balance of liberalism and *censitaire* élitism, was to become their 'golden age'.[79] At Nîmes the Protestant mayor, Girard, held office from 1832 to 1848 and half the members of the Protestant consistory sat on the municipal council.[80] Religious divisions were the driving-force behind political dissension: just as Protestantism found its political expression in Orleanism, so it was Catholicism which fuelled legitimist opposition.[81]

In the Gard it was Protestantism which, to a large extent, fuelled anticlerical sentiment. But since this current had already been strong before 1830, the laicization of the University was undertaken with a minimum of fuss: the College of Uzès had been in lay hands since 1825, the priests left Bagnols in 1832, and the principal of the College of Alès, abbé Reynaud, died of cholera in 1835, bringing the ecclesiastical line to an end. Deeper into the Cévennes, at Le Vigan, traditionally a point of conflict between Catholic and Protestant, the municipal council decided in 1832 to elevate the *pensionnat* of Jean-Louis Ferrier into a municipal college, endowed with public funds.[82] Moreover, because the Protestants had safeguarded the position of their own religion through the mutual schools and could not be envious of the Catholic clientele of the Frères, they had no reason to try to undermine the Frères des Écoles Chrétiennes by cutting off municipal subsidies as in the Nord. Protestant municipalities in Nîmes, Uzès, and Alès therefore continued to subsidize the Frères.[83]

---

[78] B. Fitzpatrick, op. cit., p. 198.   [79] Tudesq, 'Grands Notables', p. 124.

[80] J.-D. Roque, 'Positions et tendances des Protestants nîmois au XIXe siècle', in *Droite et gauche de 1789 à nos jours* (Montpellier, 1975), pp. 207–10.

[81] A.-J. Tudesq, 'L'Opposition légitimiste en Languedoc en 1840', *Annales du Midi*, lxviii, 1956, 394 and *Grands Notables*, p. 151.

[82] A.D. Gard 2T 16, municipal council of Le Vigan, 20 Sept. 1832.

[83] Tronchot, iii, 348–50.

The law of 28 June 1833, though engineered by the Protestant Guizot, was not designed to build up the strength of Protestants in the schools. However, the Protestants who now dominated public life in the Gard were able to exploit the legislation to their best advantage. Whereas under the system of cantonal committees the Protestants had been relegated to an appendage of the official network and in fact relied more on their consistories than on the committees to which they were entitled, the *comités d'arrondissement* included notables of both religions and allowed the Protestants a powerful voice within the official scheme. Moreover the school-inspector in the Gard, the energetic former barrister Adophe Valz, could be described in 1842 as the 'all too notorious agent of Protestantism'.[84]

Under the clauses of the Guizot law providing for the establishment of communal schools and the maintenance of *instituteurs*, the Protestants were also favoured. In 1833, 62 per cent of *instituteurs* had no official stipend, and this proportion rose to 80 per cent in the Cévenol *arrondissement* of Le Vigan.[85] But the law gave public authorities power to drag the mountainous, largely Protestant areas within the sphere of enlightenment. Those communes that failed to vote the necessary rates to set up a communal school and were taxed ex officio made up 30 per cent of all communes in the Gard, the figure rising to 70 per cent in the *arrondissement* of Alès. Mixed Protestant and Catholic communes were required under the law to set up one communal school for each religion. But in the vast majority of communes, resources simply did not exist. It was the Protestant community that tended to benefit from this inadequacy. Usually the minority religion had to fall back on private teachers, but Guizot was happy to advance credits to four communes in the Vaunage in 1833 so that Protestant *écoles communales* might be set up.[86] In addition, where Protestants dominated municipal councils they often exercised a sort of despotism,

[84] A.D. Gard 1T 647, 'Visite des écoles primaires par le clergé', 1842.

[85] A.N. F17* 105; D. Bouquet and H. Vittumi, 'Pluralisme religieux et instruction primaire: la loi Guizot et son application dans le Gard, 1833-49' (D.E.S., Montpellier, 1970), pp. 22-3.

[86] Bouquet and Vittumi, p. 35. A.D. Gard 1T 35, Rector of Nîmes to prefect of Gard, 14 Mar. and 23 April, 1835.

having scant regard for the Catholic population, generally poor and defenceless even if numerically in a majority. The installation of a new bishop of Nîmes, François Cart (1838–55) who, like his predecessor Mgr de Chaffoy, was a Franc-Comtois educated at the College of Pontarlier in the Doubs, gave new heart to the Catholics.[87] Between 1838 and 1843 a series of petitions signed by oppressed Catholic communities requested the establishment of second com-munal schools with Catholic teachers.[88] In 1841 Mgr Cart appealed to the superior-general of the Marists on behalf of 'my poor children of Anduze who are threatened with being buried under Protestantism if they remain any longer without education or go to receive it in Protestant schools'.[89] Until then, the Gard had been only scantily provided for by the weak, ill-trained congregation based at Saint-Paul Trois-Châteaux in the Drôme, which had set up schools in the northern Catholic cantons of Barjac and Pont Saint-Esprit.[90] Now this congregation fused with the Marists, and Marists arrived at Anduze, 'the Geneva of the Cévennes', financed largely by the legitimist Comte de Narbonne, and in 1847 at the mining town of Bessèges, under the patronage of the Compagnie houillère et des hauts fourneaux. Aristocratic and industrial interests, faced by the Protestant Ascendancy, rallied to the support of Catholic education.

The third aspect of the Guizot law, the training of teachers, was also speedily exploited by the Protestants. An École normale was founded at Nîmes in 1831, and in 1836 57 per cent of its pupils were Protestants, though these made up only a third of the population of the department.[91] Cantonal conferences flourished in the Protestant Cévennes (Le Vigan, Saint-Hippolyte, Alès) and Vaunage (Uchaud), though they were also tried amongst Catholic *instituteurs* at Saint-Ambroix, Pont-Saint Esprit, and in the Rhône valley at Meynes (Aramon). They did not last long: Adolphe Valz, fearing

---

[87] P. Azaïs, *Vie de Monseigneur Jean-François-Marie Cart, évêque de Nîmes* (Nîmes, 1857).

[88] A.D. Gard 1T 97, 568.

[89] A.F.M. AUB 660, Anduze, Mgr Cart to superior-general of Marists, 17 Sept. 1841.

[90] P. Zind, *Les Nouvelles Congrégations enseignantes en France, 1800–1830* (thesis, Lyon, 1969), p. 366.   [91] Bouquet and Vittumi, pp. 123-4.

their subversive potential, confessed in 1837 that 'having made every effort to found them, I, more than anyone else, have contributed to their destruction'.[92] They were replaced by *cours spéciaux de perfectionnement* at Nîmes for three weeks during the summer and by the circulation of a review, the *Journal des écoles primaires du Gard.* The advent of the teaching congregations nevertheless demanded a revision of this attitude. In 1845 the Commission de surveillance of the École normale of Nîmes asked the prefect whether the course there might be lengthened from two years to three in order to equip *normaliens* to 'sustain the struggle with the Frères de la Doctrine Chrétienne and the order of Marists'.[93] Even more significant, the Cours normal to train *institutrices* to rival the female teaching orders was set up not by the state but by the Protestant consistory of Nîmes. All attempts by Villemain, the Minister of Public Instruction, to transform it into an *école normale* for both religions, partly sponsored and partly administered by the government, were firmly rejected.[94]

The constraints put on Catholic education under the July Monarchy were no less severe in the secondary sphere, and the fight back was more arduous than in a region where piety was reinforced by provincialism, as in Flanders. Municipal colleges were everywhere in the hands of lay principals, and there was no possibility of reversing the situation. The seminary of Beaucaire had been pruned back in 1828 to being a boarding-school of cassocked seminarists, though at Nîmes Mgr Cart established a *maîtrise* or cathedral school in 1839 which, besides providing altar-boys for the cathedral, was also a second *petit séminaire* in disguise.[95] The road to exile had been taken in the past by Protestants rather than Catholics, notably towards Switzerland, but under the July Monarchy the Protestant bourgeoisie colonized the Collège

[92] *Journal des écoles primaires du Gard*, 1837, pp. 206–7.
[93] A.D. Gard 1T 97, Commission de surveillance of École normale, Nîmes, to Min. Instr. Pub., 21 July 1845.
[94] A.D. Gard 1T 564, Rector of Nîmes to Min. Instr. Pub., 26 Mar. and 29 May 1843, 6 Nov. 1844; A.C. Nîmes K34, Min. Instr. Pub. to Rector of Nîmes, 3 Aug. 1842.
[95] L.-C. Delfour, *Le Cinquantenaire de Saint-Stanislas, notice historique* (Nîmes, 1892), pp. 4–5.

royal in increasing numbers, and displaced army officers, magistrates, and officials of the Restored monarchy, together with Catholic barristers and merchants, felt increasingly uneasy there. The establishment that would see to their needs was launched by the abbé Emmanuel d'Alzon (1810-80). D'Alzon, a Catholic Cévenol who still lived *ligueurs* and *dragonnades*, the son of a deputy of the Hérault, educated at the Collège Stanislas in Paris and destined for a military career before he astonished his family by going into the Church, was active in running a *catéchisme de persévérence* for Catholics at the Collège royal of Nîmes before he was appointed vicar-general to Mgr Cart in March 1839.[96] In January 1844 he purchased a *pensionnat* from an aged priest and launched his enterprise. Three qualities distinguished this new foundation. In the first place, while d'Alzon employed priests to ensure a serious Catholic education at the school, he was also aware of the academic weakness of the clergy in general. He therefore sought to reproduce Stanislas, which employed lay *universitaires* as well as priests, at Nîmes, drawing on the talent of Catholic *professeurs* who had become disillusioned with the lay University. Louis Monnier, master at the Collège royal of Nîmes, and Germer-Durand, who had moved from the college to take up a chair at Montpellier, were his prize catches.[97] Secondly, he began to group the ecclesiastical staff at the college and other recruits into a clandestine religious order, the Augustinians of the Assumption, much to the despair of Mgr Cart.[98] Thirdly, by a combination of diplomacy and subterfuge, d'Alzon campaigned for *plein exercice*, refusing to send any but his philosophy class to the Collège royal from the autumn of 1845.[99] The Maison de l'Assomption rapidly became a centre of legitimist opposition in the department with the names of prominent local ultra-royalist families —Boyer, Baragnon, Bernis, Trinquelaque—appearing on its rolls, but also drawing on opponents of the regime further afield, such as the two sons of the former prefect of the Nord, Villeneuve-Bargemon, now in exile at Nice.[100]

[96] S. Vailhé, *Vie du P. Emmanuel d'Alzon, 1810-1880* (Vol. i, Paris, 1926).
[97] Ibid., pp. 335-8.   [98] Ibid., pp. 387-400.   [99] Ibid., pp. 358-65.
[100] A.A.A., Repertoire des Élèves, 1845-52. B.M. Nîmes 30.816, *Association des Anciens Élèves de l'Assomption*, 1871.

The Revolution of February 1848, introducing universal suffrage, enfranchised the mass of Catholics in the Gard and ended the supremacy of the Protestant oligarchy which had been based on the *régime censitaire*. Catholics found liberty in the Republic and during the brief honeymoon period Emmanuel d'Alzon launched a journal entitled *Liberté pour tous* and claimed to be a republican.[101] But the *commissaires* soon revealed the true face of republicanism and Catholics who had wavered now looked for a monarchist restoration. In the Gard, Catholics inhabiting mixed Protestant–Catholic cantons opted *en bloc* for the legitimists in the elections to the Constituent Assembly, and republicans—Protestants and the Catholic wine-growers of the Rhône valley—secured only three of the ten seats.[102] In the elections to municipal councils and Conseil général of the Gard in the summer of 1848, Catholic royalists triumphed everywhere, and at Alès the legitimist-dominated Bureau d'administration of the College dismissed two masters, one for republican activities, one for freethinking.[103] The bourgeois Republic consolidated after the June Days was to the taste of few elements in the Gard. In the presidential elections of 10 December 1848, the republican Ledru-Rollin scored heavily with 14.7 per cent, especially in Protestant areas and the Rhône valley.[104] The spring of 1849 saw a shift to extremes, the Montagnards on one side, the popular legitimists, 'Montagnards blancs' or 'Jacobins enfarinés' who supported the abbé de Genoude, on the other.[105] While the Montagnards split the republican vote and were divided amongst themselves, the death of Genoude enabled the royalists to go to the polls in May 1849 in perfect discipline and conquer all eight seats.[106]

The liquidation of the 1848 Revolution in the Gard

---

[101] S. Vailhé, p. 507.

[102] R. Huard, 'La Préhistoire des Partis: le parti républicain et l'opinion républicaine dans le Gard de 1848 à 1881' (thesis, Paris, 1977), pp. 147-53.

[103] Ibid., p. 226.

[104] R. Huard, *Le Mouvement républicain en Bas-Languedoc, 1848-1881* (Paris 1982), p. 49.

[105] R. Huard, 'Montagne rouge et Montagne blanche en Languedoc-Roussillon sous la Second République' in *Droite et gauche de 1789 à nos jours* (Montpellier, 1975), pp. 139-60.

[106] Huard, 'Parti républicain', pp. 241-62; L. Loubère. 'The Emergence of the extreme left in Lower Languedoc, 1848-51', *Am. H. R.*, no. 73(2), 1968, 1631-9.

brought not only repression of *instituteurs* but also a veritable crisis of confidence about popular instruction, notably
among the conservatives of the Conseil général. De Tarteron,
landowner and *avocat* at Sumène, an outpost of Catholicism
in the Protestant Cévennes, called for the abolition of the
École normale, proclaiming that 'instruction alone, instruction
that is not guided, moderated by a supreme rule, is a scourge
for societies . . . ignorance with its honest rusticity is far
better for the happiness of society, better for civilization
itself, than an instruction that is corrupted and fuelled by
evil passions'.[107] Intellectual training was reduced to a
minimum at the École normale of Nîmes, so that an inspector
could report proudly in 1850 that 'for about the last two
years, no *élève-maître* has been entered for the *brevet
superieur*'.[108]

On the other hand, Catholic education that had been
burrowing beneath the surface of the University monopoly
was now permitted to burst forth. There was no question
of municipal colleges being recolonized by the clergy in the
Gard, but a decree of 16 July 1849 allowed the Petit Séminaire of Beaucaire to take day-boys again and in 1851 the
maîtrise of Mgr Cart became a full-blown secondary school,
the Institution of Saint-Stanislas. Most impressive, however,
was the *plein exercice* gained by the Maison de l'Assomption
in December 1849, four months before the Falloux law, and
the campaign launched by Emmanuel d'Alzon to reduce the
Lycée to a Protestant ghetto. As an inspector-general observed
in 1850, it was d'Alzon's aim to 'batter the Lycée, and to
draw off gradually the whole Catholic population'.[109] The
attack that d'Alzon mounted on the teaching of pagan
classics must be seen in the context of his war of attrition
against the Lycée. This controversy took place, of course,
on a national scale, with the publication of the abbé Gaume's
Ver rongeur des sociétés modernes in 1851.[110] Yet in 1848
d'Alzon had himself denounced the fact that 'at the college,

---

[107] A.D. Gard, Conseil général, 7 Sept. 1848.

[108] A.D. Gard 1T 566, primary-school-inspector to Conseil général of Gard,
1 Aug. 1858.

[109] A.N. F17 7920, report of Inspector-general Braive, 15 July 1850.

[110] P. Harrigan, 'French Catholics and classical education after the Falloux
law', *F.H.S.*, viii, no. 2, 1973, 255-8.

we have been living for three centuries under the spell of paganism'. The Renaissance was an aberration, divorcing men from their Christian past; the study of the classics exalted pagan virtues at the expense of Christian morality and proposed deities like Venus and Bacchus who were no more than 'personified passions'. In the lower classes of the college, therefore, only the Christian fathers would be studied; the study of 'profane authors' would be left to the upper classes.[111] In order to publicize this crusade, d'Alzon launched in November 1851 the *Revue de l'enseignement chrétien*, the first volume of which included articles by d'Alzon and Germer-Durand attacking the 'barbarism' of the pagan classics for sacrificing the spiritual to the sensual, the supernatural to the natural, the true to the beautiful, and having no more literary merit than the Christian classics.[112] Whether it was the polemic against the University or the continuing war against Protestantism that had more effect would be difficult to judge. But the results speak for themselves. In the autumn of 1843, the Lycée of Nîmes had 423 pupils, Assumption, 40; in the autumn of 1853, the figures were respectively 311 and 202.[113]

Catholic education also made advances at the primary level in the 1850s, but again, as far as the Gard was concerned, within certain limits. Its territory was not fertile ground for female congregations, but the southern Ardèche, bordering on the north of the Gard, was deeply Catholic. It was from this Massif that the Gard was colonized: the Sœurs de la Présentation from Bourg Saint-Andéol before 1848, the Sœurs de Saint-Joseph of Les Vans, of Ruoms, and of Vagnas, together with the Sœurs de Saint-François Régis of Aubenas, subsequently.[114] The only significant male teaching congregation, outside the Frères des Écoles

[111] B.M. Nîmes, *Maison de l'Assomption. Rapport de M. l'abbé d'Alzon, directeur, 1849*, 23-4.
[112] *Revue de l'enseignement chrétien*, vol. i, Nov. 1851. See also A.A.A. A23, *Rapport sur la Maison de l'Assomption*, 1852.
[113] A.A.A. DN2, 'Évolution de la population scolaire de la Maison de l'Assomption, 1843-80'; A.D. Gard 2758 and 2753, Statistics of Collège royal and Lycée of Nîmes.
[114] H. Brun, 'Les Ordres religieux du diocèse de Nîmes', *Bulletin du Comité de l'Art Chrétien de Nîmes*, ix, no. 57, 1907, 233-8.

Chrétiennes in the major centres, were the Marists, who found patrons among the nobility and industrialists. The Baronne de Merlet, who had strongly supported the foundation of Catholic schools at Anduze, left money in her will for the foundation of two church schools in the commune of Notre-Dame de la Rouvière, another Catholic outpost against the Protestant-dominated cantons of the Hautes Cévennes.[115] In the coal-basin of Alès, on the other hand, the great patrons were the mining and iron-founding companies, anxious to set up schools in the working-class towns and villages expanding so rapidly in the middle years of the century, in order to moralize and to discipline their labour forces. The Compagnie des Forges of La Grand' Combe founded a school of Frères des Écoles Chrétiennes in 1849, and this was the congregation favoured by the Compagnie des Fonderies et Forges of Alès in 1857 for the Alès suburb of Tamaris. On the other hand, the Compagnie des Forges et Mines of the Loire, which had absorbed the iron-founding part of the Bessèges concern, brought in the Marists in 1854, while the mining company of Bessèges, which survived the take-over, combined with the mining company of Trelys, near Alès, to found a Marist school in 1850 in 'the new-born locality of Rochessadoule'.[116]

Despite its disadvantages, the diocese of Nîmes was not deprived at this crucial period of a fighting bishop. Henri Plantier, a Savoyard and one-time professor of Hebrew at the Theology Faculty of Lyon, was appointed to Nîmes in 1855 largely because of his Gallican loyalties.[117] But like Mgr Cart before him, he soon came under the spell of his fiery vicar-general and latter-day crusader, Emmanuel d'Alzon. By all accounts, the Wars of Religion were not over, as the Protestants celebrated the tricentenary of their Synod of 1559 and Mgr Plantier saw fit to publish a *Lettre aux Protestants du Gard* exposing the contradictions between their 'liberty of thought' and the stake at which Servet was burned by Calvin, and equating their doctrines with anarchy

---

[115] A.F.M. AUB 660, N-D de la Rouvière, Isidore de Christol, executor of will, to superior-general, 27 Oct. 1851.

[116] A.F.M. AUB 660, Rochessadoule-Alès, Calas, manager of mines of Alès to Frère supérieur, 2 Oct. 1860.

[117] J. Clastron, *Vie de Sa Grandeur Monseigneur Plantier, évêque de Nîmes* (2 vols., Nîmes, 1882).

and revolution.[118] A sharp controversy followed, during which the Protestant chaplain of the Lycée of Nîmes, Cazaux, inquired what had happened to the liberal, Gallican Plantier of Lyon, while Frédéric Desmons, pastor of Saint-Géniès-de-Malgoirès (Saint-Chaptes), warned that 'Catholicism would always remain as he had known it; an intolerant, fanatical, dogmatic religion, a permanent threat to liberty, an obstacle to human progress and to intellectual emancipation.'[119] Protestantism was gathering its forces for a full-scale attack on the Catholic Church.

In the Gard, as in the Nord, the struggle over education reflected and reinforced local conflicts. For the Catholic Church the control of education was a means of conversion, or at least a means of saving souls from heresy, and of making more secure the alliance between throne and altar. For the state, when it was in the hands of men faithful to the principles of 1789, Protestantism could be mobilized to reinforce currents of anticlericalism in order to undermine the Catholic Church, and to secure a firm anchorage for the regime.

The struggle was embedded deep in the historical conflict between Catholic and Protestant that had been fought along the Cévennes for centuries. It was not simply a question of memories of Ligueurs and Camisards; Catholics and Protestants were at each others' throats from 1790 until the White Terror of 1815, and even 1830 saw outbreaks of communal violence. Within colleges like Nîmes and Alès, old battles were still being fought out.

It is clear that the forces of the Revolution had the upper hand in the Gard. The University was effectively dominated by the laity and no Catholic response was forthcoming until the Restoration. Then, just as Protestants from outside the department, eventually with British backing, promoted mutual education, so the Catholic response came from outside, notably from Franche-Comté, that Catholic province that provided the diocese's first two bishops after the Revolution, or from the Massif Central, that sent priests to teach

---

[118] H. Plantier, *Instructions, lettres pastorales et mandements* (Paris, 1867), ii, 57–80.
[119] D. Ligou, *Frédéric Desmons et la franc-maçonnerie sous la Troisième République* (Paris, 1966), pp. 47–8.

at Bagnols. The July Days of 1830 promoted the Protestants of the Gard to political office and virtually handed the public system of education over to them. The Collège royal became to all intents and purposes Protestant, the College of Alès had a Reformed majority. Protestant municipalities ensured that the *écoles communales* had Protestant teachers, and Catholic minorities had to fall back on private instruction. Outside the towns there was a shortage of congregations and in the Cévennes Catholics feared for the faith of their children. The École normale was overwhelmingly Protestant and that founded for girls was not public but the private creation of the Protestant Consistory of Nîmes.

It was not until after the Revolution of 1848 that the Catholics began to regain some ground. In Nîmes the College of Assumption took wing to protect Catholic and legitimist interests and inflicted heavy losses on the Lycée. In the countryside female religious congregations from the Ardèche moved into position. The management of the coal and iron companies of the Alès basin, concerned about working-class discontent, did not stint in order to build up a knot of Catholic schools in the mining villages. By 1860 Protestants and other partisans of *laïcité* had a good deal of ground to make up.

(c) Ille-et-Vilaine: Blue and White

Of our three departments, Ille-et-Vilaine, forming a large part of Upper Brittany, was by far the most reactionary. The mentality of its eastern flank, the Vitré–Fougères bloc, had been forged by the Ligueurs and the Chouans. In the west, it was marked less by counter-revolution than by Counter-Reformation: the pattern of its intensely Catholic parishes was the shadow of the old diocese of Saint-Malo. And yet there were pockets of resistance to this vigorous traditionalism, centres from which the beacon of revolution might be carried. These were above all the large towns but also, in the countryside, certain knots of communes to the north-east and south-east of Rennes. The seaboard may be considered an 'intermediate' zone, rallying to the principles of 1789 but also staunch in its defence of the Catholic religion.

In the language of contemporaries, the struggle was between the 'whites' or supporters of the Bourbons and the 'blues', the shock troops of revolution.[120]

Though tensions in Ille-et-Vilaine could be traced back to the Wars of Religion, and the Jesuits had been expelled from the College of Rennes in 1762,[121] it was the revolutionary decade that cast political and religious attitudes in an enduring mould. With the exception of the parishes around Liffré and Retiers, the clergy overwhelmingly refused the oath to the Civil Constitution of the Clergy. The only master at the College of Rennes, then in the hands of secular priests, to take the oath, was the abbé Auguste Germé, rhetoric master, who was subsequently appointed *professeur des belles-lettres* at the École Centrale. Continuity at the École Centrale, even in modern subjects, was important, as at Lille: the mathematics master, Thébault, had given public municipal lectures in mathematics since 1756.[122] But in its conception the École Centrale did mark a break with the past: boycotted by the old landowning and *parlementaire* families, it was the school of the revolutionary administrative élite. When Bonaparte instituted the *lycée* regime in 1802 he was challenged by the prefect of Ille-et-Vilaine, Mounier, formerly a magistrate of Grenoble, organizer of the Estates of Dauphiné at Vizille, and moderate member of the Constituent Assembly. At the solemn opening of the Lycée of Rennes, Mounier spoke as one who regretted the passing of the *écoles centrales* with their ideology of enlightenment, tolerance, and virtue, fearing that the Catholic education of the *lycées* would perpetuate ignorance, intolerance, and empty ceremonial.[123]

---

[120] For a fuller account, see R. Gildea and M. Lagrée, 'The historical geography of the west of France: the case of Ille-et-Vilaine', *E.H.R.*, xciv, 1979, 830–47.

[121] G. Durtelle de Saint-Sauveur, 'Le Collège de Rennes depuis sa fondation jusqu'au départ des Jésuites, 1536–1762', *Bulletin de la Société archéologique d'Ille-et-Vilaine*, xlvi, 1918; P. Ricordel, 'Le Collège de Rennes après le départ des Jésuites et l'École Centrale d'Ille-et-Vilaine, 1762–1803', *Annales de Bretagne*, xliii, 1936, xliv, 1937.

[122] L. Benaerts, *Le Régime consulaire en Bretagne: le département d'Ille-et-Vilaine durant le Consulat, 1799–1804* (Paris, 1914), pp. 172–3).

[123] A. Aulard, ed., 'Un discours de l'ex-Constituant Mounier sur l'instruction publique en l'An XII', *La Révolution française*, 55, July–Dec. 1908, pp. 271–81; Benaerts, p. 318.

The thin blue line of the revolutionary tradition was continued in Ille-et-Vilaine with the appointment in 1808 as Rector of the Academy of Rennes of Auguste Germé. But the tide was flowing the other way following the Concordat, and every attempt was being made to reconstitute the Catholic structure of education of the Ancien Régime. The Eudists, who had run the seminaries of the diocese of Rennes since the end of the seventeenth century, returned to organize a seminary in 1800 under the abbé Blanchard (1755-1830), Eudist superior of the Petit Séminaire of Rennes in 1782, now acting as the clandestine representative of the last bishop of Rennes before the Revolution, Bareau de Girac.[124] In the diocese of Saint-Malo, the original seminary of 1645, entrusted to the Lazarists, had been situated at Saint-Méen, on the border with the Côtes-du-Nord. A municipal college was set up in 1802, but in 1818 direction was given to Grardel, Lazarist superior of the seminary before the Revolution.[125] In the meantime, an *école ecclésiastique* was founded for the diocese at Saint-Malo itself in 1802 by a trio of nonjurors. One of them, Jean-Marie de la Mennais (1780-1860), son of a rich shipowner of the port, had been made subdeacon shortly after the Concordat by Cortois de Pressigny, the much venerated last bishop of Saint-Malo.[126]

Not only were seminaries reconstituted more rapidly in the diocese of Rennes than in those of Cambrai and Nîmes, they also exploited far more powerfully the privileges of exemption from university tax to become Catholic schools, replying to the aspirations of a devout province that posed a threat to the very existence of the official *lycées* and colleges. At Rennes, the seminary of Blanchard had 157 pupils in 1810, as against 218 at the Lycée, and the matter was complicated by political factors. As the prefect said, 'M. Blanchard is known to be little enamoured of our new institutions. There is a noticeable tendency in his teaching

---

[124] J. Dauphin, *Les Eudistes dans le diocèse de Rennes* (Rennes–Paris, 1910), pp. 278-90.

[125] A.D. I-et-V. 1T *Collège et Petit Séminaire de Saint-Méen*, notables of Saint-Méen to prefect, 20 May 1802; Grardel to sub-prefect, 30 May 1821.

[126] E. Herpin, H. Hervot, J. Mathurin, G. Saint Mleux, *Histoire du Collège de Saint-Malo* (Ploërmel, 1902), pp. 34-57; A. Laveille, *Jean-Marie de la Mennais, 1780-1860* (2 vols., Paris, 1903), i, 11, 14, 37.

to stir up old memories.'[127] It was the danger of these Catholic institutions, undermining the University, that the decree of 11 November 1811 was designed to eliminate. The limitation of one seminary to each department obliged that of Saint-Malo to close. The Petit Séminaire of Rennes was to send its pupils during the day to attend the classes of the Lycée and Blanchard, who refused to countenance this savage legislation, was eased out and his seminary entrusted to the bishop, Mgr Enoch, a former Oratorian, superior of the seminary of Grenoble before the Revolution, a *constitutionnel* appointed to Rennes in 1805 by the good offices of the prefect, Mounier.[128]

The aim of seminaries was not only to train priests but also, as the only exception to the University monopoly and supervised by the episcopate, to take as day-boys sons of those traditional families who would settle for nothing but a genuinely Catholic education. For them, the moral and religious education offered by the *lycées* was not enough, and at Rennes their worst fears seemed to be borne out during the Hundred Days when, on three nights in a week, about a hundred pupils of the Lycée attacked the seminary with sticks, stones, iron spikes, and knives.[129] On the other hand, the University could not tolerate the development of an alternative Catholic system of education which, under the guise of training for the priesthood, would draw away its clientele. Once the monarchy had been restored there seemed no reason why the University, not only Catholicized but also clericalized, could not provide an education religious enough even for the most particular of *dévots*. Auguste Germé was dismissed as Rector in 1814 and replaced by an *émigré*, the abbé Le Priol. The Lycée underwent metamorphosis and became the Collège royal of Rennes, its *proviseur* none other than the abbé Blanchard.

And yet conflict did continue. The moderate royalists had decided to maintain the University and restrict seminaries to the training of priests. Under the ordinance of 15 August

---

[127] A.N. F17 10213, prefect of Ille-et-Vilaine to Min. Int. 12 Dec. 1810.

[128] Lagrée, *Mentalités*, pp. 196, 215.

[129] A.N. F17 8823, superior of seminary to Rector of Academy, Rennes, 10, 11, 13 April 1815.

1815, seminarists were to wear ecclesiastical habit after two
years' study and seminaries were not to accept day-boys: the
collèges royaux existed for their purposes. After the elections
of September 1816 had weakened the ultra-royalists, the new
Minister of the Interior, Lainé, was ready to support the
campaign of the prefect of Ille-et-Vilaine and the *procureur
général* to oblige Mgr Enoch to send day-boys away from the
seminary. The constellation of interests was curious: the University was now defended by non-jurors and *émigrés*, the
seminary by an Oratorian and *constitutionnel.* Mgr Enoch
argued that the seminary was too small to take all its pupils
as boarders, and that future priests would be sent to a college
that in no sense could offer the same guarantees: 'we are still
a long way from those times when public education could
serve those studying for holy orders'.[130] The prefect, the
Comte d'Allonville, riposted, urging that the Collège royal
under Blanchard offered an 'education . . . entirely religious
and royal',[131] and here he was supported by the Rector, Le
Priol, who argued that the education of priests in isolation
from that of future soldiers, magistrates, and merchants,
instead of side by side, as in the Ancien Régime, served only
to 'perpetuate the effects of the Revolution'.[132] The *procureur général* took the superior of the Petit Séminaire, the
abbé Desrieux, to court, but the Catholic and royalist magis-
tracy of Rennes acquitted him, and the government had to
resort to administrative measures to oblige the bishop to send
away his day-boys.[133]

In a similar way, small towns on the traditionalist flanks
of the department kept up a campaign to transform their
municipal colleges into seminaries. So long as 'ministerialists'
like Royer-Collard were in charge of the Commission de
l'Instruction publique such changes would be blocked. But
the triumph of the ultra-royalists in the elections of November 1820 and the inauguration of the Villèle–Corbière ministry

---

[130] A.D. I-et-V. 3T 5 and A.N. F17 8823, Bishop Enoch to prefect of Ille-et-
Vilaine, 7 January 1817.

[131] A.N. F17 8823, prefect of Ille-et-Vilaine to bishop Enoch, 14 Jan. 1817.

[132] A.D. I-et-V. 3T 5, Rector Le Priol to prefect, 10 Jan. 1817.

[133] A.N. F17 8823, Le Priol to Royer-Collard, 8 May 1817. On this incident,
see also J. Poirier, 'L'Université provisoire', *Revue d'histoire moderne*, 2, 1927,
3–35.

opened new avenues for reaction. Once the defenders of the University had been toppled in Paris, the Catholic provinces could secure the form of education they desired. The town of Saint-Méen ceded its college to the bishopric in 1823, to become a seminary, and this was entrusted by the incoming bishop, Charles-Louis de Lesquen (1825–41), son of a Breton seigneur and an *émigré* who had fought in Condé's army, to the new congregation set up by Jean-Marie de la Mennais, the Pères de l'Immaculée Conception de Saint-Méen.[134] On the other side of the department, the municipality of Vitré was also able to transform its college into a seminary, which Mgr de Lesquen moved from Rennes. Despite the legal problems involved, the prefect urged that the frustration of this project 'would not fail to provoke a kind of struggle in which the interests of religion and education would be seen to succumb'. The clerical Grand-Maître, Frayssinous, gave his approval.[135] On the coast, where Catholic sentiment was solid but not reactionary, the municipal colleges fell in 1823 under the control of secular priests, that of Saint-Malo becoming part of La Mennais's empire.[136]

In primary education, conflict was likewise between a very thin revolutionary tradition and the massed forces of the clergy and notables. Until the Restoration, as much as and more than elsewhere, the initiative was abandoned to charity and speculation. In the towns, teachers were as often as not ex-convicts, married priests, and poor spinsters or widows; in the countryside, catechism was given by parish clergy on Sundays and feast-days during the summer months, but the prefect estimated in 1817 that 75 per cent of the population was illiterate.[137] The only positive signs were the charity schools founded during the Empire in the large towns by reconstituted congregations of Ursulines and Sisters of Saint-Thomas de Villeneuve and the *tiers ordres*. The relationship between the member of a *tiers ordre* and a Breton *recteur* like Gabriel Deshayes, a native of Beignon in the diocese of

[134] A.P.I.C. 558-A-3, P. Émile Feildel, 'Annales de Saint-Méen' (M.S. 1834), pp. 18–25; Laveille, *La Mennais*, i, 425–6, Lagrée, *Mentalités*, pp. 201–3, 222–4.

[135] A.D. I-et-V. 1T *Collège de Vitré*, prefect of Ille-et-Vilaine to Min. Int., 2 Dec. 1825; note of Frayssinous, 14 Dec. 1825.

[136] J. Haize, *Histoire du Collège de Saint-Servan* (Saint-Servan, 1908), p. 55; E. Herpin *et al.*, op. cit., pp. 104–7.

Saint-Malo, and celebrating a clandestine cult in the region in the 1790s before returning there as a curate in 1803, was at the origin of a new teaching congregation, the Sœurs de Saint-Gildas-des-Bois.[138]

No real progress was made in Brittany until the Restoration, and then it was by reaction against outside influences rather than because of them: the clergy would fight tooth and nail to retain the monopoly of elementary education they had enjoyed in the Ancien Régime. The *comités cantonaux* set up under the ordinance of 29 February 1816 were rarely convened by the clergy who had charge of them, and an enlightened, moderate royalist mayor like that of Saint-Servan was powerless to do anything about it.[139] Mutual education penetrated Ille-et-Vilaine under the auspices of the prefect, the Comte d'Allonville, who obtained a missionary, Lambert, from the S.I.E. in Paris. But mutual schools were confined to Rennes and to the seaports of Saint-Malo and Saint-Servan, and the main result of the intrusion was to stir up a veritable hornet's nest among the clergy, whose reaction was nothing short of fanatical. The *ultras* let fly against the *constitutionnels*, and Mgr Enoch, who had approved the regulations of Lambert's school at Rennes, was obliged to keep his head down. The prefect remarked that the lower clergy were no longer obedient to the bishop, as they had been under Napoleon, and that those who had preached the great mission at Rennes early in 1817 had taught them to 'fronder le gouvernement'.[140] The curate of Saint-Servan, the abbé Sauvage, used every device in his power to inhibit recruitments to the mutual school. He put pressure on mothers in the confessional; he threatened to exclude its pupils from First Communion; he mounted the pulpit and declared that the mutual school was a 'den

---

[137] A.N. F17 10213, prefect of Ille-et-Vilaine to Min. Int., 31 May 1817; R. Sancier, 'L'Enseignement primaire en Bretagne de 1815 à 1850', *Bulletin de la Société d'Histoire et d'Archéologie de Bretagne*, xxxii, 1952, 63, 67.

[138] F. Baudu, *Les Origines de la Congrégation des Sœurs de l'Instruction Chrétienne de Saint-Gildas-des-Bois: la Fondation et les Fondateurs* (Vannes, 1948), pp. 21–39.

[139] A.D. I-et-V. 1T *Comités cantonaux d'instruction primaire*, mayor of Saint-Servan to prefect, 14 Aug. 1817.

[140] A.D. I-et-V. 1T *Enseignement mutuel*, prefect to Min. Int., 15 Nov. 1817.

of impiety, a diabolical work, and that on the first day the Prince of Darkness will make his appearance among the assembled children'.[141] Further along the coast, at Saint-Brieuc, Jean-Marie de la Mennais, now vicar-general of that diocese, preached a retreat to his clergy. He told them that mutual education was a scheme by the regicide Carnot and 'the new missionaries of philosophy and Protestantism' to destroy the Catholic religion. The schools had a 'veritable republican constitution', with children teaching each other and pupils manœuvring with the military discipline of a parade ground. In addition, the pedagogy was 'material' and 'mechanical', isolated from the soul, replacing Christian education by mere instruction.[142]

The fury of the Breton priests cannot be underestimated, but it also had the propaganda value of building up support for the schools that they were now promoting, those of the Frères des Écoles Chrétiennes. In 1816 the municipal council of Rennes voted 11,000 francs to establish the Frères des Écoles Chrétiennes, against a mere 600 francs for the mutual school. One mutual teacher, Lambert, left in disgust for Nantes in May 1818 and the school of a second, Landry, was attacked by pupils of the Frères on St. Nicholas's Day, 1819.[143] But in a poor and backward region like Brittany, calling in the Frères des Écoles Chrétiennes was not an ideal solution. Their requirement of schoolhouse, lodgings, salaries, travelling expenses, and a contribution towards the noviciate, their stipulation that a minimum of three lay brothers be employed in any establishment and that schooling be free, confined them to towns that could afford large sacrifices. That rural communes would ever be provided for seemed a hopeless dream. But the Catholic Church could be flexible and responsive to local needs. It was to confront this challenge that Jean-Marie de la Mennais, vicar-general of Saint-Brieuc, developed his scheme for 'frères solitaires'. Peasant-brothers, on their own instead of in threes, would

---

[141] A.D. I-et-V. 1T *Enseignement mutuel*, prefect of Ille-et-Vilaine to Min. Int., 11 July 1817; mayor of Saint-Servan to prefect, 12 July 1817.

[142] A.F.P. 100, Sermon of La Mennais, 1817. His pamphlet, *De l'enseignement mutuel* (Saint-Brieuc, 1819), developed the same themes.

[143] A.N. F17 11778, Rector of Rennes to Min. Int., 30 Dec. 1819; Tronchot, ii, 267–8.

go into rural communes and lodge in the vicarage of the parish priest. This would serve the dual purpose of surveillance, the priest ensuring that the zeal of the teaching brother did not flag, and of economy, since overheads were reduced to a minimum, the maintenance of the brother coming essentially from school fees. The school could be set up in a room or a barn; when the inhabitants grew tired of this, they would provide wood, stone, and cartage, and construct a schoolhouse. The burden on the poor communes of Brittany would be next to nothing.[144] By 1828 La Mennais had set up twenty-six schools in Ille-et-Vilaine, half of them in the old diocese of Saint-Malo. At the same time the foundations were being laid for the almost total monopoly of girls' education by the teaching congregations. Gabriel Deshayes was responsible not only for the Sœurs de Saint-Gildas-des-Bois, who founded schools in the west of the department but also, ending a period of difficult collaboration in the launching of the Frères de Ploërmel with La Mennais, and moving to Saint-Laurent-sur-Sève (Sarthe), for the Filles de la Sagesse, who built up schools in the Janzé area.[145] The first Filles de la Divine Providence, organized by the *recteur* of Créhen in the old diocese of Saint-Malo, were trained by the Filles de la Sagesse,[146] and the first superior of the Sœurs de l'Immaculée Conception de Saint-Méen was trained in a congregation set up by La Mennais while he was vicar-general, the Sœurs de la Providence of Saint-Brieuc.[147] The first two superiors of the Sœurs de la Providence de Ruillé (Sarthe), who colonized heavily in the south-east and west of the department, were Breton women,[148] while in the north-east the Sœurs Adoratrices de la Justice de Dieu of Rillé-Fougères originated in the relationship between a

---

[144] A.F.P. 102, La Mennais to president of Conseil royal, 14 Nov. 1821, and A.F.P. 80, La Mennais to Min. Instr. Pub., 22 Nov. 1831.

[145] F. Baudu, op. cit., pp. 96–7, M.-T. Le Moign-Klipffel, *Les Filles de la Sagesse* (Paris, 1947), pp. 126–9.

[146] L. Louvière, *Tel l'ajonc sous la neige* (Châteaulin, 1963), p. 59.

[147] A.P.I.C. St. Méen, 8A II 00, 'Livre de paroisse, Saint-Méen'; P. H. Fouqueray, *La Mère Saint Félix, fondatrice des sœurs de l'Immaculée Conception de Saint-Méen* (Saint-Méen, 1924), pp. 23–4. See also Y. Citté, *Les Filles de la Providence de Saint-Brieuc* (Saint-Brieuc, 1940).

[148] T. Catta, *Le Père Dujarié* (Paris–Montréal, 1958), pp. 86–90.

peasant girl, Anne Boivent, and a young priest, Jean-Baptiste Le Taillandier, curate of Saint-Georges-de-Reintembault on the Norman border.[149]

The Catholic renaissance was in full spate in Brittany during the 1820s. But the 'Trois Glorieuses' of 1830 put the liberals in power and made possible, through the University and elementary-school system, the conquest of Brittany in the name of the Constitutional Monarchy. The first priority was clearly the laicization of the University. Since 1814, the Collège royal had been in the hands of the abbé Blanchard who, since being deprived of his seminary, had set up a *pensionnat* at the gates of Rennes at the Port Saint-Martin. Though he claimed that this was only a boarding-house, 'an annexe, a branch of the Collège royal', he also admitted that 'this establishment is nothing other than the continuation of the ecclesiastical school that I ran before the Revolution'.[150] This ambiguity was intolerable. He was replaced as *proviseur* by an authentic *bleu de Bretagne*, Louis-Antoine Dufilhol, formerly college principal at Lorient and *proviseur* at Nantes. Even so, the clerical challenge was only displaced: Blanchard's protégé of *hobereau* extraction, Louïs de la Morinière, was not only rhetoric master at the Collège royal but also director of the *pensionnat* Saint-Martin and, after Blanchard's death later in 1830, superior-general of the Eudists. Within the old husk was the new seed of Catholic education.[151]

In Ille-et-Vilaine, the liberals were an exposed column of vanguard-fighters. They were a bourgeoisie deeply divided from a nobility of mixed *épée* and robe origins not only by the events of 1830 but also by the 'bloody memories of the Civil Wars of '93 and 1815' and the 'fanatical pride of most aristocratic families'.[152] The last bastion that might be expected to bow to their authority was the *ligueur arrondissement* of Vitré, and establishments of education

---

[149] J.-B. Le Taillandier, *Vie des fondateurs et annales de la Congrégation des Religieuses Adoratrices de la Justice de Dieu* (Rennes, 1899), pp. 53–78; G. Bernoville, *Terre de Bretagne: les Sœurs de Rillé* (Paris, 1957), pp. 23–29.

[150] A.N. F17 8957, Blanchard to Grand-Maître, 18 Feb. 1828.

[151] See J. Dauphin, *Le R. P. Louïs de la Morinière* (Rennes–Paris, 1899).

[152] A.N. F19 2567, Letourneux, Procureur général du Roi at Douai to Min. Just. Cultes, 6 Dec. 1840.

were no exception. At Vitré in November 1830 the seminarists decked themselves with white cockades and greeted peasants with shouts of 'Vive Charles Dix!', 'A bas les patauds!'. The sub-prefect feared that the seminary could become 'the seat and mainstay of an insurrection'[153] and the prefect and Rector of the Academy exploited the irregularity of the surrender of the municipal college to the bishop in 1825 as justification to restore the status quo. The municipal council voted to re-establish the college under lay direction and National Guardsmen expelled seminarists and masters under the authority of a prefectoral *arrêté*. Vitré was also a place of trial for the Frères de Ploërmel, who suffered not only from the ambitions of the new municipality to develop lay primary education, but also from the tense political and military situation in the area. Though La Mennais had opened a school there only in 1829, at the invitation of the old municipality, the subsidy of 1,200 francs was suppressed early in 1831. La Mennais organized a subscription to open private schools, but in May 1832 a group of frustrated *châtelains* and ex-officers of the Garde royale staged the last Chouannerie in the country around Vitré, linked with the activities of the Duchesse de Berry in the Vendée,[154] and the new municipality exploited the threat to have these schools closed by the military authorities in Vitré and the prefect the following June.[155] It took La Mennais fifteen months to get the school reopened, struggling not only against 'a score of absurd republicans' on the municipal council of Vitré but also against liberal deputies in the Chamber of Deputies, petitioned by the municipality to have the Frères banned from teaching completely, who voiced the opinion that his congregation was 'affiliated to the Jesuits' and had fomented the Carlist insurrection of 1832.[156]

In Ille-et-Vilaine, the Guizot law served unquestionably as an instrument of colonization. Balzac had described

---

[153] A.D. I-et-V. 1T *Collège de Vitré*, commander of National Guard at Vitré, 19 Nov. 1830, and sub-prefect of Vitré to prefect of Ille-et-Vilaine, 20 Nov. 1830.

[154] See A. de Courson, *La Division de Vitré en 1832* (Vannes, 1899), pp. 8–26.

[155] A.N. F17 10298 and A.F.P. 102, La Mennais to Min. Instr. Pub., 2 Sept. 1832; Tronchot, iii, 382–4.

[156] A.N. F17 12474 and A.F.P. 103, La Mennais to Min. Instr. Pub., 10 Mar. 1834.

Brittany in 1829 as 'a frosty coal that would remain dull and black in the centre of a brilliant hearth'.[157] Two years later Pierre Legrand, another *bleu de Bretagne* and, like Renan, from Tréguier, now the energetic young Rector of the Academy of Rennes, warned Louis-Philippe that His Majesty 'would reign only over Brittany and not over the Bretons, so long as that population had not been touched by the benefits of education'.[158] The organization of supervisory committees on an *arrondissement* basis under the law of 16 October 1830 shook off the nefarious influence of the parish clergy in the cantons and placed confidence in the hands of an enlightened, urban élite. In the four *arrondissements* for which records survive, 49 per cent of the membership in 1830 came from the liberal professions and 5 per cent from trade, while 35 per cent of those serving on the *comités d'arrondissement* of 1833, more 'official' in makeup, had been active since 1830.[159] The only observers who saw fit to praise the *comités locaux* and denigrate the *comités d'arrondissement* were those who regretted the erosion of the influence of the priesthood by the bureaucracy. 'When something is done', scoffed La Mennais 'it is the sub-prefect and his clerks who decide everything.'[160]

A combination of poverty and recalcitrance made the provision of a school in every commune of Ille-et-Vilaine a very arduous task. No less than 55 per cent of communes had to be taxed ex officio, and this figure rose to 87 per cent in the *arrondissement* of Vitré and to 90 per cent in that of Redon, their backwardness matched only by their conservative and Catholic mentality. Some communes tried to raise resources by the sale of common land and waste, but for Fresneau, the sub-prefect of Redon, the only chance of enlightening his desperate region, where only one commune actually owned a schoolhouse, was a massive injection

[157] H. de Balzac, *Les Chouans* (first edition, 1829, *La Comédie humaine*, Pléiade edn., Paris, 1977, viii, 918).

[158] A.N. F17 10214, Rector of Rennes to Louis-Philippe, 2 Mar. 1831.

[159] A.D. I-et-V. 1T *Comités cantonaux, Comités d'arrondissement d'instruction primaire*: J. Trevet, 'L'Instruction primaire dans l'arrondissement de Fougères sous le régime de la loi du 28 juin 1833', *Annales de Bretagne*, xxix, 1913-14, 388.

[160] A.F.P. 100, La Mennais to Commission parlementaire pédagogique, April 1849.

of government funds. The appeal of the Frères de Ploërmel now becomes clear: they were cheap and they could guarantee religion. Again for Fresneau, 'if one day Brittany is enlightened, it will be principally by the good offices of the clergy'.[161]

In Brittany the first teacher-training centre was the noviciate of the Frères de l'Instruction Chrétienne established at Ploërmel in 1824, and the task of the new liberal regime was to reply to that initiative. The Academy of Rennes set up its own École normale in October 1831, based at Rennes and designed to serve the whole of Brittany. When it became clear that La Mennais was himself seeking to found a second noviciate at Rennes the prefect of Ille-et-Vilaine warned Guizot that 'to allow the Frères de l'Instruction Chrétienne to obtain the monopoly of elementary education would be to surrender the children of our countryside and towns into the hands of priests, to perpetuate the backward mentality that keeps the departments of the west in an exceptional position'.[162] The abolition of the third-degree *brevet*, reckoned to be of too low a standard, under the Guizot law, set a difficult challenge for the peasant-novices. But La Mennais was not to be beaten, and responded on three fronts. He warned against the over-education of elementary teachers, asking why a qualified *instituteur* should settle in a humble country school if he were equipped to become a clerk or *petit fonctionnaire*.[163] Ploërmel, with its tailor's shop, woodwork shop, and forge—appearing as 'a Fourierist phalanstery bound by ties of religion and devotion'[164]—not only equipped the Institute with its material needs but also engrained attitudes of the craftsman rather than those of the intellectual. Again, La Mennais set up a crash course at Ploërmel to train his lay brothers for the *brevet* so many of them lacked[165]—while Ille-et-Vilaine went without any cantonal conferences for its *instituteurs*.

---

[161] A.D. I-et-V. 1T *Comités d'arrondissement*, sub-prefect of Redon to prefect, 5 Feb. 1834.

[162] A.F.P. 100, prefect of Ille-et-Vilaine to Min. Instr. Pub., 3 May 1833.

[163] A.F.P. 104 and A.N. F17 12474, La Mennais to Min. Instr. Pub., 7 Nov. 1837.

[164] A.N. F17 12474, report of *Inspecteur supérieur*, 1 Oct. 1847.

[165] H.-C. Rulon, *Chronique des Frères de l'Instruction Chrétienne de Ploërmel*, no. 199, July 1954, p. 637.

Most skilfully he used the fact that his Institute was not providing missionaries for the liberated slaves of the West Indies
to argue that he had to send his most experienced and
qualified teachers there, so that certain irregularities must
be tolerated at home, notably the provisional authorization of lay brothers who did not have the *brevet*, if the
education of the Breton peasantry were to go ahead.[166]
In the end, of course, La Mennais had history on his side.
The old diocese of Saint-Malo, the basin of the Vilaine,
and the Vitré–Fougères fringe produced novices for his
order, not for the École normale of Rennes, as the registers
of the institution show.[167]

Despite the anticlerical constraints of the July Monarchy,
the Catholics were building the foundations of their secondary-
school system as surely as they were those of their elementary
schools. The municipal college of Vitré was recovered by
secular clergy in 1839. In the same year, the abbé Blanchard's
successor as head of the Eudist order, abbé Louïs, purchased
the old Benedictine abbey of Redon, which then housed a
small *pensionnat* run by priests.[168] Despite the need for a
front man with the right academic qualifications, the real
superior was Gaudaire, a Breton of the Morbihan trained
at Saint-Sulpice (1805-70), who would build up the establishment over three decades. But for the moment the campaign for *plein exercice*, launched in 1844 with the support
of sub-prefect Fresneau, was frustrated by Rector Dufilhol.
Of more moment was the opening of a *pension* at Rennes by
the new bishop of Rennes, Godefroy Brossays Saint-Marc
(1803-78), in 1842. Brossays Saint-Marc, member of an
old bourgeois family of Rennes, was elevated to the see in
1841 as a 'new model' bishop who would defend the views
of the government. For one adviser, he was 'of all the clergy
of Rennes the only ecclesiastic who was a sincere friend of
the constitutional government and heartily devoted to the
person of His Majesty Louis-Philippe'.[169] He was, however,

---

[166] A.F.P. 104, La Mennais to Min. Instr. Pub., 15 Sept. 1837.
[167] A.F.P. Noviciat de Ploërmel, Registres d'entrée, 1823-56.
[168] H. Le Gouvello, *Histoire du Collège Saint-Sauveur de Redon* (Redon,
1913), pp. 24-42.
[169] A.N. F19 2567, Letourneaux, Procureur général du Roi, Douai, to Min.
Just. Cultes, 6 Dec. 1840.

an ambiguous personage, not prepared to sacrifice the interests of the Church in any way. For Michel Lagrée, 'there was in him something of the prince-bishop of the style of the sixteenth century, but in the age of the railway and liberal ideas'.[170] The *pension* that he entrusted to Prosper Brécha (1814-63), a sea-captain's son and now one of the Prêtres de Saint-Méen, was originally only a boarding-house for boys attending the Collège royal during the day. But this refusal of autonomy to Catholic education was anathema to Brossays, the more so because the young philosophy master, Charles Zévort, was teaching the eclectic doctrines of Victor Cousin, which Brossays believed to be 'irreconcilable with Christian faith'. Brossays demanded of the Minister, Villemain, that Zévort be dismissed, and when Villemain refused, withdrew his chaplain from the College, which amounted to placing an interdiction on it. The intransigent line taken by Brossays was in part to give weight to arguments for *plein exercice*, the complete separation of the *pension* from the Collège royal. His most telling motive was nevertheless the need to place himself at the head of the clerical and legitimist sentiment of his diocese, and destroy any suspicions that he was a prefect in purple. In a confidential letter to the Garde des Sceaux, Martin du Nord, he explained that 'if in our old Brittany a bishop is still so powerful, he owes it to the faith of his followers, and woe betide the pastor if the least suspicion plays on this vigilance to guard the flock, esteem, authority, confidence, everything is lost for him'.[171] His critics were not slow to recognize this. The emissaries of Villemain noted that the bishop 'submits to a powerful influence, and that influence is that of the clergy to whom he wishes to pledge himself and that of the legitimist party'.[172] Rector Dufilhol recorded that Brossays had 'two interests, to undermine the University and to win over a clergy that begrudged his elevation to the see'.[173]

---

[170] Lagrée, *Mentalités*, p. 233.

[171] A.A. Rennes C1, Brossays Saint-Marc to Martin du Nord, Garde des Sceaux, Min. Just. Cultes, 24 Nov. 1843.

[172] A.N. F17 7985, delegates Beudaut and Dutrey to Min. Instr. Pub., 27 April 1844.

[173] Rector of Rennes to Paul-François Dubois, cit. A. Rebillon, 'L'Université et l'Église à Rennes au temps de Louis-Philippe', *Annales de Bretagne*, lii, 1945, p. 135.

In the end Brossays Saint-Marc was victorious. Zévort was moved on in October 1844, Dufilhol himself in September 1847. The episode illustrates the influence of the Breton *terroir* on the politics of its bishops. And this is not the only case we shall have reason to investigate of a liberal bishop obliged to conform to the clerical sentiments of his diocese. In the diocese of Rennes, one source permits the plotting of the geography of militant Catholicism: a petition against the University monopoly in favour of *La Liberté de l'enseignement* which collected 13,055 signatures in March 1845 (Map 4).[174] The context was the introduction of Villemain's second bill on secondary education in February 1844 which gave away very little to the Catholics and the formation of a Comité électoral pour la défense de la liberté religieuse, headed by Montalembert and Eugène Veuillot. In Ille-et-Vilaine the struggle set ministerial Orleanists and the republican left on one side and, on the other, the *comités pour la liberté de l'enseignement*, headed by the legitimist deputy for Vitré–Fougères and the legitimist *Journal de Rennes*. The petition of 1845 clearly demonstrated the bastions of fervent Catholicism in the department: the west, notably the cantons of Bécherel, Montauban, and Saint-Méen, and the eastern border, the Vitré–Fougères bloc. It was a pattern familiar from the past, and one that would recur.

In Ille-et-Vilaine, where legitimist landowners and the clergy were so powerful, universal suffrage in 1848 did very little for the republican cause. Brossays Saint-Marc played the grand elector in April 1848, boasting that he had '60,000 votes in his pocket',[175] and the clerical list triumphed in twenty-six of the forty-three cantons of the department. In the presidential elections of 10 December 1848 Ledru-Rollin polled a mere 0.5 per cent of the vote, and in May 1849 Brossays Saint-Marc was again on the war-path, working through the Comité des Amis de l'Ordre. Using the electoral power of the Breton clergy and the conservative *Journal de Rennes*, he isolated both republicans and die-hard legitimists

---

[174] *Journal de Rennes*, 29 Nov. 1845; Lagrée, *Mentalitiés*, pp. 57–60.

[175] J. Meyer, 'De la Révolution politique aux débuts du monde industriel, 1789–1880', in J. Delumeau, ed., *Histoire de la Bretagne* (Toulouse, 1969) p. 426.

one signatory in less than 100 inhabitants

Map 4. Petition for *La Liberté de l'enseignement* in Ille-et-Vilaine, 1845.

and obtained two-thirds of the vote for his list.[176] Committed to order, the bishop did not take long to rally to the

[176] H. Goallou, 'Les Élections à l'Assemblée legislative.en Ille-et-Vilaine, 13 Mai 1849', *Annales de Bretagne*, lxxx, no. 2, June 1973, 359-402.

Empire, urging that it was the only guarantee against 'universal cataclysm'.[177]

Those *instituteurs* who had manifested the slightest penchant for republicanism were quickly brought into line. The Comité d'arrondissement of Fougères reminded *instituteurs* at the end of 1848, 'you are exercising in your commune a sort of priesthood on behalf of civilization' and that 'we would never suffer that you transform yourselves into apostles of barbarism'.[178] The attempt by Hippolyte Carnot to subject nuns to examination before granting the *brevet* was welcomed by the new republican Rector of Rennes, Théry,[179] but deplored by La Mennais. In his opinion, the *brevet* for girls was 'a misfortune . . . only too often they lose their simplicity, their piety; they become haughty, cunning, greedy for money'.[180]

Now, of course, there was nothing to stop the flood-tide in favour of Catholic education. At Redon, Gaudaire's *pension* had been accorded *plein exercice* as the Collège Saint-Sauveur as early as October 1847. In April 1849 the chrysalid of the *pension* Brécha at Rennes opened to reveal the Collège Saint-Vincent, and the warnings of the Rector of Rennes that this would deliver a 'fatal blow' to the Lycée were rapidly realized.[181] The aristocratic image of the Catholic college outweighed the attractions of higher academic standards at the Lycée, and the sons of magistrats, high civil servants, and faculty professors were followed to Saint-Vincent by the sons of tradesmen who had a clientele in legitimist and clerical circles.[182] Moreover, whereas about 150 pupils of the *pensionnat* Saint-Martin run by the Eudists had attended the Lycée as day-boys before 1849, now, despite a social inferiority which often exposed them to ridicule, they went to Saint-Vincent.[183] But Brossays Saint-

---

[177] H. Goallou, 'L'Évolution politique d'Ille-et-Vilaine, 1851-79' (thesis Rennes, 1971), i, 173.

[178] Trevet, art. cit., p. 625.

[179] A.N. F17 12478, Rector of Rennes to Min. Instr. Pub., 4 Aug. 1848.

[180] A.F.P. 100, La Mennais to Commission parlementaire pédagogique, April 1849.

[181] A.N. F17 8959, Rector of Rennes to Min. Instr. Pub., 16 Feb. 1849.

[182] A.N. F17 7990, general inspections, 24 June 1858, 31 May 1860.

[183] A.N. F17 7985, Rector of Rennes to Min. Instr. Pub., 14 Jan. 1851; general inspection, 1857; A.E. M.F. 28, report of P. Coyer, director of Institution Saint-Martin, 14 Feb. 1887.

Marc did not stop there. Taking advantage of the Falloux law, which permitted municipal councils to hand over their colleges to private persons yet continue to subsidize them, he rapidly established a network of Catholic colleges which would channel their older pupils towards Saint-Vincent or the seminary of Saint-Méen. The municipal council of Saint-Malo entrusted its college to the bishop in September 1853, that of Vitré did so in August 1853. In both colleges Brossays Saint-Marc installed the diocesan congregation of secular priests who ran Saint-Vincent, the Prêtres de Saint-Méen. In order to ensure that the humanities classes would be concentrated at Saint-Vincent, he ruthlessly suppressed the rhetoric class at Saint-Malo, and telescoped the second into the third class; at Vitré he refused to maintain any class above the fourth (age thirteen).

The clerical upsurge was equally pronounced in elementary schooling in the 1850s. In the administration of education, the bureaucracy and University lost ground to the clergy and conservative notables. It was these who dominated the Conseil départemental de l'instruction publique and its examining commission, with the result that the pass rate for the *brevet* of teaching brothers was noticeably higher than that of lay *instituteurs*, even before the chairmanship of the commission fell to Brossays Saint-Marc's vicar-general in 1858.[184] In the *délégations cantonales* of 1854 the proportion of liberal professions among the membership was under 15 per cent, while *curés* made up 26 per cent of the delegations and manœuvred them at will.[185] In Ille-et-Vilaine Catholic leaders firmly believed that *laïcité* was imposed by outside authority and that grass-roots opinion was Catholic. This was demonstrated in the summer of 1850 at Vitré, where the municipal council interpreted the Falloux law—albeit falsely—in the sense that municipalities now had the right to appoint schoolteachers, and proceeded to oust the lay *instituteurs* authorized by the *comité d'arrondissement*, replacing them with Frères de Ploërmel.[186]

---

[184] A.N. F17 10332, affairs of examining commission of Conseil départmental de l'Instruction publique of Ille-et-Vilaine, 1858.
[185] A.D. I-et-V. 1T, Conseil départemental de l'Instruction publique, 1854.
[186] A.N. F17 10316, dossier on *affaire* of Vitré, June–Oct. 1850.

Liberty of education, the right of any qualified and decent individual to open a school, was a mixed blessing at the elementary level. As the primary-school-inspector of Fougères observed in 1856, 'between secondary private education and primary private education, there is no comparison: liberty means life for one, death for the other'.[187] The struggle in the parishes to obtain appointment to the *école communale* was therefore fierce, and the Frères de Ploërmel, using every form of pressure at their disposal, made important gains, especially in the Redonnais. But the education of girls was overwhelmingly in private hands, two-thirds of girls' schools in Ille-et-Vilaine in 1851, as against 8 per cent of boys' schools, and the Inspector of the Academy reflected that 'for too many *institutrices*, liberty means abandonment'.[188] A distinction must nevertheless be made. Whereas private lay *institutrices* had scarcely the wherewithal to survive, the female congregations which were now multiplying across the department were generally in a comfortable position, for they were funded by voluntary subscriptions collected by the *curés*, the endowments brought by aspiring nuns of rural or urban bourgeois backgrounds, and the legacies of the legitimist nobility.

Finally, the École normale of Rennes, like so many others, was refashioned under the decree of 24 March 1851 to satisfy as far as possible the requirements of the clergy. A rustic honesty and morality was preferred above intelligence, and the director of the École normale vowed that 'were I responsible for awarding bursaries, I would eliminate without pity those who had not been educated in a rural primary school'.[189] Within the training schools, the syllabus was confined to a minimum of skills. The future *instituteur*, as La Mennais had once remarked, had to be preserved from the perils of over-education. 'How, after that', now asked an inspector-general, 'could we prevent that ill-digested encyclopedia from troubling his head and giving him ideas

---

[187] A.N. F17 9326, report of primary-school-inspector of Fougères, 9 April 1856.
  [188] A.D. I-et-V 1N Conseil général. Insp. Acad. Rennes to Conseil académique. 16 Aug. 1851, pp. 200–1.
  [189] A.N. F17 9658, Campion, director of École normale of Rennes to Commission de surveillance, 15 July 1852.

above the humble situation we have assigned to him?'[190] The religious atmosphere of the *internat* was now so intense that Brossays Saint-Marc himself could say, in 1857, 'the moral condition of the École normale is that of a *petit séminaire*'.[191]

From the point of view of liberal, secular government, Ille-et-Vilaine was the most intractable of the three departments. Breton separatism, protected by the fiefs of a former marchland together with the intense Catholicism of the old diocese of Saint-Malo, meant that attempts to colonize the area for revolutionary regimes would meet with concerted resistance, and the battle of blue against white would continue in electoral campaigns, pulpits, and schools even after guerrilla warfare had ceased in the *bocage*. Only the towns and a few knots of communes in the north-east and south-east of the department could be relied on by the men of 1789. Strife over education was sharpened by the strife between patriots and loyalists and the impossibility of compromise was illustrated by the change in attitude of Brossays Saint-Marc once he was promoted to the see of Rennes: there could be no such thing in Brittany as a liberal bishop.

The revolutionary period crystallized the rivalry of the two camps, but continuity across the Revolution was still important. The Eudist and Lazarist orders reappeared while the brothers La Mennais were the tradition of the saintly diocese of Saint-Malo incarnate. The 'blue' tradition that emerged essentially from the Revolution required the support of progressive ministries in Paris if it were to make any headway.

There was no question that either the Napoleonic system or the July Monarchy could dispense with the Catholic Church, dominating Breton life as it did. The University and its colleges was controlled by Catholics after 1814, but even then the Church fought for a whole network of seminaries outside the colleges. Saint-Malo, Saint-Méen, Vitré, and Rennes all acquired seminaries at one point or another between 1800 and 1830, while Blanchard, Rector of the Academy and *proviseur* of the Collège royal of Rennes,

---

[190] A.N. F17 9313, inspector-general to Min. Instr. Pub., 28 Sept. 1849.
[191] A.N. F17 9656, Campion to Commission de surveillance, 2 July 1857.

admitted to running his pre-revolutionary seminary alongside the college.

The liberal bid to laicize the university in 1830 was part and parcel of the campaign of the Orleanist regime, which recognized the *tricolore*, to extend its authority into such bastions of legitimism as the west of France. Struggles over the seminary and Breton *frères* at Vitré were given bitter significance by the last episode of Chouannerie put down in the region in 1832. But the upshot was that Catholic secondary education expanded outside the University, sinking roots under the July Monarchy, flowering after 1848 and casting long shadows over the University. At the elementary level it was clear that the education of the people must be the privilege of the congregations. Mutual education had been swept away by the fanatical Breton clergy and new congregations were geared to missions not only in towns but also throughout the countryside, the old diocese of Saint-Malo being particularly fertile in enterprises. After 1850 the congregations surged from their power-base in the private sector into the *écoles communales* and asserted a near monopoly over the education of girls, those sanctuaries of the Catholic tradition.

3

The Anticlerical Offensive, 1860–1914

The year 1860 was a watershed not only in the history of the Second Empire, but also in that of the nineteenth century as a whole. Napoleon III's alliance with Piedmont against Austria resulted in an invasion of the Papal States and the alliance between the Empire and the Catholic Church that had formed the basis of the restoration of order after 1848 broke apart. The Church, supported by legitimist notables, could not stomach revolutionary nationalism in Italy, let alone the rape of the Temporal Power of the Pope. The authoritarian Empire now strove to consolidate its authority at the expense of the ultramontane clergy, although the elections of 1863 demonstrated the limited effectiveness of the system of official candidatures and the ability of Catholic notables to defend their position. As a result, the regime inaugurated an anticlerical offensive, presided over by the Minister of Education, Victor Duruy (1863-9), and began to erode the influence of the Church in the school system. An attempt was made to purge the clergy from the municipal colleges they still held and to set up secondary secular education for middle-class girls, though the protests of the episcopate reduced the latter initiative to ashes.[1] At the elementary level a start was made to turn back the invasion of *écoles communales* by the Catholic congregations that had reached its peak in the 1850s. A law of 14 June 1854 which transferred the right to appoint *instituteurs* from Rectors to prefects could be used to great effect by the imperial bureaucracy. The demands of the Ligue de l'Enseignement, founded in 1866 by 'men of '48' who wanted to back up universal suffrage by compulsory free education, was taken up in part by the government under the law of 10 April 1867. Providing subsidies to communes

---

[1] F. Mayeur, 'Les Évêques français et Victor Duruy: les cours secondaires de jeunes filles', *R.H.E.F.*, lvii, no. 159 (1971), p. 271. In general, F. Mayeur, *L'Éducation des filles en France au XIXe siècle* (Paris, 1979), pp. 113-38.

which themselves voted additional rates to establish free
public education, the law was seen by the Catholics as
the mobilization of funds behind official lay education,
and a Société générale d'éducation et d'enseignement was
set up to defend the interests of Catholic education. Lastly,
some municipalities took advantage of the unqualified
status of many teaching brothers to expel them from
municipal schools. The last ten years of the Empire in
many ways foreshadow the offensive of the republicans
after 1879.

For most of the 1870s nevertheless the campaign against
the Church's influence in education had to be suspended.
The Empire fell in 1870, and though a Republic was declared,
the republicans found themselves as a result of the events
of 1870-1 castigated as warmongers, dictators, and social
revolutionaries. The elections of February 1871 resulted in
a royalist landslide, although the popular vote was less for
the restoration of the monarchy than for peace, liberty,
and social order. The Paris Commune of March 1871, a
revolutionary-patriotic protest against those conservative
notables who were collaborating with Bismarck, using the
full force of the European order to safeguard property,
even at the price of national humiliation, provoked a revenge
of militant Catholicism that threatened to swamp the very
principle of *laïcité*. Argument raged on both sides. For the
republicans, defeat, the failure of democracy, and social
war were explained by lack of popular enlightenment. The
propagation of popular education would, by contrast, build
up patriotic strength, secure the Republic on the firm basis
of civic virtue, and, ending the gulf between the educated
élite and the illiterate mass, promote the 'fusion of classes'.
For the Catholics, on the other hand, it was France's
abandonment of the ways of true religion that had provoked
at one and the same time the divine chastisement of defeat,
the incurable disease of revolution, and the atheistic mobs
that had burned and pillaged Paris and murdered its archbishop. Only an act of national repentance and a return
to the Church could restore France to her providential role
as eldest daughter of the Church and defender of the Papacy,
sanctify the Christian monarchy that was a gift of God

where other regimes were acclaimed by numbers, and provide security for the family and private property.

For two years Adolphe Thiers steered a middle path between radicals and reactionaries, trying to entrench the 'Republic without republicans'. But for the Right he was an inadequate guarantee against the forces of revolution, and in May 1873 monarchists and Bonapartists overthrew him and established the Moral Order regime under the presidency of Marshal MacMahon. In the period 1873-7 France came within a hair's breadth of monarchist restoration. Whereas the years of uncertainty in 1871-2 had seen a battle for influence between the Ligue de l'Enseignement and the Société générale d'éducation et d'enseignement over free, compulsory education, after 1873 it was Catholic education that received a new lease of life. The circular of Jules Simon, Thiers' very moderate Minister of Public Instruction, dated 28 October 1871, which permitted municipal councils to opt between lay teachers and congregations for their *écoles communales* in instances other than the death, resignation, or dismissal of the incumbent, gave *carte blanche* to conservative municipalities to 'clericalize' their schools. At the secondary level, municipal colleges which remained in the hands of priests were safe, private Catholic institutions, including those of the Jesuits, flourished, and recruitment to the priesthood and religious congregations was stepped up by the foundation of *maîtrises*, *alumnats*, and *juvénats* which would take boys after their First Communion, cut them off from worldly temptations, douse the fires of puberty, and reduce wastage to a minimum. Finally, the Moral Order period saw the fruition of the Catholic campaign for higher education. Catholic faculties now existed as an alternative to lay faculties for the graduates of Catholic colleges, even though the law of 12 July 1875 gave them less than was hoped for: Catholic universities were not allowed to issue degrees, nor did they have freedom to devise their own syllabuses. As at the Restoration, the hegemony of the University remained in the last instance intact.

The second wave of the anticlerical assault began after the triumph of the republicans in the elections of 1876 and 1877. By then they had managed to shake off the image of warmongers, dictators, and social revolutionaries. Instead, they were able to portray the Right as quixotic crusaders, prepared to risk war with Prussia in order to restore the Temporal Power of the Pope, as counter-revolutionaries who would overturn the Constitution of 1875 and use the authoritarian methods of the Seize Mai to keep a republican majority out of power, and as partisans of the Ancien Régime who wanted nothing better than to bring back serfdom and tithes. By contrast, the republicans claimed that they themselves were now men of peace, democracy, and private property. The main concern of the republicans in power was not to undertake social reform but to guard against clerico-royalist restoration by training new generations of patriotic citizens, prepared to defend the Republic against its enemies within and without.

Although this was the most vigorous anticlerical campaign for ninety years, the republicans were not able simply to bulldoze Catholic education out of existence. They too had to contend with the historical geography of France, the pattern of attitudes shaped in the provinces by decades of conflict. The reaction of the provinces will be dealt in the next chapter. For the moment it is worthwhile defining the stages of the war.

The first step in the republican bid to complete its control of secondary education was to undermine the private Catholic colleges that had grown since the Falloux law by exploiting the fact that many of the congregations that taught in them were strictly, under revolutionary-Napoleonic legislation, unauthorized. The Jesuits and the Assumptionists were prime targets. Article 7 of the bill introduced by Jules Ferry, who would dominate the republican campaign, called for the dissolution of these congregations. The bill was defeated by the Senate and the expulsion of the unauthorized congregations therefore took place under the decrees of 29 March 1880. But where it mattered these Catholic colleges could survive, both by entrusting their property to *sociétés civiles* and by transferring direction to secular priests or claiming that the congregation was now 'secularized', free from all vows of community. Secondly, the republican government sought to revive the short-lived *cours secondaires* for girls of Victor

Duruy under the Camille Sée law of 21 December 1880, although in many provincial towns the initiative had been taken some time before. Even so, the entrenched position of the female religious orders was difficult to assail, the lay *pensionnats* provided facilities of continuous study, supervision, and enclosure that the *cours secondaires* did not have until they became ladies' colleges in the later 1880s and 1890s, and it was never certain whether the courses aimed at attracting the daughters of the bourgeoisie or, alternatively, those of the *classes moyennes*. But under the auspices of the Republic, the secular secondary education of girls was there to stay. In the third place, the triumph of republicans at the municipal elections of January 1881 spelled the end of priestly control in those municipal colleges where it still existed. But often the college premises belonged to the secular clergy, so that it was up to the public authorities to build new schools while the clergy continued their work in private institutions now independent of the University.

It is clear that changes of circumstance at the local level were as important as national legislation in the process of laicization. This was equally true at the level of elementary schooling, and again it was the 'town-hall revolutions' favouring the republicans in 1878 and 1881 that were significant. Following the elections there was a series of 'spontaneous' laicizations, often in revenge for forced 'clericalization' under the Moral Order regime. Naturally, the Catholic clergy and notables were able to salvage Catholic education where they wanted it: subscriptions were opened and endowments solicited to continue in private schools, as after the Revolution of 1830. It was only after the legislative elections of August 1881 which consolidated the republican majority in Parliament that measures were taken on a national scale to eliminate the influence of Catholicism in *écoles communales*. The law of 28 March 1882 banned the teaching of catechism in those schools, partly because of the association of altar with throne, partly because a theological morality that preached resignation in this world and striving for personal salvation in the next was seen to be anti-social, and partly because it was a fruitless way to discipline populations that remained beyond the scope of organized religion.

On the other hand, the revolutionary–romantic insistence on individual rights also had to be qualified for the sake of bourgeois order. The 'moral and civic education' now introduced into the public schools was a morality independent of religion and yet social in commitment. Starting from the scientific fact of the interdependence of men in society according to the division of labour, it elevated the fact of the subordination of the individual to society into a morality based on *devoir*. The protests of the Catholic camp against the 'école sans Dieu' were vociferous, since for them there could be no morality without religion, no social order without fear of eternal damnation. But primary schools were not suddenly swept by a cynical, ersatz code of ethics. Though Jules Ferry reassured *instituteurs* in his letter of 17 November 1883 that they would not have to become 'improvised philosophers or theologians', there was small danger that they would, given the fact that so many were rooted in traditional rural backgrounds and that such pains had been taken since the Falloux law to guard against their over-education. Again, schoolteachers in firmly Catholic areas would have to think twice before adopting an aggressively laic stance: that was the easiest way to lose their clientele to the private schools. And so, especially in the girls' public schools, crucifixes remained on the walls, prayers were said, and the pupils were taken to mass, as in the old days. Lastly, even if catechism were banned from the classroom, that did not prevent the parish clergy, now calling for the co-operation of honest families, to increase the dose of religious instruction outside it.

'Défense religieuse', along with protection, was a strong argument for the Right in the elections of 1885 and they gained ground heavily at the first ballot. Unfortunately, the prospect of monarchist restoration was whispered between the two rounds and the republican party, hitherto divided between moderates and radicals, now pulled itself together and recovered the advantage. A fresh blow against clerical influence in the schools was now struck by the government of 'republican concentration' which included the radicals: the law of 30 October 1886 required that teaching congregations must now evacuate the *écoles communales*. Those bastions of reaction where notables and clergy combined to keep the population in subjection were to be colonized for the Republic. The application of the law would nevertheless be difficult in strongly Catholic areas. There was a shortage of lay teachers who might replace the congregations, partly because of the pitiful salary they were paid, partly because the *écoles normales d'institutrices*, mandatory in every department under a law of 9 August 1879, were only beginning to turn out their graduates. Property rights constituted a second obstacle to laicizations. Where the *école communale* was owned by the commune, as was the case with most boys' schools, there was no barrier to laicization. But few rural communes had the resources to build a second school, so that the girls' school was often owned by a private benefactor, the parish priest, the vestry, the *bureau de bienfaisance*, or the congregation itself, and put at the disposal of the commune on condition that it install a teaching congregation. Such property could not simply be confiscated, and the commune was now obliged to build or acquire a girls' school of its own before laicization could go through. For these reasons, while the law of 1886 stipulated that *frères* should be out of the public schools within five years, no such rigid time-limit could be imposed in the case of nuns. Lastly, even when laicization did go through, Catholic parishes could reply by founding a private shool, and continue to employ religious congregations. Financing such schemes might pose acute problems since the law of 1886, reinforced by a Conseil d'État ruling of July 1888, prohibited communes from using public funds for private schools. But private patronage and voluntary subscriptions might overcome such problems where Catholic education was held to be indispensable, and the parish clergy could then go to work populating the church school at the expense of 'la laïque'. Pupils of the lay school were refused benches in church, ploughed in catechism classes, and turned away from First Communion. Their parents were subjected to exhortation from the pulpit and in the confessional and even refused absolution. In a deChristianized area, such tactics would remain without effect, but a pious population would fall quickly into line.

The anticlerical assault of the Ferry era had the effect of dividing the ruling classes of France. Necessary in order to consolidate the Republic against clerical–royalist reaction, it was a luxury that could not be afforded when danger threatened from another quarter with the rise of socialism after 1890. The propertied classes now sank their ideological differences and combined forces behind the policies of protection, imperialism, and social reform. This was the period of the Ralliement, during which the moderate Right (the *ralliés*) abandoned counter-revolution and sought to create a Tory Republic, and the moderate Left (the progressists) abandoned anticlericalism and succumbed to the 'esprit nouveau'. The covert conquest of power by the conservatives was guaranteed by the ministry of Méline in 1896, but the elections of May 1898 ended the conservative ascendancy and exposed France to the dangers of revision. The Dreyfusards, campaigning against the iniquities of Church and army, forced the Right to fall back on a counter-revolutionary position, denouncing the Republic of Jews, Protestants, and Freemasons and looking to an eternal France symbolized if not by the monarchy then by the Catholic Church, the secular arm of which was provided by France. The *ralliés* now joined the 'intransigents' who had not accepted the compromise with the Republic, the progressists joined the radicals and socialists in defence of the Republic. The Jacobin alliance was reconstituted, and the Waldeck-Rousseau cabinet of June 1899 adopted a policy of clipping the wings of Church and army as its first priority.

The final phase of the anticlerical assault was aimed, beyond the *écoles communales*, at the private schools of the teaching congregations, mostly pre-1886 in the case of girls' schools, mostly post-1886 in the case of boys' schools. The instrument of Waldeck-Rousseau's policy, as much a step forward for trade-union rights as a blow to the congregations, was the Associations law of 1 July 1901. Waldeck-Rousseau was a man of some subtlety who, as a young lawyer and one of the founding fathers of the republican party at Rennes in the 1870s, understood the need to play down anticlericalism and freemasonry if republicanism was going to make any headway in the Breton countryside.[2] His original bill made the authorization of each congregation a matter for administrative decree and made possible the authorization of congregations in one area while prohibiting them in another, according to political expediency and the wishes of local inhabitants.[3] These provisions explain why in 1902 a large-scale inquiry was conducted, and municipal councils of communes where religious congregations were active were asked to comment on the suitability of those congregations. Since the only communes that had teaching congregations in 1902 were either a small minority where the *école communale* for girls had not been laicized, or those in which, after laicization, private schools had been built at the expense of the local inhabitants, it might be imagined that all municipalities wanted to keep their private schools run by congregations. However, in the first place, those private schools had sometimes been established by a clerical party against the wishes of the municipality and, in the second, many municipalities had changed hands in the elections of May 1900. The vote of the municipal councils is thus some gauge of anticlericalism in the localities, albeit bearing in mind that their view might only be that of the majority in the commune.

The Associations law was not, however, destined to be enforced with any discretion. A parliamentary committee decided that authorizations would have to be decided in the Assembly, not by the administration, and requests from congregations would be accepted or refused *en bloc*.[4] In addition, the elections of April–May 1902 undercut the conservatives and gave undisputed victory to a Bloc des Gauches composed of radicals, radical-socialists, and ministerial socialists. Waldeck-Rousseau resigned as Président du conseil in June and was replaced by Émile Combes. The difference between the two was the difference between a *bleu de Bretagne* who wanted to control but conciliate

[2] P. Sorlin, *Waldeck-Rousseau* (Paris, 1966), pp. 158–9. Waldeck-Rousseau was elected deputy for Rennes in 1879.

[3] M. O. Partin, *Waldeck-Rousseau, Combes and the Church: the politics of anticlericalism, 1899–1905* (Duke University, 1969), p. 27; M. Larkin, *Church and State after the Dreyfus Affair* (London, 1974), p. 89.

[4] M. O. Partin, op. cit., p. 31; M. Larkin, op. cit., p. 90.

Catholics, and a meridional who had not only received an ecclesiastical education but had also taught between 1857 and 1860 in d'Alzon's Assumptionist College at Nîmes, and, in revolt against the congregations that had shaped him, sought to eliminate them from the face of France.

In the hands of Combes the Associations law would be enforced with a new ruthlessness. His decree of 27 June 1902, of little significance, required the closure of schools opened by unauthorized congregations since the passage of the law, and concerned about 120 establishments in the whole of France. Much more severe was the circular of 9 July, which ordered the closure within a week of all communities that had not requested authorization. Confusion, ignorance, and misplaced trust put many schools in this category. Then, from the spring of 1903, came the wholesale rejection of petitions for the authorization of congregations and more expulsions. Lastly, a law of 7 July 1904 prohibited from teaching even those congregations that had been organized, including the Frères des Écoles Chrétiennes.

Even this brutal policy could not be wholly enforced in the provinces. Catholic resistance was vigorous, reinforced in some areas by a regionalist opposition that challenged the imperialism of Paris. There were instances of republican municipalities and deputies in Catholic areas obtaining a stay of execution on behalf of the Church schools, so as not to forfeit narrow majorities at forthcoming elections. Property rights were again a problem. Schools owned by dissolved congregations were forfeit, but the school might be taken over by a *société civile* or tontine which harboured the congregation. On occasions, the tontine itself leased from the congregation, providing only a front for illicit proprietorship. Where the private school was owned by the vestry or a private patron, the expulsion of the congregation would have to await the building or rental of premises sufficient to accommodate all those pupils suddenly deprived of teachers. Even the dissolution of congregations might not be catastrophic from the point of view of Catholic education. Members might continue to teach, either by slipping over a nearby border into exile or, more commonly, by 'secularizing' themselves, divesting themselves of their religious habit, and obtaining 'letters of secularization' from their superior-general and diocesan bishop, certifying that they were no longer a member of a congregation. The only hitch was the prohibition of 'secularization *sur place*', the reopening of a private school by the lay brother or nun who had taught there previously, but this could be solved by neighbouring staffs changing places. Of course, the drama of dissolution did produce much wastage from the congregations, but the Catholic Church was coming to rely increasingly on a devout lay staff, and could watch with some wry pleasure the difficulties experienced by academic authorities in the recruitment of their own teachers. Lastly, many female congregations survived because they were not uniquely dedicated to teaching but 'mixed', having charge of hospitals and pharmacies, orphanages and nurseries, or taking catechism classes. In these capacities they could either undertake propaganda to recruit to Catholic schools, or function as clandestine Catholic schools in their own right.

The last episode of the story is one of Catholic revenge, firmly demonstrating the limits of official anticlericalism in regions that were historically Catholic and determined to maintain some form of Catholic education. Catholic schools were now those of secularized congregations or a pious laity, standing against the public schools that proclaimed an official neutrality in matters of religion, and each was striving to maximize its clientele. Whereas with the introduction of 'moral and civic education' in 1882 the Catholics had made capital out of the slogan 'l'école sans Dieu' in order to populate the private schools of the congregations, now they sought to populate the private secularized schools by making a *cause célèbre* of the use of unpatriotic history textbooks. The context was an attempt by radical deputies to prosecute parents or tutors who prevented their children or wards from attending the whole of the course made compulsory under the law of 28 March 1882, or from reading the books prescribed by the departmental commission. In reply, the bishops of France published a letter on 14 September 1909 denouncing the usurpation by the state of the imprescriptible right of families under natural and divine law to educate their children. The school was an extension of the family, and Catholics were obliged to send their children to Christian schools on pain of non-absolution. Only in exceptional circumstances might they use official schools, and then families and priests must ensure that religious instruction was given outside the class-room. In addition, the conscience of the child must be kept whole and priests and parents, urged to form parents' associations, must increase their vigilance as to the practices of schools. Above all, they must verify that certain history textbooks now placed on the Index were not used, apply pressure to the lay teacher to withdraw them if they were, and, if nothing was done, remove their children from the school.[5]

A reading of these texts and a comparison with texts used in private schools reveals that there were indeed republican and Catholic interpretations of the history of France.[6] For the Catholics, the Middle Ages symbolized a chivalric, Christian, and paternalistic world, while for the republicans it was characterized by feudal strife, serfdom, and intolerance, relieved only by the emergence of the towns, *Tiers État*, and Estates General. For Catholics, the Reformation was the revolt of arrogant individualism against order, authority, and tradition that could only result in anarchy; for the republicans, it represented the triumph of the liberty of conscience. For Catholics, the French Revolution was the rule of sects and the Terror; for the republicans, it marked the assertion of the sovereignty of the people over divine-right monarchy. Having said that, it should be emphasized that some of the texts had been used in the *écoles communales* for ten or twenty years, and that the standards of history teaching were notoriously low, so that it was the rhetoric rather than the substance of the debate that was important. The controversy witnessed an attempt by the

---

[5] *Lettre pastorale des Cardinaux, Archevêques et Évêques de France sur les droits et devoirs des parents relativement à l'école*, 14 Sept. 1909.

[6] The texts in question were A. Aulard and A. Debidour, *La Première Année d'histoire de France* (1894), C. Calvet, *Histoire de France* (1898), G. Devinat, *Histoire de France* (1898), Gauthier and Deschamps, *Cours élémentaire d'histoire de France* (1904), Guiot and Mane, *Histoire de France* (1905), L.-G. Rogie and P. Despiques, *Histoire de France* (1907). On the texts used in private schools, see J. Freyssinet-Dominjon, *Les Manuels d'histoire de l'école libre, 1882-1959* (Paris, 1969).

Catholic clergy to recover an influence in elementary schools that it had not enjoyed since the Revolution. A population in the grip of the clergy, and in some cases of the traditional alliance of priesthood and squirearchy, would in the first instance send a delegation to the schoolteacher asking for the withdrawal of the offending book, then instruct their children not to recite their history lessons, and finally, *in extremis*, boycott the school, all of which created an atmosphere in which the difficult birth of a private school was made more likely. At the provincial level, the extent to which local populations accepted or refused the clergy's offensive provides a reasonable estimate of the geography of clericalism and anticlericalism on the eve of the First World War, and makes it possible to measure the changes that had taken place after over a century of conflict over the school.

Anticlericalism was a useful weapon for the republicans only in certain circumstances. While the monarchist and Bonapartist Right remained a danger and the working classes were still relatively unorganized, the cult of patriotism and civic virtue could be employed as a force for social integration, mobilizing popular elements in a Jacobin alliance against counter-revolution. But in the years after 1906 class conflict grew more severe and war with Germany became ever more menacing. The threat to society was no longer the Right but the labour movement and socialism, ready to use the strike weapon, more powerful in Parliament, and antimilitarist if not antipatriotic in the name of the international working class. By 1913 the issue of clericalism and anticlericalism was dead, and republican notables now looked for an alliance with the Right. In this second Ralliement, a Union nationale was constructed by Poincaré around the campaign to restore the three-year military-service law abrogated by the Bloc des Gauches. The enemy was the SFIO and CGT, together with the Radical Party reorganized under Caillaux. The teachers' unions, which on this question were out of line with the national orthodoxy, were dealt with severely. Lay education was in some respects forsaken in the interests of the Union Sacrée.

The Response in the Provinces (2)

The effect of successive waves of anticlerical legislation on the provinces would be determined largely by a geography of attitudes that in 1860 was already firmly engrained. Where the influence of the Catholic Church was insignificant or strongly contested, anticlerical measures would be enforced with no difficulty, even welcomed as liberation from the grip of the clergy. Where, on the other hand, Catholicism was solidly entrenched, the lay laws would be fiercely resisted. Our three departments demonstrate a graded scale of results. In the Nord, clearly divided between Catholic Flanders and a more irreligious Cambrésis and Avesnois, the anticlerical legislation reinforced the division between north and south. In the Gard the hold of Catholicism was more tenuous, challenged not only by Protestantism but also by its most recent avatar and ally, *laïcité*, but it made up for quantitative weakness by its militancy. Though the presence of Catholicism was certainly undermined by the end of the period, in its strongholds along the flank of the Cévennes it remained impregnable. The most significant case was nevertheless Ille-et-Vilaine, the last rampart of the feudal nobility of Maine and Anjou and the vestibule of pious, saintly, superstitious Brittany. There the laws of Ferry and Combes remained without effect, as if shipwrecked on its granite base.

(a) The Nord: Compromise

In the Nord, the anticlerical offensive clearly aimed at the erosion of the clergy's bastion of power in Flanders, both Flemish and French-speaking. The alliance between the Empire and the Church against revolution had been sealed by the election of Kolb-Bernard as an official candidate at Lille in August 1859, but the Italian question snapped that alliance and in the elections of June 1863 the official candidate was defeated at Hazebrouck by Ignace Plichon, a convert from Orleanism, for whom the Flemish clergy had campaigned fiercely.[1] Dehaene, principal of the college of Hazebrouck, had not been foreign to the candidature of Plichon, and Victor Duruy, anxious to eliminate not only priest-principals but also the practice whereby a principal of the University openly funded rival ecclesiastical institutions and appointed their staffs, seized the opportunity to dismiss Dehaene on 6 March 1865. On the 8th, known locally as the 'journée de la débâcle', masters and pupils scuttled the college, smashing windows and tearing up paving-stones, leaving only four lay masters and a few sons of *fonctionnaires.*[2] Mgr Regnier proposed at once to Dehaene that he open a private ecclesiastical institution that would serve as 'the *petit séminaire* of Flanders and the north of the diocese'.[3] The inauguration of the Institution Saint-François d'Assise in October 1865 symbolized the victory of a Catholic province over the centralizing state.

Resistance was also mobilized in Flanders against Duruy's experiment of *cours secondaires* for girls. Mgr Regnier followed the initiative of Dupanloup, bishop of Orleans, issuing a pastoral letter from Cambrai that was published in the legitimist *Émancipateur* and read from the pulpits. The 'haute bourgeoisie of Lille' was said to have boycotted the scheme, which succeeded in the Nord only at the College of Valenciennes, under the patronage of a young *avocat*, Louis Legrand, who demanded to know 'in the name of what principle is the intelligence of women being kept in darkness?'[4] But Valenciennes was an industrial town in the barely Catholic south of the department.

The decisive battle in the Nord nevertheless took place as the result of a local initiative. Since 1848 the Frères des Écoles Chrétiennes had found their way back into the municipal schools of Lille, and now directed nine. But the

---

[1] J. Lemire, p. 283, 'La Flandre a fait Plichon et Plichon a fait la Flandre'. M. Emerit, 'Du Saint-Simonisme au Catholicisme: Ignace Plichon, député du Nord', *Revue du Nord*, lvi, no. 220, 1974, 29–42.

[2] A.D. Nord 1T 52/2, sub-prefect of Hazebrouck to prefect of Nord, 23 Mar. 1865; Lemire, pp. 290–5.

[3] Mgr Regnier to abbé Dehaene, 10 Mar. 1865, cit. Lemire, p. 351.

[4] A.N. F17 8755, Rector of Douai to Min. Instr. Pub., 26 Feb. 1868; *Écho de la Frontière*, 14 Nov. 1868.

over-expansion of teaching congregations meant that while the directors of primary schools would have the *brevet*, scarcely trained novices employed as assistants in the 'small' or 'alphabet' classes rarely did. It was on this weakness that the anticlerical municipal council of Lille focused. It had been campaigning since 1863 to have every teaching brother in its schools, assistant as well as director, equipped with the *brevet*, but it was the law of 10 April 1867 which gave it the authority to cut the salaries of non-qualified *frères*.[5] Incensed, Frère Philippe, superior-general of the Frères des Écoles Chrétiennes, threatened to withdraw all forty-three of his teachers unless the municipality reconsidered its position. But the municipality wanted nothing better than laicization. On 21 June 1868 the teaching brothers duly moved out of the municipal schools of Lille, and lay *instituteurs*, assistants, and sufficient *élèves maîtres* from the École normale moved in at 8 a.m. on 22 June.[6] The clerical aristocracy and bourgeoisie of Lille, headed by the Comte de Melun and the *avocat* Théry, with the vicar-general, Charles Bernard, acting as liaison with Cambrai, were equal to the task of collecting enough funds to reopen privately the schools of the Frères des Écoles Chrétiennes. A Comité des écoles libres was set up to undertake the task, a body which interlocked with the branch of the Société d'éducation et d'enseignement founded at Lille by the Comte de Melun in 1869.[7] But because pupils now had to pay school pence, the Frères tended to lose the mass of the working class and fall back on the labour aristocracy and *classes moyennes*.[8]

The debates which followed the defeat of 1870 revealed once again in the Nord the tension between Flanders and the interior. Louis Legrand, representing Valenciennes, introduced a motion on compulsory free education into the

---

[5] A.D. Nord 1T 90/2, municipal council of Lille, sessions of 22 Dec. 1865, 13 July 1866, 4 and 30 Dec. 1867. See R. Hemeryck, 'La Laïcization des écoles des Frères à Lille en 1868', *Actes du 95e Congrès national des sociétés savantes*, Reims, 1970 (Paris, 1974), 869–75.

[6] A.D. Nord 1T 84/3, Insp. Acad. Douai to Min. Instr. Pub., 1 July 1868.

[7] A.D. Nord M 222/773, prefect of Nord to Min. Int., 13 Mar. 1868; Comte de Melun to prefect of Nord, 4 Mar. 1869.

[8] Member of Comité des écoles libres, Lille, to superior-general, 12 Nov. 1874, cited by Hemeryck, 'La Laïcization des écoles des Frères à Lille', 894.

Conseil général in October 1871, arguing that 'ignorance is the principal cause of our political abasement and of our misfortunes' and that 'a wide diffusion of enlightenment is the necessary corollary and natural auxiliary of republican institutions'.[9] He was supported by seven councillors from the 'deChristianized' south of the department and one for Dunkerque. In the following spring the petition of the Ligue de l'Enseignement obtained 46,000 signatories, 1.4 per cent of the population, notably in the *arrondissements* of Douai, Cambrai, and Avesnes, with some support on the coast at Dunkerque and Gravelines and in the working-class town of Roubaix. On the other hand, 116,000 signatures were collected by the *comités catholiques*, representing 4 per cent of the population, far in excess of the 1.2 per cent recorded for France as a whole. Flanders provided the basis of the support, industrial French-speaking Flanders just as much as rural *Flandre flamingante*.[10]

The period of Moral Order permitted the Catholics not only to strengthen their position in Flanders but also to make gains further south. The circular of 28 October 1871 on municipal options between lay teachers and congregations played into the hands of the congregations, notably in the *arrondissements* of Dunkerque, Hazebrouck, and Lille. At the secondary level, the abbé Dehaene transformed the Institution Saint-François d'Assise into what it really was, a *petit séminaire*. Cardinal Regnier was happy to shoulder the additional burden in order to draw on the Flemish *terre sainte* as much as possible for his priests.[11] Flemish municipal colleges run by priests and independent ecclesiastical colleges also prospered. Indeed, for the first time since the Revolution the Jesuits returned to Lille to open a college, taking over the day-school of Saint-Joseph that the Société Saint-Bertin ran in conjunction with its boarding-school at Marcq. The Jesuit College opened under Père Pillon in 1872, growing so rapidly that it had 500 pupils when it

---

[9] A.D. Nord 1N 113, Conseil général, 25 Oct. 1871, p. 680.

[10] B. Ménager, 'Prélude à la bataille scolaire: les pétitions de 1872 concernant l'enseignement primaire dans le département du Nord', *Revue du Nord*, lv, no. 217, 1973, 135-43.

[11] Lemire, p. 515. The decree is of 12 December 1873.

transferred to new premises of magnificent ugliness in the rue Solférino in 1876.[12] Lastly, the Catholic *patronat* of the Nord, Kolb-Bernard, the *négociant* Henri Bernard, the linen-spinner Philibert Vrau, together with the abbé Bernard, vicar-general to Cardinal Regnier and the Comte de Caulaincourt, a leading social Catholic, were at the forefront of the campaign for Catholic higher education. By 1877 Faculties of Law, Letters, Science, and Medicine had been set up at Lille, making it one of the Catholic universities alongside Paris, Angers, Lyon, and Toulouse.[13]

Change came with the elections of February 1876, at which the republicans gained eleven of the eighteen seats in the Nord. Authoritarian practices after the Seize Mai *coup* took three seats from the republicans in 1877, but the electoral geography remained consistent: the monarchists strong in Flanders, the republicans having the advantage in the large towns and in the south, especially around Valenciennes.[14] How far the republicans could now undermine the electoral influence of the clergy and the monarchists was now the central issue. By opening their first salvo against the Jesuits, the republicans hoped to mobilize mass support for the anticlerical campaign. Cardinal Regnier noted that 'the attacks levelled against Catholicism are masked under the name "Jesuitism". "Jesuitism", there is the bugbear of the moment.'[15] Yet the Catholic élite always had the means to obtain for itself the education it wanted. Père Pillon was brought before the Conseil académique of Douai in December 1880 on the charge of 'reconstituting a congregation' and was duly suspended.[16] But the Collège Saint-Joseph continued under the abbé Baunard, a professor at the Catholic University

---

[12] P. Delattre, op. cit., vol. ii (Enghien, 1953), 1332-5.

[13] On the Catholic University of Lille, see Mgr Baunard, *Les Deux Frères: cinquante années de l'Action Catholique à Lille: Philibert Vrau, Camille Feron-Vrau* (Paris, 1911), i, 191-219; J. Gadille, *La Pensée et l'action politique des évêques français au début de la Troisième République, 1870-1883* (Paris, 1967), i. 144-5, 340-1; P. Pierrard, 'Les origines de l'enseignement supérieur catholique à Lille', *Ensemble d'écoles supérieures et de Facultés Catholiques* (32e ann. no. 1, 1975), 3-33.

[14] Y.-M. Hilaire, *Atlas électoral Nord, Pas-de-Calais, 1876-1936* (Lille, 1977), pp. 44-7.

[15] Cit., Orhand, p. 310.

[16] A.D. Nord 6V 57, Conseil académique of Douai, 10 Dec. 1880.

of Lille, and in 1884 the Jesuits returned to open a so-called annexe or Petit Collège of Saint-Louis de Gonzague.[17]

The establishment of *cours secondaires* for girls was an assault on one of the main strongholds of Catholic education. At Douai, the Inspector of the Academy reported in 1879 that 'the secondary education of girls is almost exclusively in the hands of the clergy. Confronted by the powerful institutions of the Dames de Flines and those of the Sainte-Union, where the bourgeois families of Douai, even the liberal ones, send their daughters, only one lay *pension*, that of the Mlles Théry, has been able to survive.'[18] The municipality of Lille had founded an E.P.S. in 1870, and a second, paying girls' school, the Institut Fénelon, in 1877. Dunkerque set up a *cours secondaire* in 1878, Roubaix, Valenciennes, and Douai courses in 1879, Cambrai in 1880, and, after the Camille Sée law, Armentières in 1881 and Maubeuge in 1882. But these were little more than municipal lecture courses, hoping to draw on the ladies' boarding-schools for an audience, and the system that had developed at Douai and Valenciennes whereby academic standards were raised, yet the virtues of enclosure and education preserved by the giving of private lessons in lay *pensions* by *lycée* masters, had proved itself a viable alternative. The clientele of the *cours secondaires* therefore remained humble, composed at Valenciennes of Protestants, the daughters of *universitaires*, and girls from the *écoles communales*. And as the mathematics master who was its organizing genius observed, 'it is not with pleasure that many parents would see their daughters sitting on the same bench as girls from the *école communale*'.[19] In 1888, the *cours secondaire* at Valenciennes was elevated into a college, but with only 52 pupils against 250 in private lay *pensions* and 390 in convents.[20] Only the introduction of the *internat* would make possible an increase in numbers and an *embourgeoisement* of the intake.

[17] P. Delattre, op. cit., vol. ii, 1341–3, 1352; H. Beylard, art. cit., p. 81.

[18] A.D. Nord 2T 917², Inspec. Acad. to Rector of Douai, 31 May 1879.

[19] A.D. Nord 2T 917⁶, *professeurs* of Lycée of Valenciennes to Rector of Douai, 8 Jan. 1885; Cannac, maths. master at Valenciennes to Insp. Acad., 8 Dec. 1879.

[20] A.D. Nord 2T 939, *directrice* of College of Valenciennes to Rector of Lille, 22 Jan. 1889. (The seat of the Academy was moved from Douai to Lille in 1887.)

The importance of local attitudes was again clear in the question of the municipal colleges—now confined to Flanders—that were still in the hands of the secular clergy. Clerical elements could not be tolerated in the University after the republican victory, but nothing could be done so long as the municipalities maintaining the colleges remained in Catholic-royalist hands. After the elections of January 1881 the republicans secured a majority on the municipal council of Tourcoing and negotiations began with the Academy about state aid for a *lycée* of special education, adapted to the needs of the industrial towns of Tourcoing and Roubaix.[21] Meanwhile the abbé Leblanc, principal of the college, was hauled before the Rector and told that his college was 'not part of the University except in name . . . that in reality, it was nothing but a *petit séminaire*'; to continue, it must 'become lay from head to foot'.[22] The municipal council broke its links with the college, but Tourcoing was not only industrial, it was also deeply Catholic. While the authorities constructed a gigantic *lycée*, abbé Leblanc opened the Institution Libre du Sacré-Cœur and soon had over 400 pupils. In clerical Flanders, Leblanc could count on forces to fill his school: as the *proviseur* of the Lycée complained in 1891, the clergy dominated the industrialists, and the industrialists dominated the small shopkeepers, the commercial clerks, and the foremen who risked losing their clients or their jobs if they sent their sons anywhere but the Sacré-Cœur.[23]

The campaigns of laicization at the level of elementary schools must be set in the context of the geography of education in the department. The initiative left to communes to choose between lay teachers and congregations and the liberty to found private schools tended to give each region the type of schooling it aspired to or deserved. In 1879–80 the division was clear between Flanders and the interior, boys' schools run by the teaching congregations largely

---

[21] A.D. Nord 2T 858, Zévort, Directeur de l'Enseignement secondaire, to Rector of Douai, 7 July 1881.

[22] H.-J. Leblanc, *Le Collège communal de Tourcoing, 1858-83* (Lille, 1885), p. 248.

[23] A.D. Nord 2T 308⁴/³, *proviseur* of Lycée of Tourcoing to Insp. Acad. 10 Oct. 1891.

confined to Flanders, with the exception of Saint-Amand, nuns' schools spreading further south but thinning out in the Avesnois and parts of the Cambrésis.[24] This pattern is reinforced by the geographies of recruitment in the nineteenth century of the Catholic clergy on the one hand and of the lay *instituteurs* trained in the *écoles normales* on the other. In the Nord it was Flanders, and above all *Flandre flamingante*, without the rival attractions of industrial employment, its Catholicism protected against new ideas by a barrier of language, that produced priests (Map 5).[25] By contrast, it was the coastal fringe, and especially the southern half of the department, that bred *instituteurs* (Map 6).[26]

The 'spontaneous' laicizations following the municipal elections of 1878 and 1881 made even closer the correspondence between schooling and mentalities. The Flemish *arrondissements* were not stirred, but the teaching congregations suffered important losses in the *arrondissements* of Douai, Valenciennes, and Avesnes.[27] Roubaix followed Lille's initiative of 1868 in 1882, led by the radical Émile Moreau, who protested against the 'ceaseless invasion of the Ignorantins'.[28] But the Catholic camp was prepared, and not just the clergy but the lay élite, led by Philbert Vrau and the industrialist, Jonglez de Ligne, which was busy raising subscriptions to fund private schools run by the congregations.

The fighting spirit of northern Catholicism was again demonstrated over the application of the law of 28 March 1882 on compulsory lay education. The Catholic community was dead set against the abolition of the catechism in schools, and forced their opinions on to its new archbishop, Mgr Duquesnay (1814-84), elevated to Cambrai as a prelate who would accept their new political regime. But though as bishop of Limoges he had drunk a toast on 13 February 1881 to the Republic, the following August at Lille, supported if not pressurized by the notables of the Comité catholique, he proclaimed, 'legality is an adversary

---

[24] A.N. F17, 10558, État de situation, 1879-80.
[25] E. Masure, *Le Clergé du diocèse de Cambrai, 1802-1913* (Roubaix, 1913).
[26] École normale of Douai, Registres d'inscription, 1856-1913.
[27] Ménager, *Laïcisation*, p. 83.
[28] A.F.E.C. NC 788, municipal council of Roubaix, 17 Jan. 1882.

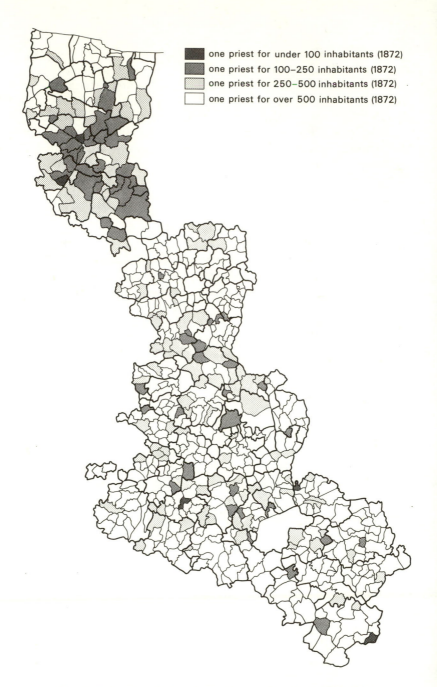

Map 5. Ordination of priests in the diocese of Cambrai, 1802–1913

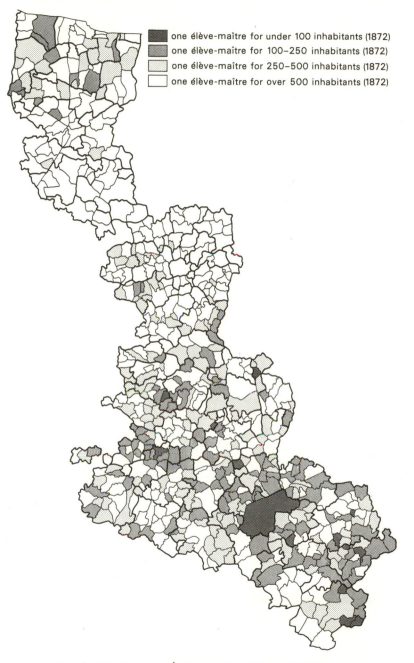

one élève-maître for under 100 inhabitants (1872)
one élève-maître for 100–250 inhabitants (1872)
one élève-maître for 250–500 inhabitants (1872)
one élève-maître for over 500 inhabitants (1872)

Map 6. Recruitment to École normale of Douai, 1856-1913

with which we have to contend'.[29] The republican administration cried in despair that 'Duquesnay has swindled us',
and the Conseil général of the Nord voted by 29 votes to
22 to suppress the 'budget des cultes' in retaliation.[30] The
Comité catholique of Lille revealed its own extremism over
the law of 28 March, launching on 21 April 1882 a Catholic
Ligue de résistance against 'compulsory atheism', all the
members of which swore on oath not to send their children
to the 'école sans Dieu'.[31] Duquesnay nevertheless struggled
to assert his independence of this lay élite, founding in May
1882 a Comité diocésain d'enseignement libre under an
ecclesiastic in order to loosen the coils of the Comité catholique.

In a region as anticlerical as that of Avesnes, the law of
28 March was enforced without any difficulty. But in the
arrondissements of Lille and Hazebrouck parish priests, who
dominated many of the school commissions set up under
the same law to enforce compulsory lay education, connived
at absenteeism on the part of opponents of the 'école sans
Dieu'.[32] Besides, there were always difficulties in applying
'independent morality' in practice. Old, tired, and incapable
instituteurs were only gradually replaced by young men with
'the greenness of age, but also the necessary knowledge and
pedagogic vigour'.[33] Recruitment to the École normale of
Douai was upgraded in the years 1893-1914, in that while
10 per cent of trainees were from the ranks of *cultivateurs*,
15.2 per cent sons of wage-earners and the same proportion
from the *artisanat*, 10.8 per cent came from the world of
commerce, 20.9 per cent were sons of *employés* and *petits
fonctionnaires*, and 19 per cent were second-generation
instituteurs.[34] The only problem was that the establishment

[29] J. Gadille, 'Monseigneur Duquesnay et la République, 1872-1884', *Revue
du Nord*, xlv, no. 178, 198.

[30] Gadille, p. 205; A.D. Nord 1N 125, Conseil général, 25 Aug. 1881.

[31] A.D. Nord 1T 68/1, report of *commissaire spécial*, 21 April 1882; Ménager,
Laïcisation, pp. 54-5.

[32] Ménager, pp. 67-8, 75.

[33] A.D. Nord 1N 136, Conseil général, 1892, report of Directeur départemental, p. 557.

[34] École normale of Douai, Registres d'inscription, 1893-1913. The sample
is of 1093 cases.

of a hierarchy of pay-scales as early as 1860 had been accompanied by the rigidity of promotion by seniority, so that in the large towns an *instituteur* might be thirty-five before he had the direction of a municipal school.[35] Secondly, a teacher who tried to preach the new gospel in a deeply Catholic area would see numbers diminish rapidly, especially if a private school run by a congregation existed nearby. As the *institutrice* of Hellemmes, near Lille, confessed in 1900, 'if I suppress all religious practices, the nuns will carry off half my pupils'.[36] At the turn of the century the socialist *Réveil du Nord* ran a campaign to 'laïciser les écoles laïques'. Lastly, catechism was still taught outside the school, and resistance to modern ideas, the vehicle of which was Flemish, could be reinforced by use of Flemish, which Mgr Duquesnay learned rapidly to call 'the language of Heaven'.[37] For the Directeur départmental the Flemish language was archaic and impervious and 'must be replaced by French in order to open up the country to the influence of the *siècle*'.[38] But the revenge of the Catholics came in August 1885, when the Nord fell to the conservative list. For Jules Cambon, the prefect, this was a manifestation of the ' "Flemish soul", the same temperament which in the sixteenth century, that golden age of their history, drove them to break with Orange at the same time that they were resisting the Duke of Alba'.[39]

The state of affairs was clearly intolerable and the tough prefect Saisset-Schneider, a converted satrap of the Moral Order regime who replaced Cambon in 1887, was to use the legislation of 30 October 1886 finally to break the spell of the teaching congregations. But still problems rose. Laicizations set up increased demands for staff, and the stipends of the teaching body were so meagre that the authorities spoke

---

[35] A.D. Nord 1N 140, Conseil général, report of Directeur départemental, 27 July 1896, p. 248.

[36] *Réveil du Nord*, 20 Dec. 1900, cit. Ménager, p. 266.

[37] A.D. Nord 2V 76, sub-prefect of Hazebrouck to prefect of Nord, 16 Oct. 1882.

[38] A.D. Nord 1N 134, Conseil général, report of Directeur départemental, 1 Aug. 1890, p. 581. ('Directeur départemental' was the new title of the Inspector of the Academy in the Nord.)

[39] A.N. F19 5900, prefect of Nord to Min. Cultes, 23 Dec. 1885.

of a 'crise des instituteurs',[40] while on the other hand the nuns were being expelled so slowly from the *écoles communales* that not all qualified *institutrices* could find posts. At the same time, voluntary subscriptions and the patronage and pressure of landowners and industrialists multiplied the private schools in which the congregations could teach. Between 1879-80 and 1899-1900 the proportion of the school population that attended the schools of the congregations (public or private) fell in the Nord from 24.7 per cent to 18.8 per cent in the case of boys and from 62.4 per cent to 45.2 per cent in that of girls. This represented a loss of only about a quarter of the clientele, and though the teaching congregations were forced back geographically to their strongholds, the old pattern of education was ineradicable. Congregations were eliminated from boys' schools south of Lille and from girls' schools in the Pévèle (cantons of La Bassée, Seclin, Pont-à-Marcq, and Cysoing) and south of the Escaut, while their hold on the coal-basin in between was weakened. But their positions in *Flandre flamingante* and industrial Flanders remained impregnable. It has been shown that Catholicism still had a strong hold on many textile workers of the Lys valley and the Roubaix-Tourcoing area, often only recently uprooted from the Catholic country-side of Béthune or Flanders on either side of the Belgian frontier.[41] And a refusal to countenance laicization was demonstrated at the polls, for laicization did not produce the electoral advantage sought by the republicans. Flanders rallied to Boulanger in the by-elections of August 1888 and returned a phalanx of royalists together with a Boulangist in the *arrondissement* of Dunkerque in the general elections of September-October 1889.[42]

After 1890 the anticlerical offensive was in any case suspended because the threat of socialism encouraged the ruling groups, republican or royalist, to sink their differences and rally in defence of property. The threat in the Nord was

[40] A.D. Nord 1N 138, Conseil général, report of Directeur départemental, 31 July 1894, p. 790.

[41] Y.-M. Hilaire, 'Les ouvriers de la région du Nord devant l'Église Catholique, XIXe-XXe siècles' *Mouvement social*, no. 57, Oct-Dec. 1966, 181-201.

[42] J. Néré, 'Les Élections de Boulanger dans le département du Nord' (thesis, Paris, 1959), p. 197; Hilaire, *Atlas électoral*, pp. 72-4.

particularly clear, as Marx's son-in-law Paul Lafargue was elected to the Chamber of Deputies for Lille in November 1891, profiting from indignation over the 'fusillade de Fourmies' at the other end of the department, as Guesde's Parti Ouvrier took the industrial centres of Roubaix and Caudry in the municipal elections of 1892, and as Guesde was himself elected deputy for Roubaix in 1893.[43] In response, there was a shift to the right, away from socialists and radicals, on the part of moderate republicans grouped around the *Écho du Nord*, and a Ralliement to the republican regime of most royalist deputies. Royalism in the Nord never had a backbone of legitimist *châtelains* to support it, and the Orleanist landowners of the Cambrésis, magistrates of Douai, and Catholic *patronat* of Flanders were now preprared to use the Republic to obtain protective tariffs and measures of social defence, while campaigning to make it more religious, more conservative, and more national.[44] The Ralliement left the legitimist rump isolated and achieved some success in the legislative elections of May 1898, when Guesde was defeated at Roubaix by the textile magnate Eugène Motte.

The displacement of the conservatives in France as a whole in those elections and their exposure to the Dreyfusard campaign for revision, which threw together *ralliés* and intransigents and provoked republican concentration in the face of counter-revolution, had a curious parallel at Lille, its own Dreyfus affair. The Flamidien case appealed vividly to the anticlerical imagination, depicting as it did the perverse nature of celibacy. On 8 February 1899 the body of a young boy, missing for three days after attending the *patronage* or youth club of the Frères des Écoles Chrétiennes in their school complex at the former Hôtel de la Monnaie, was discovered in a little-frequented part of the building. When traces of blood and semen were found on the sheets of a Flemish brother, Flamidien, rumours spread like wildfire, and anticlerical riots broke out.[45] Bands roamed the

[43] C. Willard, *Les Guesdistes. Le mouvement socialiste en France, 1893–1905* (Paris, 1965), p. 224.

[44] A. Bonnafous, 'Les Royalistes du Nord et le Ralliement', *Revue du Nord*, xlvii, no. 184, 1965, 29–41.

[45] A.D. Nord 1T 123/12, Commissaire spécial to prefect du Nord, 8 Feb. 1899.

city shouting 'A bas la calotte!' and insulting priests. The school was closed by the prefect, but crowds chanting the Internationale pelted stones at the windows of the former Jesuit College of Saint-Joseph and the offices of *La Croix*, so that the horse-gendarmerie had to be brought up to protect them.[46] The day after the boy's funeral, the socialist leader Paule Mink lectured at the Maison du Peuple, denouncing clerical education and the clergy's manipulation of women: the confessional had not only been used after 1848 and 1870 to extract denunciation of their husbands' revolutionary activities, but it had also degenerated into 'a permanent orgy. Across the narrow partition, mouth against mouth, confessor and penitent exchange amorous confidences, deflowering by their avowals the secret of restless nights.'[47] Paule Mink's speech provoked a new wave of rioting. An *Appel à la Population*, urging families to withdraw their children from the schools of the congregations, the signatures headed by that of the socialist mayor of Lille, Gustave Delory (1857-1925), who well remembered his education at the hands of the Frères until the age of eleven, was posted on the city walls and anticlericalism became the theme of carnival proceedings at Lille and Roubaix.[48]

The Flamidien affair transformed the pattern of political alignments in the Nord. The moderate republicans of the *Écho du Nord* broke away from the conservatives and joined forces not only with the radicals but also with the socialists against clericalism. There were now 'on one side the Catholics, liberal conservatives, all convinced partisans of the innocence of brother Flamidien; on the other, all the republicans, from moderates to collectivists, upholding with no less conviction the guilt of Flamidien'.[49] The Catholic party campaigned to shift guilt for the murder on to a 'syndicate' of *agents provocateurs*, while on the republican side it was the socialists— in defiance of Guesde's warnings that anticlericalism was a blind alley, a means by which the bourgeois republicans

[46] A.D. Nord 1T 123/12, prefect of Nord to Min. Int., 17 Feb. 1899.

[47] Ibid., lecture of Paule Mink reported by *Progrès du Nord*, 13 Feb. 1899.

[48] Ibid., Commissaire central, Roubaix, to prefect of Nord, 10 Mar. and 7 April 1899.

[49] Ibid., Commissaire spécial, Lille, to prefect of Nord, 26 Feb. 1899.

could head off consideration of social questions—who led the attack on the Church in the Conseil général. In April 1899 they pressed for support for a bill that would disqualify from teaching all those who had taken a vow of chastity.[50] And in August 1900, after victory in the municipal elections, they introduced motions for the abolition of all clerical garb, for the secularization of property held 'in mortmain', and for the prompt application of the decrees of 1880 for, as one of them said, twenty years after their expulsion 'it was not possible to take a single step in the streets of Lille—or anywhere else—without treading on a Jesuit'.[51]

The geography of anticlericalism in the Nord was made clear again as the official attack on the private schools of the congregations gathered momentum. The inquiry of 1902 shows that in communes around Lille and along the Escaut, in industrial centres like Vieux-Condé, Denain, and Bouchain, a desire to be rid of the congregations was pronounced. But in Flanders, both around Hazebrouck and Bailleul and in the textile conurbation between Lille and the Belgian frontier, the congregations were hotly defended.[52] The polarization of opinion was clear in the legislative elections of 1902. The socialist vote declined, but radicalism strengthened its position in the Avesnois and at Tourcoing, while the Right was stronger in the *arrondissements* of Dunkerque, Hazebrouck, and Lille, with 35 per cent of the vote in the Nord as against 23 per cent in 1898.[53] Indeed, one dimension of the response to the anticlerical offensive was becoming increasingly important: the assertion of Flemish regionalism in defence of Catholic integrity. In forty-three northern communes in 1900 the catechism was still taught in Flemish and the *curé* of Killem (Hondschoote) was hauled over the coals for refusing to teach the *instituteur's* son the catechism in French. Waldeck-Rousseau informed the archbishop that from 1 January 1901 the stipends of

---

[50] A.D. Nord 1N 143, Conseil général, 10 April 1899.

[51] A.D. Nord 1N 144, Conseil général, 21 Aug. 1900, p. 104; 30 Aug. 1900, pp. 580-1. See in general, J. Bruhat, 'Anticléricalisme et mouvement ouvrier en France avant 1914', *Mouvement social*, no. 57, 1966, 61-100.

[52] A.D. Nord 6V 64, 88, 91, 114-16, 165-6, 183-5, 197-9.

[53] Y.-M. Hilaire, *Atlas électoral*, pp. 118-25.

the parish clergy who refused to catechize in French would be suspended. But the official view at Cambrai was that Flemish was the mother tongue in both French and Belgian Flanders, and that 'while people study French at school, at home, in the street, at games, at work, they instinctively speak Flemish'.[54] When attempts were made to apply the Associations law under Combes's circular of 9 July 1902, *Le Figaro* reported that '*Flandre flamingante* is seething.' There was turmoil in the cantons of Dunkerque, Hondschoote, Bergues, and Wormhoudt and it was claimed that 'Belgian newspapers of the frontier towns are exploiting these disturbances to bring home to the Flemish population of France the advantages of a separatist movement that would assure them respect for their traditions and for their religious faith'.[55]

Even with the full force of the Bloc des Gauches mobilized, there were limits to the success of radical persecution. Republican municipalities in Catholic areas were placed in a difficult position on the eve of the 1904 municipal elections and the mayor of Merville, a small Flemish town on the Lys and the only republican cantonal capital in the *arrondissement* of Hazebrouck, warned the sub-prefect that a closure of the school of the Filles de l'Enfant-Jésus before the elections would be suicidal.[56] Industrial clericalism was also an obstacle. Where the local employer such as the mining company of Aniche had built the girls' private school and the municipality had no resources to build another public school, the nuns could expect a stay of execution until premises had been found for pupils who would otherwise be turned on to the street.[57] In the Nord, exile was always a possibility, and this time it was the *classes moyennes* rather than the aristocracy who profited from the proximity of the Belgian frontier. The Marists moved their noviciate at Beaucamps to Pommereul, between Condé and Mons, although

---

[54] A.D. Nord 123/5, Waldeck-Rousseau to archbishop of Cambrai, 18 Dec. 1900; notes made by ? vicar-general Lobbedy, archdeacon of Dunkerque-Hazebrouck, for meeting with prefect of Nord, 11 Dec. 1901.

[55] A.N. F19 6083, cutting from *Le Figaro*, dated Dunkerque, 25 July 1902.

[56] A.D. Nord 6V 116, mayor of Merville to sub-prefect of Hazebrouck, 23 July 1903.

[57] A.D. Nord 6V 198, telegram of Min. Cultes, 3 Aug. 1903.

Hainaut would provide considerably fewer priests than Flanders, and in 1906 opened a *pensionnat* destined to replace those of Lille, Haubourdin, and Beaucamps at Péruwelz, some miles to the north of Pommereul.[58] The Frères des Écoles Chrétiennes of Passy in Paris built a massive *pensionnat* at Froyennes which also accommodated the pupils of the Pensionnat Saint-Pierre of Lille.[59] Lastly, many female congregations which did not confine their activities to teaching were permitted to survive and often continued to teach in a clandestine way. The Filles de Saint-Vincent de Paul were allowed to stay at Denain because the rapidly expanding steel town had no hospital or hospice other than that they provided, even though the director of the steel company at Denain–Anzin which employed them was reported to have 'transformed his factory into a religious establishment where a kind of White Terror reigns'.[60]

One further picture of the historical geography of the Nord is provided by the clerical offensive of 1909 against republican textbooks, designed to build up a clientele for the private Catholic schools no longer run—at least nominally —by the congregations. In the south of the department the campaign met little success. The municipal council of Petite-Forêt (Saint-Amand Rive Droite) attacked the 'exorbitant pretensions of M.M. les prêtres' and requested the government to take every measure possible to 'shelter the lay school from the Jesuitical attacks of the clerical brood'.[61] In the Avesnois the clergy were completely powerless. The *curé* of Taisnières-en-Thiérache was able to prevent only one pupil taking his Gauthier and Deschamps to school, and when an anxious mother inquired whether her son might postpone his history lessons until after the First Communion, the *instituteur* replied bluntly that in that case he would not be permitted to sit the *certificat d'études*.[62] But in *Flandre flamingante*, the campaign in favour of the private schools

[58] A.F.M. Ch. Fr. Norbert, 'Historique de la Province de Beaucamps, 1838–1944' (typed M.S. 1944), pp. 230–7; A.F.M. BEL 660, Péruwelz.

[59] A.F.E.C. NG 237.

[60] A.D. Nord 6V 198, Commissaire de police to sub-prefect of Valenciennes, 1 Aug. 1903.

[61] A.N. F17 9125[6], municipal council of Petite-Forêt, 23 Oct. 1909.

[62] Ibid., Directeur départemental to Min. Instr. Pub., 5 Nov. 1909.

was triumphant. Tension was high in the cantons of Grave-lines, Hondschoote, Wormhoudt, and Bergues, at breaking-point in those of Bailleul, Merville, Hazebrouck, and Cassel. The clergy led the struggle from the pulpit, refused absolu-tion to parents and First Communion to children who patronized the official schools, while at Morbecque and Nieppe it was reported that 'children are buried without ceremony, their coffin undraped, with a scandalous ill will, simply because they attended the public schools'.[63] Lay *instituteurs* were as unwelcome and uninfluential in rural Flanders as the *curés constitutionnels* had been, over a century before.

On the eve of war militant *laïcité* was forced into a dif-ficult position because a vanguard of *instituteurs* was caught up with the socialist and CGT opposition to rearmament while the President Raymond Poincaré was seeking a 'Union nationale' with the Right in order to bring back the three-year military-service law abolished in 1905. The Syndicat des instituteurs et des institutrices laïques du Nord dated back to February 1906, but following the Congress of teachers' unions at Chambéry in August 1912 the Right saw a last chance for victory over lay education by denounc-ing the *instituteurs* as antimilitarist and antipatriotic. The circular of the Minister of Public Education dated 23 August 1912 thus condemned teachers' unions as 'seats of national disintegration' and ordered them to dissolve by 10 Septem-ber. Though the syndicate dissolved, it was immediately reconstituted as *l'Émancipation* and affiliated to the Bourse du Travail of Lille.[64] It drew on a wave of support from other unions, municipalities, and the socialist contingent in the Conseil général du Nord, who argued in favour of teachers' unions that they were antimilitarist, against the army as a strike-breaker, against expenditure on guns instead of butter, and against wars of colonial plunder, but that they were in no way antipatriotic.[65] Even at this juncture the old geography

[63] A.N. F17 9125[6], Directeur départemental du Nord to Min. Instr. Pub., 22, 26, 28 Oct., 16 Nov. 1909.
[64] A.D. Nord M 595/77, assembly of Syndicat des instituteurs, 8, 10, 24 Oct. 1912.
[65] A.D. Nord 1N 157, Conseil général, 23 Sept. and 2 Oct. 1912. The motion

was evident, for support for the campaign came from Lille, Roubaix, and the industrial area based on Valenciennes, including Denain and Vieux-Condé. The north of the department saw the advantage of appropriating nationalism as a weapon to hit back at lay education.

Catholic education demonstrated that in the Nord it was a vigorous and combative force. While the alliance of the Catholic Church with legitimist royalism was dispensable, its alliance with Flemish separation was symbiotic, each reinforcing the other. For that reason, the decline of the royalist cause did nothing to undermine its challenge; rather, in the face of the mounting social threat and the German menace, Catholicism could rally the support of property-owners and nationalists. There was not necessarily any contradiction between regionalism and nationalism: Catholics distinguished between the anticlerical Republic and eternal France, the eldest daughter of the Church. Moreover, in the nationalism of a thinker like Maurice Barrès, based on the cult of *la terre et les morts*, family, region, and nation were interdependent manifestations of solidarity, one growing out of another.

The battle over schools in the Nord was deeply influenced by historical conditions. Mgr Duquesnay, like Brossays Saint-Marc, was obliged to take up an uncompromising stance in order to retain the support of his flock. At the same time, persecution clarified the historical geography of the Nord, defined more boldly the areas that were prepared to stand by the Church and those that were not. Catholicism and indifference, clericalism and anticlericalism were not determined by class. There were anticlerical workers in Lille and Roubaix, but there were Catholic workers in Tourcoing and Armentières. There were anticlerical peasants in the Avesnois, fanatical peasants in *Flandre flamingante*. Flanders was prepared to build private Catholic schools in response to laicization, its clergy continued to teach the catechism in Flemish outside school, and even lay teachers were obliged to give a Catholic flavour to their teaching in order to keep the classroom populated. In the south of the department, on the

that the government end its persecution of teachers' unions was defeated 16:36.

other hand, there was little that clerical fanaticism could achieve, whether the clergy were fighting the Ligue de l'Enseignement, the state education of girls, or so-called antipatriotic history textbooks. The bastion of Catholic strength was Flanders, both Flemish- and French-speaking, on both the French and the Belgian sides of the border. While the south of the department was consolidated by anticlericals, Catholics dug themselves into the north. A sort of truce, if not deadlock, had been achieved.

(b) The Gard: Progress

In the Gard the exacerbation of the school question after the break with Rome was articulated along the line of conflict between Catholic and Protestant. While the Protestants were generally satisfied with Napoleon III's support for Piedmont in Italy, the clergy led by Emmanuel d'Alzon and the Catholic laity of the Sociétés de Saint-Vincent de Paul, especially where they were confronted by Protestant populations as at Nîmes and Alès, campaigned for the papacy.[66] Moreover, the Catholic hierarchy, no longer supported by the imperial regime, fell back on a counter-revolutionary and ultramontane position, denouncing the rising tide of revolution which, in the Gard, might easily be identified with Protestantism, the inevitable result of free thought and the revolt against God. This was certainly the tenor of Mgr Plantier's pastoral letter of January 1866, which extolled Pius IX as the 'defender and avenger of true civilization', standing for revealed truth against rationalism, authority against revolution, property against communism.[67]

The secondary secular education of girls was for Mgr Plantier only another aspect of revolutionary subversion. In reply to Dupanloup's initiative he let loose a stinging letter from Nîmes, warning mothers of the dangers of exposing their daughters to public lectures and to teachers who, their morality apart, were 'rationalists, pantheists, Protestants,

[66] A.D. Gard 6M 341, Commissaire central of Nîmes to prefect, 14 Jan. 1860; sub-prefect of Alès to prefect, 9 Feb. 1860.

[67] H. Plantier, *Instructions, lettres pastorales et mandements*, vol. iii (Paris, 1867), 407-48.

Jews and goodness knows what else'. Instead of women who 'shelter their intellectual beauty like their physical beauty behind the veil of Christian modesty', these courses would turn out a race of *'esprits forts* and free thinkers'.[68] Courses were set up early in 1868 at Le Vigan and at Nîmes, but the *lycée* master who organized the lectures at Nîmes reported that 'they were only really attended by young Protestant girls'.[69] Victor Duruy's attempt to make elementary education free for those communes that were prepared to contribute also had the result of confirming religious divisions. Fifty-five communes petitioned for free education, and though the Conseil général pleaded lack of funds and stonewalled on many of the requests, twelve of the original eighteen grants went to the poor, Cévenol, and largely Protestant *arrondissement* of Le Vigan.[70] And while the founders of the Bibliothèque populaire of Nîmes, especially the Protestant banker Ali Margarot, and the Cercle anduzien de la Ligue de l'Enseignement, established in April 1869, were ardent supporters of free education, the Catholic and royalist *Opinion du Midi* was fiercely hostile to it.[71]

The issue of free, compulsory, lay education was at the heart of the school question in 1866-72. Emmanuel d'Alzon denounced 'revolution, and its mother, free thought' only weeks before the collapse of the Empire.[72] On the other hand, the *Gard républicain* noted early in 1872 that 'instructed, the people would be less pliable and the country squires who are trying to drag back as much as they can of the good old times would see a large number of their candidates left in the field at elections'.[73] In the Gard, unlike in the Nord, it was the progressive forces that had the upper hand. In the Conseil général in April 1872, a motion for

[68] H. Plantier, *Lettre motivée d'adhésion et de félicitation adressée par Mgr l'Évêque de Nîmes à Mgr l'Évêque d'Orléans à l'occasion de sa brochure intitulée: M. Duruy et l'éducation des filles* (Nîmes–Paris, 1868).

[69] A.D. Gard 2T 156, Courcière, master of Lycée of Nîmes, to Insp. Acad. 27 April 1868.

[70] A.D. Gard 1T 720, report of Insp. Acad. Nîmes, 1867.

[71] R. Huard, *La Bataille pour l'école primaire dans le Gard, 1866-72* (Nîmes, 1966), pp. 26-31; *Mouvement républicain*, pp. 179-80.

[72] A.A.A. A23, *Discours de la distribution des prix, Collège de l'Assomption*, 30 July 1870.

[73] *Gard républicain*, 5 Mar. 1872, cit. Huard, *La Bataille*, p. 35.

compulsory elementary education was carried by a resounding 29:6, a vote on free education by 26:8. Opponents of the scheme were limited to the mountainous west, the north at Saint-Ambroix and the coalfields of Alès, and parts of the Rhône valley.[74] The petitioning movements of 1872 are even more revealing. Support for the Catholic campaign was sparse, confined to the backward, reactionary cantons of Alzon and Sumène in the west, communes such as Saint-Ambroix and Saint-Florent in the north and Saint-Gilles in the south. Mixed Protestant and Catholic cantons like Le Vigan and Saint-Hippolyte, forming a 'front line' for both communities, were deeply divided. But the triumph of the Ligue de l'Enseignement was incontestable. Over 29,000 signatures were collected, representing 6.4 per cent of the population, a proportion passed only by the Seine and the Ardennes. Heavily based in the Protestant Gardonnenque and Vaunage, support was nevertheless very strong in the Rhône Valley from Beaucaire to Pont Saint-Esprit, indicative of a growing anticlericalism on the Catholic plain.[75]

Under the Moral Order regime it was possible for the Catholics to regain some ground, although their activities always remained circumscribed within certain limits. Catholic expansion under the circular of 28 October 1871 was centred on the coal-basin of Alès. As hamlets suddenly became substantial mining villages, a demand was set up for teaching by the congregations, coming not only from mining companies like that of Bessèges but also from Catholic municipalities. Frère Louis-Régis, director of the Marist School at Bessèges, became something of an ambassador for the Petits Frères de Marie in the area, reporting to his superior-general in 1874 that 'it seems to me important that we should gain control of this coal-basin which will soon be crossed by a railway line and will develop along with the mining and chemical industries. In addition, this region is very religious and could become a centre of vocations for our community.'[76] The

---

[74] A.D. Gard, Conseil général, 13 April 1872; Huard, *La Bataille*, p. 43.

[75] Petitions in A.N. C 4131. The figure of 29,356 is from Huard, *Mouvement républicain*, p. 302. He presents a map on p. 182.

[76] A.F.M. AUB 660, Saint-Florent, Fr. Louis-Régis to superior-general, 29 Aug. 1874.

question of vocations was posed with particular acuteness in departments like the Gard, where *fonctionnarisme* was strong and Catholicism, outside the mixed Protestant–Catholic areas, was relatively weak. The new bishop at Nîmes after 1875, Louis Besson (1821–88), another in the series of prelates imported from the Franche-Comté, the son of an *avocat* of Baume-les-Dames in the deeply Catholic part of the Doubs,[77] was as alert to the problems of recruitment as Mgr Cart who had founded the *maîtrise* that became the Collège Saint-Stanislas. By 1877 Mgr Besson had founded *maîtrises* in his episcopal palace at Nîmes, at Alès, and, to draw on the expanding population of Catholic peasant-workers in the Cévennes, at Bessèges and La Grand'Combe. But these were long-term projects; as a short-term solution he resorted to bringing in young priests from the mountains of the diocese of Besançon, since 'this excellent soil remains more fertile than ever amid the universal sterility'.[78] Lastly, there was the issue of Catholic higher education. Emmanuel d'Alzon launched the second series of his *Revue de l'enseigne-ment chrétien* at Nîmes in March 1871 under the slogan 'Delenda est Carthago', the modern Carthage being the University, its monopoly of degrees being the last citadel to fall. Despite his efforts, Nîmes did not manage to acquire a Catholic university, although from 1875 *conférences catholiques* were active, the law taught by Catholic barristers, ecclesiastical history, mathematics, and the arts by masters of the Assumptionist college, including d'Alzon.[79]

As elsewhere, the situation in the Gard was transformed by the elections of 1876-7. In 1876 the republicans won four of the six seats, and though they lost Uzès as a result of re-actionary pressure in 1877, the pattern of voting remained much the same. Royalism remained strong in the west of the department, in the cantons of Trèves, Alzon, and Sumène, in the north, and east of a line drawn from Barjac to Saint Gilles. But republicanism was dominant in the Protestant

[77] See L. Bascoul, *Vie de Mgr Besson, évêque de Nîmes, Uzès et Alais* (2 vols., Arras–Paris, 1902).
[78] L. Besson, *Lettres pastorales*, 10 Nov. 1877.
[79] A.D. Gard 2T 200, 'Declaration d'ouverture' of Conférences Catholiques of Nîmes, n.d.

Cévennes, Gardonnenque, and Vaunage—all thirteen rural cantons with a Protestant majority voted republican in 1877 —and it was beginning to challenge the Catholic conservatives in the Rhône valley, especially around Beaucaire.[80] Change at the base resulted in change at the top: Paul Dumarest, the prefect appointed in March 1879, was an uncompromising anticlerical. As Raymond Huard has said, 'under his hard fist, the republican conquest would be brought to its conclusion'.[81]

Catholic secondary education was not at its strongest in the Gard, but defence of what existed was still fierce. The working-class Catholic quarter of the Enclos Rey in Nîmes reacted with hostility to the expulsion of the Recollets as an unauthorized order in October 1880,[82] but the death of the director of the Assumptionist College, Germer-Durand, made it possible to clear out the Assumptionist staff with little trouble the following December. On the other hand, the college was made over to a *société civile* under the presidency of a magistrate, Numa Baragnon, and though the staff was now ostensibly made up of secular priests, an anonymous informant asked the prefect in 1888 'why and how do you consent to thirteen monks!!! at the College of the Assumption?'[83] The secondary education of girls was placed on a firmer footing than in 1868, at Nîmes in 1880, at Alès in 1881. But while at Alès a Catholic headmistress was appointed in order to win over the Catholic as well as the Protestant population and the *cours secondaire* was elevated into a college in 1887, at Nîmes it was said in 1883 that 'the clientele has not been taken from the Catholic convents or even from the large Protestant *pensionnats*; it has been made up almost entirely of girls requiring to pursue a career who until now would not have thought of starting secondary education'.[84] Indeed, when an E.P.S. for girls opened, most of them, especially those with teaching in mind, went there

[80] Huard, *Mouvement républicain*, pp. 404–13; S. Schram, *Protestantism and Politics in France* (Alençon, 1954), pp. 89, 93.

[81] Huard, *Mouvement républicain*, p. 413.

[82] A.N. F19 6072, prefect to Min. Int., 31 Oct. 1880.

[83] A.D. Gard, Fonds du Cabinet du Préfet, 1, 'un citoyen qui vous aime' to prefect, Feb. 1888.

[84] A.D. Gard 2T 156, Insp. Acad. to Rector of Montpellier, 3 Aug. 1883.

as a quicker, cheaper, and more practical alternative. It was only in 1900, after the application of considerable pressure on the municipality by the republican organizations,[85] that Nîmes acquired a ladies' college.

The magnitude of the task of laicizing elementary schools is best examined geographically. It is clear that Protestantism was an important component of *laïcité*, for the schools of the congregations in 1879–80 were squeezed east of the Barjac–Saint-Gilles line that marked the limit of the Protestant presence. The congregations hung on in the Causses and Catholic Cévennes, overlapping with lay schools both around Le Vigan and in the southern part of the department. Here was dry tinder for conflict (Map 7).[86] Over the century as a whole, the Gard provided very few priests, but the Catholic east, the mountainous border with the Vivarais in the north, and the Causses, at the limit of the Rouergue, in the west, did make some contribution.[87] *Instituteurs* were recruited above all from the Protestant areas, the Cévennes and the Vaunage, but conflict in the Protestant–Catholic frontier zones around Le Vigan and in the south was a stimulus to the lay teaching mission, while anticlericalism in the Catholic parishes of Nîmes also produced recruits.[88]

Spontaneous laicization following the 1878 and 1881 municipal elections, albeit with the firm support of the prefect, was very important in the Gard, a wave of anti-clericalism from below. Between 1878 and 1886 the proportion of *écoles communales* in the hands of the congregations fell by 71 per cent in the case of boys, by 33 per cent in that of girls, where the figures for the Nord were respectively 47 per cent and 24 per cent. Again, the movement started earlier, with 1878 being the decisive election leading to laicizations in the Rhône Valley and in the borderlands, in such towns as Saint-Hippolyte and Alès. At Alès, historically the Catholic frontier post in the Cévennes, on the occasion of the prize-giving at the schools of the Frères des Écoles Chrétiennes in

[85] A.M. Nîmes, R-1-270, petition of 'cercles et chambres républicains' to municipality of Nîmes, 31 Mar. 1894.

[86] A.N. F17 10554, État de situation, 1879–80.

[87] A. Ep. Nîmes, Register of ordinations, 1817–1913.

[88] École normale of Nîmes, Registres d'inscription, 1842–1913.

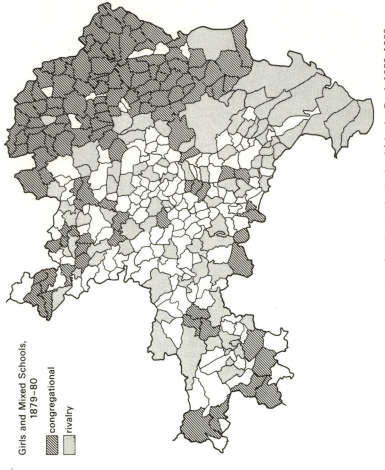

Girls and Mixed Schools,
1879–80

congregational

rivalry

Map 7. Lay teachers and congregations (girls' and mixed schools) in the Gard, 1879–1880

the rue Taisson and the suburb of Rochebelle in August 1879, the books presented by the new republican municipality were rejected by the *curé* of Rochebelle as hostile to the Church and its teachings, and replaced by approved texts furnished by the local mining company.[89] The municipality seized on this affront to vote the laicization of the Frères' schools by twelve votes to eight on 8 October 1879. The Frères nevertheless claimed that the schoolhouse had been made over to the town by the abbé Taisson in 1821 on condition that it be entrusted to their Institute, and refused to move. As a result, the gendarmerie was sent in to expel them by force and the Baron de Larcy, the local senator, challenged the Minister of the Interior to defend the spoliation of Church property ordered by the prefect.[90] Following the municipal elections of 1881 there were further laicizations in mixed communities including Nîmes, where the republican municipality was led by the Protestant banker Ali Margarot. The main argument was that only in non-Church schools could the 'fusion of religions' properly take place, but it was one which conveniently suited the purposes of Protestant opponents of the congregations.

The elimination of religious instruction from the public schools was a natural extension of this approach, and implemented after the monarchists were routed in the legislative elections of August 1881, forced back to the first *arrondissement* of Nîmes. Reactions to the idea of the 'école neutre' varied according to religious affiliation. 'From the beginning' reported the school-inspector of Uzès, 'the Protestant clergy, followed by the Protestant population, have accepted the new teaching.'[91] This was partly because liberal, as opposed to orthodox, Protestantism which emphasized perfectibility rather than sin, the Gospels rather than the Cross, duty and virtue rather than faith, was itself a major component of the lay ideology, partly because Protestants had a sense of *civisme* explained by the fact that only the

[89] A.D. Gard 1T 513, *curé* of Rochebelle to mayor of Alès, 22 Aug. 1879; primary-school-inspector of Alès to Insp. Acad. Nîmes, 20 Sept. 1879. See also Bruyère, *Alès*, pp. 583-7.

[90] *Journal Officiel*, 10 Dec. 1879, pp. 10853-68.

[91] A.D. Gard 1T 811, primary-school-inspector of Uzès to Insp. Acad., 3 Mar. 1889.

granting of civil rights in 1789 had secured them properly against religious intolerance and persecution. One of the leading campaigners for *laïcité* in the Gard, Frédéric Desmons, naturally combined these tendencies: not only was he the liberal Protestant pastor of Saint-Géniès de Malgoirès but he was also a notable in the masonic movement and a radical politician, elected deputy for Alès in 1881.[92] The Catholics on the other hand immediately took up a position of Ligueurs. Mgr Besson warned his flock vaguely against the dangers of the morality of Rousseau, but real influence was in the hands of the fanatical lower clergy, whether the notorious abbé Chapot who held forth in the cafés of the popular quarter of the Enclos Rey at Nîmes, where the clientele still drank to Saint Bartholomew's Day,[93] or the clergy of the canton of Saint-Ambroix, who denounced the Republic as 'a government of filth and corruption', run by 'demolishers and bomb-throwers'.[94]

It was still possible for the Catholics to fall back on the teaching of catechism outside the school. For Mgr Besson, catechizing by the parish priest had long been a priority.[95] Even the Protestant Consistory of Nîmes urged parents to take up the responsibility that *instituteurs* had laid down, calling on 'the sons of Huguenots' to transmit to their children 'this Christian and Protestant faith for which your fathers suffered so much'.[96] Again, it was likely that many *instituteurs* would not be capable of propagating a moral and civic education to great effect. The director of the École normale of Nîmes was complaining in 1879 that 'élite minds are rare, especially in the milieu where we recruit', most of them familiar only with their village patois before they came to Nîmes.[97] Yet an analysis of the schools' registers before and after 1880 shows that the pattern of recruitment was

---

[92] See D. Ligou, *Frédéric Desmons*, p. 104.
[93] A.D. Gard V4, mayor of Nîmes to prefect of Gard, 3 July 1882.
[94] A.D. Gard V4, Commissaire de police of Saint-Ambroix to prefect, 29 April 1880.
[95] L. Besson, *Lettres pastorales*, 21 Nov. 1877.
[96] A.C. Nimes K 28, president of Consistory of Nîmes to parents of Église réformée, 15 Oct. 1881.
[97] A.D. Gard 1T 666, director of École normale to Commission de surveillance, 20 July 1879.

changing: the proportion from artisan or wage-earning backgrounds combined remained constant at just under a quarter,
but pupil-teachers from peasant backgrounds, including
(usually small) proprietors with *cultivateurs*, fell from 52.8
per cent to 30.7 per cent, while those with white-collar backgrounds leaped from 2.8 per cent to 15.1 per cent and
second-generation *instituteurs* rose from 7.3 per cent to
16.4 per cent.[98] These petty-bourgeois *instituteurs* were
more likely to be the redoubtable 'black hussars of the
Republic' of whom Péguy spoke.

Fig. 2. Social Origins of *Élèves-maîtres* at École normale, Nîmes

	1842-79	1880-1914
proprietors	25.7	15.1
liberal professions	1.0	0.8
cadres	1.0	1.8
industrialists, entrepreneurs	–	1.3
négociants, marchands	1.7	1.6
shopkeepers	7.3	7.3
artisans	23.3	16.7
employés, petits fonctionnaires	2.8	15.1
instituteurs	7.3	16.4
cultivateurs	27.1	15.6
wage-earners	0.7	5.7
servants	2.1	2.6

Despite agitation on the Catholic side, the republican
movement in the Gard was strong enough to take full control
under the *scrutin de liste* arrangement that operated in the
elections of August 1885. Laicization was now to be completed under the law of 30 October 1886. There were, of
course, certain obstacles in the way of a quick victory.
Where *écoles communales* were not owned by the commune,
the commune had to build its own municipal schools before
laicization could go through. But at La Grand'Combe, the
building of the new lay-school complex was financed by the
mining company which also owned the schools, public before
laicization in 1892, then private, in which the congregations
taught. The Inspector of the Academy condemned the way in
which 'the company is substituting itself for the municipality'

[98] École normale of Nîmes, Registres d'inscription. The samples are of 288
and 384 cases.

and saw 'serious difficulties' ahead.[99] The usual response to the laicization of *écoles communales* in Catholic areas was the building of private Catholic schools. Once the private Catholic school had been built, all manner of means could be used to populate it at the expense of 'la laïque'. At Rivières in the canton of Barjac the sub-prefect of Alès reported that 'under pressure from landowners, priests, fanatical elements in the population and a prejudiced municipality, a *de facto* monopoly of private schooling, at the expense of public education, has been created'.[100] The lay *école communale* had been reduced to a single pupil, and he from a neighbouring commune.

The Catholic Cévennes was a privileged area in the Gard. Looked at as a whole, the progress of lay education was substantial. Between 1879–80 and 1899–1900 the proportion of boys in schools of the congregations declined from 46.2 per cent to 29.7 per cent, a loss of over a third, while the proportion of girls taught by nuns fell from 62.5 per cent to 46.3 per cent, a loss of nearly a quarter. But a glance at Map 8 shows that in the case of girls' education at least, the geographical foundations of the strength of the congregations remained intact. They could not be shifted from the mountainous west and north, nor from east of the Barjac-Saint Gilles line. The pattern of resistance was clear in the elections of 1889. While an opportunist now captured the constituency of Uzès, the Orleanist Comte de Bernis remained at the first constituency of Nîmes and Fernand de Ramel, candidate of the companies, was returned for the second constituency of Alès.[101]

The Catholic *patronat* was largely successful in its war against socialism. In 1898 the socialist candidate obtained 57 per cent of the vote at Bessèges but only 29 per cent at La Grand'Combe.[102] For this reason, there was very little pressure for Ralliement with the moderate republicans. In any case, the republican family remained divided from

---

[99] A.D. Gard 1T 956, Insp. Acad. to prefect, 5 Nov. 1890.

[100] A.D. Gard V 343, sub-prefect of Alès to prefect of Gard, 10 Sept. 1902.

[101] D. Ligou, *Frédéric Desmons*, pp. 156–7.

[102] J.-M. Gaillard, 'La Pénétration du socialisme dans le bassin houiller du Gard' in *Droite et gauche* (Montpellier, 1975), p. 284.

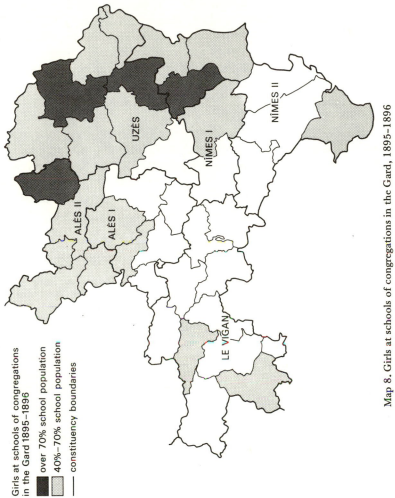

Girls at schools of congregations
in the Gard 1895–1896

■ over 70% school population

▨ 40%–70% school population

— constituency boundaries

ALÈS II

ALÈS I

UZÈS

NÎMES I

NÎMES II

LE VIGAN

Map 8. Girls at schools of congregations in the Gard, 1895–1896

royalists because of the legacy of the Wars of Religion, and the strength of the Montagnard tradition of 1849 made easy a 'red shift' towards socialism: the Protestant-dominated *arrondissements* of Alès and Le Vigan returned socialist deputies, albeit ones who later broke with the P.O.F., in 1898. In contrast with the Nord, there was no Ralliement in the Gard and no significant realignment of forces as a result of the Dreyfus Affair: Catholic royalism and a republicanism both Protestant and anticlerical continued to confront each other, as in the French Revolution.

Indirectly, the Gard did feel the shock waves of the Dreyfus Affair, because the Assumptionists, who had run the College at Nîmes until 1880 and had been at the head of the antiDreyfusard campaign, were now officially dissolved. The fact that the staff at the College were now secularized as ordinary priests did little to impress the anticlerical authorities: in 1899, 1903, and again in 1912 they were accused of reconstituting a congregation, while in 1909 they were ejected from d'Alzon's magnificent edifice in the avenue Feuchères on the grounds that the *société civile* had only a fictitious existence, and that the congregation was the real proprietor.[103] But it was the fate of the congregations that taught in the primary schools that had most significance. The geography of support for them or opposition to them is evident from the inquiry of 1902.[104] Anticlericalism was strong in the Protestant regions, especially those bordering on Catholic populations, as around Saint-Hippolyte, Sauve, and Quissac, but hostility was increasingly evident in the wine-growing plain, at Saint-Gilles, Beaucaire, and Aramon. On the other hand, clericalism was articulate in the east behind the Rhône Valley, in the cantons of Marguerittes, Lussan, and Remoulins, in the industrial north, where the municipalities were largely composed of the managers and engineers of the great companies, and in the mountainous Causses region around Alzon, Le Vigan, and Sumène. But such preferences would have difficulties in weathering the storm that swept the Bloc des Gauches to power in the spring

[103] *La Maison de l'Assomption à Nîmes, passé, spoliation, survivance* (Nîmes, 1922).
[104] A.D. Gard V 325, 328-39, 343.

of 1902. The socialists had already taken Nîmes I from the royalist deputy in a by-election of 1901, and they retained it along with Le Vigan and Alès I. Radicals held Nîmes II in the person of the Protestant Doumergue and conquered Uzès, leaving Fernand de Rame isolated in the Catholic-industrial constituency of Alès II. The only consolation was that the Protestant bourgeoisie of Nîmes, frightened by the progress of a socialist in 1901, had voted with the Right at the second ballot.[105]

Under Émile Combes, sometime master at the Assumptionist College of Nîmes, the Associations law would be applied mercilessly to destroy the congregations. But the Catholic party was game to defend the position it still held. Mgr Béguinot, a native of the Cher, appointed to the see of Nîmes in 1906 as a stranger to the counter-revolutionary traditions of the Midi, soon came under the influence of the young, ultramontane clergy accustomed to look to the legitimist prelate of Montpellier, Mgr de Cabrières, for a sign. Béguinot reacted to the circular of 9 July 1902 by protesting that most of the sixty schools due for closure were 'situated in Protestant country'.[106] Popular protest was angriest in the Catholic–industrial region. On 23 July 1902 women invaded the schoolhouse of Saint-Ambroix to prevent the departure of the nuns and miners from Molières caused uproar in the town before reinforcements of gendarmerie arrived from Bessèges and Nîmes.[107] The extent to which crowds were manipulated by the industrial companies is difficult to gauge. But when the schools financed by the coal and steel companies and run by the congregations in the industrial area around Bessèges, Robiac, and Saint-Ambroix were closed in 1902–3, over a third of the school population was lost to lay public schools. This was not the case in the rural Catholic cantons of Génolhac and Barjac, where most of the pupils remained in private schools.[108]

Where the efficacy of the industrial companies was not in

---

[105] D. Roque, art. cit., p. 229.
[106] A.N. F19 6275, Mgr Béguinot to president of Republic, 20 July 1902.
[107] A.N. F19 6080, commander of gendarmerie of Alès to prefect of Gard, 23 July 1902.
[108] A.D. Gard 17 883, 884, Enquêtes of Nov. 1902 and Nov. 1903.

doubt was in questions of property-ownership. The lack of municipal school premises would hold up the closure of private Church schools. But La Grand'Combe was not only an extreme version of a 'ville patronale', it was also, as one school-inspector put it, 'a state within a state'.[109] It maintained not only close on two thousand children in the private schools of the Frères des Écoles Chrétiennes and the Sœurs de Saint-Vincent de Paul at La Grand'Combe, its faubourgs of La Levade and Champclauson, and in the neighbouring commune of Laval, but it had also built, at its own expense, the three lay municipal schools. In addition, it maintained the town hall, police station, post office, Catholic and Protestant churches and vicarages, the medical services and the cemeteries, the roads and street-lighting, the water-supply, wash-house and sewerage, the public conveniences and refuse disposal.[110] In 1904, the sub-prefect of Alès, calculating that it would cost 250,000 francs to buy the schoolhouses owned by the company, hoped that he might negotiate a staggered closure of the Church schools which would be taken over and leased by the municipality. 'It is reasonable to assume', he reflected, 'that the company will not hold to ransom the commune it administers.'[111] Eight years later, when nothing had been achieved, the prefecture reported that even if the funds were available to build *écoles communales* for 2,000 children, there remained the problem of a site. 'The mining company owns everything in the commune, even the soil', and besides, at La Grand'Combe, perched on a hillside overlooking the River Gardon, there was no room to build. If lay municipal schools ever were constructed, the company would secularize its schools the day they opened, and the municipal schools would remain empty.[112] In July 1914 there were 220 private schools in the Gard, eight run by religious congregations, seven of which were at La Grand'Combe. And they survived, because the

---

[109] A.D. Gard 1T 882, primary-school-inspector of Alès to Insp. Acad. Nîmes, 15 Oct. 1903.

[110] A.D. Gard 18J 801-5, Cie houillière de la Grand'Combe: registre de dépenses de prévoyance patronale, 1910–14.

[111] A.D. Gard V 326, sub-prefect of Alès to prefect of Gard, 10 Sept. 1904.

[112] A.D. Gard Fonds du Cabinet du Préfet 81, *secrétaire général* of prefecture to Min. Int., 6 Nov. 1912.

Minister of the Interior's decision of 2 August 1914 suspended the operation of the laws of 1901 and 1904.[113]

Elsewhere in the Catholic industrial region, 'secularization' was a response to closure. But the sub-prefect of Alès reported that these secularized schools 'have nothing "lay" about them but the name. All operate under the patronage of the same "Catholic" committees; the clergy has ultimate control; teaching is done according to the syllabus of the congregations. Finally, it is a secret to no one that the schools have been reopened with the *arrière-pensée* of restoring them purely and simply to the nuns who ran them previously, the day the congregations receive the authorization they are requesting.'[114] It is clear that the process of secularization took its toll of the congregations. A survey carried out in 1908 on the fate of the Frères des Écoles Chrétiennes who in 1904 had been teaching in the district of Avignon, which included the Gard, recorded that 10.5 per cent were still active in France, 15.3 per cent were secularized, and 29.3 per cent working in exile abroad, but that 18.5 per cent had retired and 26.3 per cent had left the congregation.[115] Increasingly, private Catholic schools had to be staffed by lay teachers, and the ecclesiastical college of Sommières, where the Latin contingent had never been very large, was transformed into a private *école normale*.[116]

One last loophole exploited by the female congregations was the fact that it was only their teaching activities that were prohibited. There could be no real objection to the Sœurs de Saint-Joseph des Vesseaux at Les Mages in the Catholic canton of Saint-Ambroix, who protested that they were only minding the children of the agricultural labourers during the period of the wine-harvest, especially as they were doing no school exercises but only knitting, sewing, and crochet work. Except that the *directrice* expelled in 1902 was then discovered in a hayloft with girls poring over their *cahiers* and textbooks.[117]

---

[113] A.D. Gard 1T 1050, report 'pour l'Inspecteur d'Académie mobilisé' to Min. Instr. Pub., 10 April 1915.

[114] A.D. Gard V 331, sub-prefect of Alès to prefect of Gard, 7 Feb. 1907.

[115] A.F.E.C. NC 281, District of Avignon.

[116] *Directoire trimestriel de l'Immaculée Conception de Sommières*, 1 Oct. 1910.

[117] A.D. Gard V 325, Commissaire spécial to sub-prefect of Alès, 27 Nov. 1902; Insp. Acad. to prefect of Gard, 23 Dec. 1902.

The campaign by the Catholic clergy to propagate and populate Catholic schools by mounting an attack on the history
texts used in the public schools provides a final image of the
pattern of clericalism in the Gard. In a large town such as
Nîmes, enthusiasm could be whipped up among the Catholic
élite. At the end of 1909 Colonel Keller, president of the
Société générale d'éducation et d'enseignement, addressing a
meeting at Nîmes organized by the Ligue des Femmes françaises, challenged his audience: 'Our fathers organized the
Ligue to save our country from Protestantism. Are we no
longer worthy of them?'[118] In the Cévennes, resistance took
the form of a refusal to recite history lessons. More virulent
were the clerical assaults in the Causses, notably in the cantons
of Le Vigan and Alzon. At Campestre, where there were no
private schools, the boys used Guiot and Mane, the girls Gauthier and Deschamps. The *instituteur*, though a Protestant,
inclined before the *curé*'s tirades; the *institutrice*, who resisted
his threats, was attacked as a 'maîtresse jacobine' and 'blocarde
pur sang' and threatened with excommunication, while her
school was placed under an interdiction.[119] The following
spring the *curé* organized catechism classes on Tuesdays, Thurs-
days, and Sundays at 11 a.m. and, announcing that 'catechism
must take precedence over the school', warned that those who
failed to attend would be excluded from First Communion. At
his wits' end, the school-inspector urged that the case be
brought to the notice of the *procureur de la République*, to
'make this turbulent priest understand that we have had enough
of his dealings and that we are going to stand up to the
systematic sabotage that he is undertaking against our
schools'.[120] Resistance could be spontaneous as well as
official. André Chamson tells of endemic warfare between
Catholic children of the secularized Frères' school and Protestants of 'la laïque', the insults of 'Enfants de chœur
. . . Fils du Pape!', 'Républicains . . . Protestants!', 'Culs
blancs . . . Royalistes!', and 'On est des Camisards!' followed
by stone-throwing.[121] But the limits of clerical influence

---

[118] A.N. F17 9125[5], cutting from *Journal du Midi*, 28 Dec. 1909.
[119] A.D. Gard, primary-school-inspector of Le Vigan to Insp. Acad. Nîmes,
1 Nov. 1909.
[120] Ibid., 18 June 1910.
[121] A. Chamson, *Le Chiffre de nos jours* (Paris, 1954), pp. 306, 338.

were now clear, having been pushed back to the Causses and the Catholic and industrial Cévennes. In the Vaunage and increasingly along the Rhône Valley, that great conquest for anticlericalism, clerical agitation to replace proscribed books encountered resolute opposition from municipalities now in the hands of republicans.

Historical traditions made any form of compromise very difficult in the Gard. Rivalries forged by the Wars of Religion could not easily be moderated. The Catholics of the Gard were very much Ligueurs in the counter-revolutionary, ultramontane mould, while the Protestants could not escape from their past as Huguenots and Camisards. Protestantism was not the only root of Radicalism. As the century wore on it merged with the broader tradition of 1848, a Mediterranean Radicalism that was anticlerical rather than specifically Protestant, and became the politics of the wine-growing plain of the Rhône. In the Vaunage and the Cévennes it was quite different, and the determining factor was the Calvinist presence.

The school question was fought out against the background of this history, and could not but be influenced by it. The Protestant Church, unlike the Catholic Church, did not stand in the way of *laïcité*. On the contrary, Protestantism was an important component of anticlericalism and anticlericals could rely on Protestants for their support. On the Catholic side there was no room for conciliators: Mgr Béguinot found difficulty in controlling his ultramontane clergy. Equally, the struggle over schools was one means by which militant Radicalism triumphed in the Gard. Catholics had not the resources to respond effectively to laicization with private schools or to the closure of private schools with secularized teachers. And so Catholic bases were gradually eroded—in the Rhône valley and to some extent in the Causses–Cévennes region. Bishops and religious orders became increasingly reliant on recruits from the Franche-Comté and other Catholic areas. By 1914 the hub of Catholic resistance was the north of the department, both in the countryside and in the industrial centres of Alès, Bessèges, and La Grand'Combe. It would be wrong to blame the pressure of employers for this state of affairs: most of the miners and ironworkers were peasants who had come down from the Massif Central. What mattered was not class but historical geography.

(c) Ille-et-Vilaine: Defeat

The alliance of the Catholic Church and authoritarian Empire against the threat of republican and socialist revolution had provided a framework for the expansion of Catholic education to a position of hegemony in Ille-et-Vilaine by the end of the 1850s. The Frères de Ploërmel had conquered schools in the west of the department, the teaching congregations of nuns had infiltrated everywhere with the exception of last-ditch defences around Liffré and Retiers, and apart from the Lycée of Rennes, public lay secondary education was confined to three small colleges on or near the seabord, at Saint-Servan, Dol, and Fougères. But the achievement of Brossays Saint-Marc was not confined to education: as a grand elector, he had demonstrated that he could manipulate the political power of the conservatives in the department almost at will and his see was made into an archbishopric following the imperial visit of 1858. But it was a balance of forces destined to change radically from this moment. The Second Empire had a strong authoritarian populist streak that suffered only with difficulty the tutelage of clerical and conservative notables. A muscular prefect like Paul Féart, promoted from the Bonapartist fief of the Gers in 1858, would look to make Ille-et-Vilaine into another colony of the administration, relying to some extent on former Orleanists but seeking to exploit the power of universal suffrage by appealing directly to the peasantry and even to the working classes. By December 1859 he had organized the election of an official candidate, the Comte de Dalmas, in the legitimist citadel of Vitré-Fougères. The archbishop still barred his way, but the defeat of the papal zouaves at Castelfidardo consummated the rupture between Church and state. From now on the Empire would reveal itself a 'blue' regime, and its campaign against the entrenched positions of clericalism and royalism in many ways prepared the way for the success of the Third Republic in Brittany.

A war of attrition against the archbishop's near monopoly of education was clearly a priority. The law of 14 June 1854 had transferred the right of appointing *instituteurs* from Rectors to prefects, but municipal councils still had the right to opt in principle between a lay teacher or congregation on the death, dismissal, or resignation of an incumbent. The Le Sel affair of 1861–2 proved to be a trial of strength between prefect and archbishop. After the death of their lay *instituteur*, the municipal council of this cantonal capital in the Redonnais opted for a teaching brother. Féart on the other hand was intent on having another lay *instituteur*, and imposed one ex officio. Brossays Saint-Marc addressed a petition to the Senate in February 1862, demanding the invalidation of *arrêtés* of Féart, 'as contrary to *La Liberté de l'enseignement* and as constituting arbitrary and illegal acts'. On his side, Féart urged the necessity of circumscribing the archbishop's pretensions, and the Senate eventually rejected the petition.[122] Féart followed up his triumph in the legislative elections of June 1863, securing the return of official candidates in all four *arrondissements*.

Clericalism nevertheless retained its vigour and resilience and made clear what were the limits of the centralized state. There was no echo of the controversy over the secular secondary education of girls in the traditionalist diocese of Rennes. What was significant in Ille-et-Vilaine was Duruy's scheme for subsidized free education in *écoles communales*, under the law of 10 April 1867. The project was especially attractive to the poor communes of Britanny, notably to those seeking to fortify their embattled position *vis-à-vis* the teaching congregations. Those which opted for free education in Ille-et-Vilaine were concentrated in the 'blue' zones between Combourg and Liffré and around Retiers. The Catholics were not slow to react to what they saw as a war machine directed against them. Brossays Saint-Marc had already denounced such policies as 'a means of undermining private schools'.[123] A Comité de l'Enseignement libre founded

---

[122] A.N. F17 10349, Brossays Saint-Marc, *Pétition au Sénat*, 4 Feb. 1862; Féart to Min. Instr. Pub., 15 Feb. 1862. The Senate considered the matter on 30 April 1862.

[123] A.A. Rennes, C5, letter of Brossays Saint-Marc, early 1866.

by the Catholic laity in April 1868 warned that the Duruy law threatened to 'give a crushing superiority to official schooling' and to 'destroy private education'.[124] As the committee gathered funds to plough into the private sector, letters of anguish poured into the office of the Inspector of the Academy from lay teachers, overwhelmed not only in the heart of the Vitré area but also in the 'frontier' area of La Guerche and the 'blue' canton of Retiers.[125] The only problem faced by the partisans of Catholic education was the shortage of Frères de Ploërmel, who had close on exhausted themselves by building up their strength in the west of the department. The provisional solution found in the Vitré area, while awaiting the arrival of a Frère, was the opening of a school by the local curate. But when the *curé* of La Guerche opened a school, Baroche, the Minister of Justice and Worship, brought this practice to an abrupt end by telling Brossays Saint-Marc that clergy in receipt of state stipends in order to undertake the parish ministry would lose those stipends if they tried to teach as well.[126]

Defeat put an end to anticlerical imperialism, and the anticlericalism of the republicans was stifled in February 1871 by a landslide victory of legitimists and Catholics, watched over by Brossays Saint-Marc. In the controversy over free, compulsory, lay elementary education that raged over the following year, the ideological geography of the department was clearly defined. A motion on compulsory education proposed in the Conseil général in November 1871 was carried narrowly by 22:20. Supporters of the motion represented the 'blue' central band of the department, and were mainly lawyers; its opponents, mostly landowners together with an ironmaster of the forges of Paimpont, represented the Catholic–legitimist flanks.[127] At the popular level support for radical education policies seemed less certain. The campaign of the Ligue de l'Enseignement obtained only a few

---

[124] B.M. Rennes, Comité de l'Enseignement libre, *Programme et Constitution* (Rennes, 4 April 1868.)

[125] A.N. F17, 12221, letters of *instituteurs* and *institutrices* of *arrondissement* of Vitré to Insp. Acad. Rennes, Jan. 1869.

[126] A.F.P. La Guerche, abbé Desbois to superior-general, 12 June 1866; A.N. F19 3972, Baroche to Brossays Saint-Marc, 31 Aug. 1867.

[127] A.D. I-et-V. 1N Conseil général, 8 Nov. 1871.

hundred adherents in large towns like Rennes, Saint-Servan, and Saint-Malo, and in 'blue' communes in the south-east corner, Retiers and Martigné-Ferchaud. By contrast, the Catholic petition, promoted by the Comité de l'Enseignement libre, obtained over 30,000 signatures, 5.2 per cent of the population, notably in the 'pious' cantons of Bécherel, Montfort, Saint-Méen, and Maure in the west, the legitimist Vitré–Fougères block in the east, and in the north the area around Châteauneuf and Dol, more Catholic than conservative.[128]

Under the Moral Order regime it was possible for the Breton Catholics to consolidate their position, and even to make more ground. The Rector of the Academy of Rennes complained that the circular of 28 October 1871 played entirely into the hands of the congregations,[129] while the reactionary prefect Delpon de Vissec did not hesitate before lashing out to suspend *instituteurs* with republican sentiments in 'blue' communes.[130] Persecution and poverty resulted in 'numerous desertions' from the profession of *instituteur* in these years,[131] while Brossays Saint-Marc was able to console himself in his old age that he had seen congregations of nuns establish themselves in almost every commune,[132] so that the few lay *institutrices* turned out from the *cours normal* annexed to a *pensionnat* of Fougères remained 'sans emploi'.[133]

It was the excesses of the Moral Order regime that made possible the triumph of the republicans in Ille-et-Vilaine in 1876. They won six of the eight seats but could not capture Vitré, even when the conservative vote was divided between the liberal monarchist historian, Arthur de la Borderie, and a die-hard legitimist, former commander of papal zouaves, the Comte Le Gonidec de Traissan, who won.[134] In the

---

[128] A.N. C4136, Pétitions, 1870-2, Ille-et-Vilaine.

[129] A.N. F17 9262, Rector of Rennes to Min. Instr. Pub., 24 Dec. 1873.

[130] A.N. F17 11591, prefect of Ille-et-Vilaine to Min. Instr. Pub., 12 Dec. 1873; Insp. Acad. to prefect, 30 Nov. 1875.

[131] A.N. F17 9608, Insp. Acad. to Rector of Rennes, 29 April 1873.

[132] A.N. F17 10869, Caillard, Déléguée générale pour l'inspection des écoles normales, 5 Sept. 1875.

[133] A.D. I-et-V. 1N, Inspector Acad. Rennes to Conseil général, 7 Sept. 1877.

[134] Goallou, 'Évolution politique', iii, 1021 ff.

elections following the Seize Mai crisis of 1877, during which the Moral Order regime managed to cram congregations into communes as 'blue' as Liffré, the republicans were driven out of Fougères, but they demonstrated that they could make the running in the west of the department, that was intensely Catholic but prepared to consider moderate republican deputies if they did not trample on the faith.

In this sense, the anticlerical weapon was one that had to be handled with circumspection in Ille-et-Vilaine. On the one hand, the ascendancy of the clergy and squirearchy had to be broken in order to consolidate the Republic. On the other, care had to be taken not to alienate the mass of the population, which might otherwise incline to republicanism, by persecuting their religion. There were neither Jesuits nor Assumptionists to be expelled as unauthorized congregations from Ille-et-Vilaine, but the gendarmerie and troops who drew up to turn the Carmelites out of their convents in October 1880 were confronted by a crowd of 6,000 and the prefect noted 'an almost superstitious veneration for anything wearing a habit, an attitude from which a large number of citizens are far from being liberated, even though they have now rallied to the republican party'.[135] While provoking popular Catholic fervour, the dissolution of the congregations contributed to the authority of the episcopate. Brossays Saint-Marc had died early in 1878, doubtless unable to stomach the republican *raz de marée* and the Eudists, who for thirty years had bent the Institution Saint-Martin to the interests of the archiepiscopal Saint-Vincent, took advantage of the vacancy to establish *plein exercice*. But the new archbishop, Mgr Place (1814–93), a former Vatican official and bishop of Marseille, announced on his arrival that 'there is no room at Rennes for two private establishments of *plein exercice*', and promptly decapacitated the rival institution again.[136] The obligation of the Eudists after 1880 to claim that they were not a congregation but only secular priests under the direction of the episcopate strengthened Mgr Place's

---

[135] A.N. F19 6073, prefect of Ille-et-Vilaine to Min. Int. 22 Oct. 1880.

[136] A.E. MF 28, Mgr Place to Le Doré, superior-general of Eudists, 22 June 1879; J.-B. Rovolt, *Vie du T.R.P. Ange Le Doré* (Besançon, 1925), i, 278.

position, and though the Holy Congregation at Rome resolved
in favour of the Eudists in 1884, Place himself both refused
Saint-Martin *plein exercice* and kept its graduates out of the
Grand Séminaire until 1892.[137]

That two ecclesiastical colleges could and did flourish in
Rennes only demonstrates the size of the Catholic clientele.
The education of girls was dominated by the convents to an
even greater degree. A *cours secondaire* was set up in 1880,
but it had no more than fifty pupils before 1885 and Rennes
was the last university town in France to set up a college
or *lycée* for girls, as late as 1906. Similarly, though the
academic authorities considered putting pressure on the
municipality of Vitré to clear the priests from its college,
the inspector of the Academy reflected in 1881 that 'Vitré
is the *congréganiste* and clerical town *par excellence*. My conviction is that a lay college, unless we populated it with
scholars, could not at this moment be re-established in a
town which crowds together so many opponents of our
educational institutions.'[138]

The laicization of the elementary schools would be no
easier task. The 'blue' communes not overwhelmed by the
female congregations had been reduced to a vestigial few,
while the Frères de Ploërmel had colonized the old diocese
of Saint-Malo, the declining weaving districts of the Vilaine
above Rennes, and the countryside around Vitré. The areas
around Liffré, Retiers, and Bain, where there was some
overlap with lay education, were potential flash-points.[139]
The patterns of recruitment to the priesthood and the
École normale of Rennes in the nineteenth century illustrate
the same tension. Ille-et-Vilaine was generous to the Church,
but the outline of the old diocese of Saint-Malo and the
Vitré–Fougères bloc stand out (Map 9).[140] On the other
hand, the zones which promoted *instituteurs* were the
'blue' strongholds: in the north-east a wedge between Combourg, Liffré, and Saint-Georges-de-Reintembault; in the

---

[137] Rovolt, i, 283–5.

[138] A.D. I-et-V. 1T *Enseignement secondaire privé*, Insp. Acad. to Rector of
Rennes, 2 July 1881.

[139] A.N. F17 10554, État de Situation, 1879–80.

[140] A. Vaillant, 'Le Clergé séculier en Ille-et-Vilaine au XIXe siècle' (*mémoire
de maîtrise*, Rennes, 1972).

☐ one priest in under 250 population (1896)

Map 9. Ordination of priests in the diocese of Rennes, 1803-1900

south-east a knot of communes around Martigné-Ferchaud (Map 10).[141]

'Spontaneous' laicization accomplished very little in

[141] École normale, Rennes, Registres d'inscription, 1831-1913.

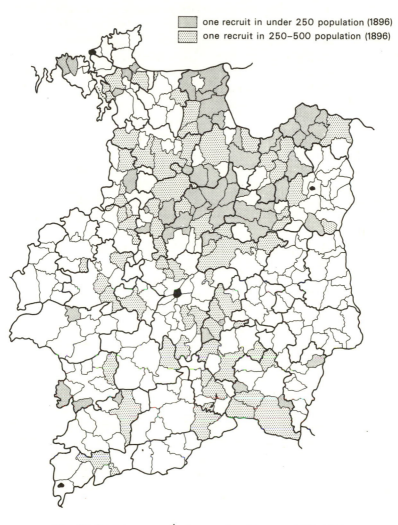

Map 10. Recruitment to École normale of Rennes, 1831–1914

Ille-et-Vilaine. After the municipal elections of 1878, communes like that of Liffré that had been 'clericalized' during
the Seize Mai tyranny reverted to lay instruction. At Martigné-
Ferchaud, hotly contested for a century, the gendarmerie
had to be brought in to turn out the Frère, and that evening

crowds gathered in front of the vicarage, chanting 'A bas les Chouans!'[142] But anticlericalism was not coterminous with republicanism. When the radicals of Rennes led by the tanner and president of the Chamber of Commerce, Edgar Le Bastard, took advantage of the death of the director of the school of the Frères des Écoles Chrétiennes to demand its laicization, the moderate republican mayor resigned and fought an election in alliance with the royalists in defence of the congregation: he was perfectly aware of the support for the Frères in the working-class suburbs affected.[143] It was not until the elections of January 1881 that Le Bastard's radical list triumphed and all the municipal schools underwent laicization. And at that point Mgr Place was ready, launching a Comité des écoles libres and appealing to 'the vigour and tenacity of Breton faith' in order to gather funds for private schools.[144] Again, the main problem was the shortage of Frères, for on top of the crisis of recruitment now came the law of 16 June 1881 on free education, which required all teachers and their deputies to have the *brevet* and meant frenzied retraining in many congregations. A violent quarrel broke out between Mgr Place, who had publicly announced the maintenance of all Church schools, and the superior-general of the Frères des Écoles Chrétiennes, who regretted that he could not provide the staff and was concentrating those he did have on the remaining *écoles communales*, which tapped municipal funds.[145]

It was clearly for their moderation in matters of education, not for their extremism, that the republicans were swept back to power in the legislative elections of 1881, which left the royalists with only the *arrondissement* of Vitré. But that victory was followed up by the law of 28 March 1882 which banned the teaching of catechism in schools. It was condemned by the archbishop of Rennes,

---

[142] *Journal de Rennes*, n.d. Cutting in A.F.E.C., NC 673.

[143] *Journal de Rennes*, 2, 4 Dec. 1879; J. Meyer, *Histoire de Rennes* (Toulouse, 1972), p. 365.

[144] Mgr Place, *Lettre à Messieurs les curés de la ville de Rennes sur l'expulsion des Frères des écoles municipales et sur la fondation d'écoles chrétiennes libres*, 9 Mar. 1881.

[145] A.F.E.C. NC 777, Mgr Place to superior-general, 13 Mar. 1881; superior-general to Mgr Place, n.d.; Magr Place to superior-general, 19 Mar. 1881.

and rendered largely ineffectual by Catholic populations. 'Here in our Breton countryside', complained a landowner near Combourg, 'the new law on education is trampled under-foot.'[146] All sorts of obstacles stood in the way of imple-menting moral and civic education in Ille-et-Vilaine. Lay *instituteurs* came usually from a coarse, rural background— 26.1 per cent of *normaliens* were sons of *cultivateurs* in the period 1880–1913, as against 15.6 per cent in the Gard and 10.0 per cent in the Nord, and it was not until 1895 that the Inspector of the Academy noted the arrival of a new genera-tion of teachers who started as assistants 'at a time when ideas of pedagogical progress were in favour'.[147] While a protagonist of 'la foi laïque' might be welcome in a 'blue' or anticlerical commune, in a reactionary or clerical one he would appear as a *curé intrus*, in the manner of 1791. Either he would have to make himself 'petit, petit', or he would provoke fanatical hostility. Of course, the teachers who proposed texts for use in the schools (the choice of the cantonal *conférences pedagogiques* being subject to the decision of a departmental commission of school-inspectors and *école normale* directors) could always opt for the most anodyne of alternatives. Indeed, the more extreme works of Steeg, Compayré, and Paul Bert were left on one side in preference for Lavisse's *Instruction morale et civique*, while the unspeakable Madame Henry Gréville was rejected for Clarisse Juranville's *Le Savoir-Faire et le Savoir-Vivre*, which urged that the mission of girls was to become 'angels of the hearth'.[148] But even then, Breton children arriving at school at the age of seven had already received a heavy dose of religious education, and after 1882 the Catholic Church made sure that this dose was increased. The official diocesan publication, the *Semaine religieuse* of Rennes, pressed mothers to exploit their early influence on the child and 'simply, frankly, to be catechists'.[149]

The school question was doubtlessly not the only issue

---

[146] A.N. F17 9125[2], M. Sannin of Meillac to Min. Instr. Pub., 10 Aug. 1885.

[147] A.D. I-et-V. 1N Conseil général, 1895, report of Insp. Acad. p. 577.

[148] C. Juranville, *Le Savoir-Faire et le Savoir-Vivre* (Paris, 1879), p. 97.

[149] *Semaine religieuse du diocèse de Rennes*, 25 Nov. 1882, 25 Dec. 1886.

in the elections of August 1885, but 'défense religieuse' was a solid platform around which conservatives could rally. Their proportion of the popular vote increased from the 1881 figure of 22.4 per cent to 39.2 per cent, and the republicans held the department under the *scrutin de liste* by only the narrowest of majorities, while the rest of Brittany passed to the Right. Clearly, it was now more necessary than ever for the republicans to destroy the electoral influence of nobility and clergy, and the law of 30 October 1886 was employed for this purpose. Much work remained to be done. Between 1878 and 1886, the congregations had lost only 25 per cent of their boys' *écoles communales*, compared with a 71 per cent loss in the Gard, while in girls' and mixed public schools they had actually improved their position by 6 per cent. And little now worked in favour of the anticlerical offensive. Though a new École normale d'institutrices was established at Rennes in 1886, the dominance of the female congregations was so overwhelming that many lay schoolmistresses still remained unemployed. Little more than a third of girls' or mixed schools were owned by the communes, and these were laicized by 1905. But where communes had leased premises from the parish priest, vestry, *bureau de bienfaisance*, or the congregation itself, new building had to take place and the laicization of these schools was not complete until 1912. School-building was the standard response of Catholic communities to the laicization of their schools. Nobility and clergy were generous with their patronage, but they were not alone. There were many cases of the bourgeoisie of Rennes or Nantes, bankers, shipowners, and lawyers, investing in rural influence by building schools. And the community participated either by voluntary subscription or by cartage and construction work.[150]

Once the private schools had been built, they had to be populated. The argument of the authorities was that laicization had been resisted and private schools organized by a conspiracy of '*curés*, curates and country squires'.[151] It is clear

---

[150] A.D. I-et-V. 1T *Congrégations, statistique (1902–6)I*, prefect to Min. Instr. Pub., 29 April 1904.

[151] A.D. I-et-V. 1T *Laïcisations et fermetures (1886–1913)*, Insp. Acad. to prefect, 18 April 1891.

that the Breton clergy did use all forms of pressure, from refusing First Communion to children to refusing absolution to parents, in order to combat the 'école sans Dieu'. In 1898 the dowager Marchioness of Kernier was said to have 'mobilized the ban and *arrière-ban* of her farmers' at Val d'Izé, near Vitré, to prevent the laicization of the school of the Sœurs de Rillé.[152] But it would be wrong to imagine that there was only a 'feudal reaction' and no popular pressure in favour of the congregations. For the Procureur général of Rennes the Bretons were 'simple, credulous, fanatical masses, cowed under the yoke for centuries, coarse and brutal in their habits, inclined to mysticism, enamoured of the supernatural, a prey to the basest superstitions and addicted to alcohol'.[153] The analysis of the mayor of Combourg was more subtle. Writing from the rural constituency of Saint-Malo, which was inclined to republicanism but was also Catholic, he noted that 'if the political leaders have succeeded in rallying a republican majority, it is thanks to their tolerance, to their moderation, to their benevolent attitude, and to the absence of any demonstration contrary to the beliefs and prejudices of the country-dwellers.'[154] But now the Republic had revealed itself of the deepest blue and aroused the Chouan that slumbered in so many Bretons. In the elections of 1889 the rural constituency of Saint-Malo returned the royalist Vicomte de Lorgeril, and the republicans lost all but one of their seats in the department.[155]

From then on, there was no question in Ille-et-Vilaine of a Ralliement that would suspend the struggle between blue and white, anticlerical and clerical. The Bretons defended their schools stoutly. Between 1879–80 and 1899–1900, the proportion of children at the schools of the congregations in Ille-et-Vilaine fell only marginally, from 49.8 per cent to 39.5 per cent in the case of boys, from 86.3 per cent to 80.4 per cent in the case of girls, the monopoly of the nuns seemingly

[152] A.D. I-et-V. 1T *Laïcisations et fermetures (1886–1913), instituteur* of Izé to school-inspector of Vitré, Nov. 1898.

[153] A.N. F19 5610, Procureur général of Rennes to Min. Cultes, 15 July 1888.

[154] A.D. I-et-V. 1T *Laïcisations et fermetures (1886–1913)*, sub-prefect of Saint-Malo to prefect of Ille-et-Vilaine, 21 Aug. 1890.

[155] J.-Y. Le Priellec, 'Les Élections législatives en Ille-et-Vilaine, 1881–1897' (*mémoire de maîtrise*, Rennes, 1974).

inviolable. Geographically, the strength of the Frères de Ploërmel was incontestable in the old diocese of Saint-Malo and the countryside around Vitré, while only in a few cantons to the north of Rennes and in those of La Guerche and Retiers could lay *institutrices* challenge the female congregations. The Dreyfus Affair might have been exploited for the purposes of anticlericalism, but there were few Dreyfusards apart from the professors of the Faculty of Rennes who organized the Ligue des Droits de l'Homme and the Association des Bleus de Bretagne. Not only the royalists and traditional Catholic clergy but also the Christian democrats around the abbé Trochu's *Ouest Éclair* and the Boulangist workers of Rennes were antiDreyfusard.[156] The survey of 1902 amply demonstrated, should there have been any need, support for the congregations in the parishes. There were a few traces of opposition in the embattled centres of anticlericalism around Liffré, Janzé, and Retiers, but the flanks of the diocese were as clerical as ever.[157] Moreover, these sympathies were confirmed at the polls. Rennes was conquered in May 1900 by an alliance of Catholic republicans and royalists and while the Bloc des Gauches swept France in 1902, the conservatives of Ille-et-Vilaine not only consolidated the *arrondissements* of Montfort, Redon, and Vitré, but also took their revenge by conquering Fougères and Saint-Malo 1. As the *curé* of Val d'Izé had told his congregation, 'when you vote, you are choosing between the candidate of God and the candidate of the Devil. With the one, you will keep your Christian schools, with the other, you will see them closed. It is up to you.'[158]

The Bloc des Gauches was nevertheless bent on completing the anticlerical offensive, and installed a hard-hitting radical prefect, Delpech, whose first thoughts were to consider cutting the stipends of clergy who had campaigned actively for the Right.[159] In reply, the Right fell back increasingly on a Breton regionalism. Cardinal Labouré called the private

---

[156] J. Meyer, *Histoire de Rennes*, p. 423.
[157] A.D. I-et-V. 1T *Fermetures d'établissements congréganistes, 1901–4, G; Demandes d'autorisation*, 1902, K', K".
[158] A.N. F19 5929, prefect to Combes, 24 May 1902.
[159] A.D. I-et-V. 1V 1507, prefect to Combes, 16 May 1902.

Catholic schools 'irrecusable evidence of Breton faith'. 'If the free-thinkers of the Midi and Centre are happy with a neutral or atheistic education', he continued, 'that is their concern.' But 'the Breton people is too religious and too proud ever to yield in resignation to the yoke of a sect'.[160] Resistance to Combes's measures was indeed forceful. Its most spectacular manifestations were in response to the expulsion of the Frères de Ploërmel following the rejection of their request for authorization by the Chamber of Deputies in April 1903. Curiously enough, very little trouble seems to have been recorded in the old diocese of Saint-Malo, except on the coast at Cancale. But where clericalism was combined with legitimism the mixture was explosive. There were crowds of several hundred to resist the gendarmerie and troops at Amanlis and Piré, in the former weaving canton of Janzé, and of close on two thousand at Vitré, where notables such as Le Gonidec de Traissan were at large among the demonstrators. Early in 1904 there was more trouble in the canton of Argentré and in the 'frontier' canton of La Guerche.[161] The social hierarchy of the Vitré region was doubtless a feature of this resistance but, like the Chouan wars, resistance was a defence of the integrity of the Catholic community against the revolutionary authorities and their troops from without.

Behind resistance was the capacity to survive. In Brittany, perhaps more than elsewhere, the means existed to cushion the blows of even the most savage Combist measures. Republican politicians faced with the realities of provincial opinion pressed the government to delay procedures that would destroy already fragile electoral support. Informed at the end of 1905 that the school of the Frères de Ploërmel at Saint-Servan was due for closure, Robert Surcouf, the republican deputy who had won the rural constituency of Saint-Malo in 1898, not least because of the buccaneering and royalist past of his family, warned that the moment, six months before the next legislative elections, was 'ill-chosen', and that Saint-Servan was his 'weak spot'. The prefect was equally anxious to

---

[160] *Semaine religieuse*, 26 July, 16 Aug. 1902.
[161] On these incidents, see A.D. I-et-V. 1T *Congrégations, interventions des liquidateurs, prise de possession des immeubles* (1903-9), T.

preserve a few republican seats in the department, the closure was postponed, and Surcouf managed to hold on to his constituency in May 1906.[162] Again, a dissolved congregation lost the right to hold property, but not only could schools be leased from a *société civile* or tontine, but that tontine might itself only be a link in the chain, itself subletting from the congregation which remained the effective proprietor.

Next, it was always possible for members of congregations to secularize themselves, and continue teaching in private schools as pious laity. There was obviously a certain amount of wastage from the congregations, even in Brittany, but the academic authorities were also having difficulty recruiting lay teachers and in 1904 an Association départementale pour la protection de l'enseignement primaire privé was set up in Ille-et-Vilaine under the presidency of the cardinal-archbishop in order to train and maintain a lay Catholic staff in the private schools.[163] An *instituteur* of Rennes suggested that the training of teachers hidden away in vestries by the parish clergy was in fact worse than suffering the congregations, for 'it has let the clergy worm its way back into the school and I know that the priest is more retrograde than the *cher frère*'.[164]

Lastly, because so many female congregations were 'mixed', nursing as well as teaching, they could survive officially as nurses and continue to have influence over education. The sub-prefect of Fougères urged that Sœurs de Ruillé and Sœurs de Rillé be authorized to continue as nurses, not only because it would be impossible to find other people to care for the sick but also because 'the population is very attached to them'.[165] On the other hand, it was said that another congregation allowed to nurse, the Filles de la Sagesse, 'use the influence they have over the population . . . to fight the *écoles laïques* and promote recruitment to private schools'.[166]

---

[162] A.D. I-et-V. 1M, Surcouf to prefect of Ille-et-Vilaine, 14 Nov. 1905; prefect to sub-prefect of Saint-Malo, 8 Jan. 1906.

[163] A.A. Rennes C5, statutes of Association départementale, 1904.

[164] A.N. F17 9125[5], S. Connier, private *instituteur* of Rennes to Min. Instr. Pub., 8 Dec. 1910.

[165] A.D. I-et-V. 1T *Congrégations enseignantes*, sub-prefect of Fougères to prefect, 18 Nov. 1911.

[166] Ibid., sub-prefect of Vitré to prefect, 5 Nov. 1911.

Far more significant in the Catholic scheme to build up private schools at the expense of the public schools was the campaign against nefarious history textbooks launched in 1909. The offensive of the clergy, broadly supported by the large landowners, was little short of an onslaught in Ille-et-Vilaine. In the *arrondissement* of Redon the sub-prefect reported 'chronic pressure on the part of the Breton clergy aimed at school attendance as well as at the use of classroom texts', reinforced by the activities of parents' associations and the witch-hunts of 'dames bien pensantes', seeking out the proscribed books.[167] Losses from *écoles communales* were reported to be high in the Redonnais, and the toll was similar in the Vitré–Fougères bloc. As in Flanders, the Catholic beliefs and practices of the majority of the population could be exploited to great effect by the clergy. 'They make our little pupils cry in confessional', reported the prefect, and 'refusal of sacraments has become a general way of fighting our schools'. Moreover, the clergy had the support of 'the *hobereaux*, large landed proprietors who have the whole rural population at their mercy'. Farmers often had the obligation to patronize the private school written into the terms of their lease; where they sent their children to the lay school, artisans risked losing their employment, tradesmen their commerce, paupers their relief.[168] There were nevertheless one or two areas where the railings of the clergy were not listened to and the republican *Avenir* kept a careful account of communes where an engrained anticlericalism triumphed, such as Liffré and Retiers.[169] The 'blue' bastions of Ille-et-Vilaine were still the same, still holding out against the offensives of *curé* and châtelain, but they were heavily under siege.

The last word must nevertheless go to the five Breton bishops of Rennes, Nantes, Vannes, Saint-Brieuc, and Quimper. In a letter to Catholic parents published in October 1913, they made very clear the distance that existed between the enactment of a law and its application, a distance that we have ourselves been at pains to emphasize. 'The founders of

---

[167] A.D. I-et-V. 1M, sub-prefect of Redon to prefect, 26 Oct. 1909.
[168] A.N. F17 10366, prefect to Min. Instr. Pub., 25 Feb. 1913.
[169] *Avenir hebdomadaire*, 19 Sept. 1909, 16 Jan. 1910.

the lay school', they said, 'forgot that the success of a law derives not only from the will and strength of legislators, but also and above all from the temperament and character of the people who must apply it. Many laws of all kinds have failed because of this. When the *laïcité* of the school was decreed, it was never considered that it would be impossible to impose it on regions as Catholic as our own.'[170]

The stubbornly held view that France was a centralized state, that policies devised in Paris could be directly implemented in the provinces, and that the provinces had to await instructions from Paris before venturing any move must therefore be modified. Schools and schoolteachers sprang up according to local needs. Chaos during the French Revolution did not mean that people were not being educated but only that the state had no control over what was going on. The work of the *maîtres de pension*, artisan teachers, and 'dame schools' of penniless spinsters, so much maligned by the reformers, provided an indispensable service at the margins of historical knowledge at the beginning of the nineteenth century. As with the medical profession, the teaching profession had to be disciplined and improved by the state, but the state did not create it. The establishment of a public system of education was a gradual process that worked with materials already to hand.

The importance of a dialogue between Paris and the provinces rather than a one-way chain of command is underlined by the politics of the school question. Reforms in education were not always initiated in Paris. Republican ministries were installed as a result of victories in the legislative elections of 1876 and 1877. Municipal elections were just as significant, for it was the 'town-hall revolutions' of 1878 and 1881 that initiated a spontaneous wave of laicizations, before the blanket measure of 1886. There were of course more laicizations in the Gard than in Ille-et-Vilaine, but that was the reflection of a stronger current of republicanism at the polls. Looking at the problem from the Paris angle, it becomes clear that provinces could not be made to swallow 'reforms' that they did not want. A trenchantly anticlerical prefect like Dumarest

---

[170] A.N. F17 9125[5], Letter of Breton bishops to Catholic parents, 24 Oct. 1913.

in the Gard could carry through sweeping changes because that was the sense of the electorate. But where a department had voted against the general trend that elected a given ministry, resistance might be concerted. In the Nord Saisset-Schneider could do little against the Catholic landslide of 1885, while in Ille-et-Vilaine the Combist Delpech was harassed all the way by a department that would have nothing to do with the policies of the Bloc des Gauches.

The relationship between governments and the Catholic Church was of course complex and a simple 'Church against state' analysis is insufficient. The Church monopolized both primary and secondary education before 1789, so that any move in the direction of state control was bound to cause some confrontation. The politics of the Catholic Church did not help. Whereas before the Revolution orders like the Doctrinaires had espoused progressive theories, the Revolution sealed the alliance between the Church and the monarchy which alone seemed able to protect its interests. Subsequent flirtations with democracy, such as that of Lamennais, were firmly reprimanded by the papacy.

Yet the attitude even of Bonapartist and liberal governments towards the Church was ambiguous. The Church was associated with reactionary regimes, but it was also a force for order and authority that could not be ignored. And a system of state education set up without the concurrence of the Church would inevitably force the Church into opposition to that system. Moral and religious instruction was therefore at the very base of the Napoleonic University and of the Guizot law, and the Catholic clergy became Rectors, *proviseurs*, principals of municipal colleges, and schoolmasters in the public sector while members of religious congregations became teachers in *écoles communales*.

The alliance was nevertheless unstable, and the royalism, ultramontanism, and 'feudalism' of the Church came to threaten both the Second Empire and the Third Republic. An anticlerical offensive became one way of anchoring such regimes against threats at home and abroad. At the same time the republicans remembered the moral influence of the Church, its dedication to the cause of social stability, and tried to evolve a code of ethics that would ensure the

same stability without, for political reasons, relying on the
sanction of religion. It was almost an impossible experiment.
But the association of the Church with royalism and counter-
revolution was not unbreakable. After the death of the
Legitimist pretender in 1883 the royalist cause looked
distinctly bleak, and the possibilities of a compromise
between Church and Republic grew brighter. Conservative
victories in the elections of 1885 were grounded not on
royalism but on tariff reform and the defence of the Catholic
Church against anticlerical legislation. When strikes and
socialism threatened the propertied classes and Germany
threatened national frontiers, there seemed no reason why
conservative republicans and moderate royalists should not
rally against the evils of international socialism, and its few
adherents among the *instituteurs.*

There was however a geography of Ralliement. It was
possible in the Nord where legitimism lacked a backbone and
socialism and Germany presented real threats, but was in-
conceivable in the Gard and Ille-et-Vilaine. These departments were economically backward, far from the frontier,
and above all divided by traditions which reinforced disputes
over religion or the constitution, and therefore disputes
over education. For, as has become clear, the conflict over
education was not everywhere the same, but coloured by the
historical geography of the provinces, the pattern of attitudes
that had been forged by generations of political and religious
conflict. The struggle between clerical and anticlerical forces
in the provinces was caught up in the web of historical
battles, expressed in the Nord in the campaign of French-
speakers to colonize Flanders, in the Gard in the struggle
of Protestants for supremacy over Catholics, and in Ille-et-
Vilaine in the bid by the thin blue line of the men of 1789
to lay to rest the partisans of the Ancien Régime. Outsiders
who came into these areas were transformed by them:
Duquesnay, archbishop of Cambrai, espoused the cause of
the fanatical Catholic camp in order to gain some influence
over it, while the sincere piety of a Franc-Comtois like Louis
Besson required some stiffening in a diocese used to the
hectoring militancy of Emmanuel d'Alzon.

But conflict over education did not just passively receive

the imprint of the historical context. The conflict over schools in the nineteenth century was another episode in the long struggle which defined the historical geography of France. It was not a class struggle, fought by the middle or working class against the feudal-ecclesiastical establishment and its dependants, but a struggle between communities, each of which had its identity and solidarity bound up with a particular cause. The anticlerical offensive altered the dimensions of those communities, but while it was possible for one camp to conquer areas the loyalty of which wavered, so that the strongholds of each side became ever more narrowly defined, those strongholds themselves remained impregnable. In the Gard, Protestant and anticlerical forces between them forced the congregations back into the corners of Nîmes, the Causses–Cévennes region, and into the north of the department, where Catholicism secured its position. In the Nord an anticlericalism that dated back to before the Revolution consolidated its grip but could make no progress in Flanders, its stalwart Catholicism protected in part by the Flemish language and in general by the separatist traditions of the Flemish 'nation'. In Ille-et-Vilaine, which fought bitterly against the ideological currents of the Third Republic, anticlericalism received a thrashing. Resistance in the marchland around Fougères and Vitré and in the *terre sainte* of the diocese of Saint-Malo might have been predicted, but in this case territory of undecided loyalty was won over to the Catholic opposition. These populations would have been content with a Catholic Republic, but the Radical campaign to break the back of the clergy and squirearchy alienated them completely. For not only did it challenge their Breton faith, it also violated the liberties of the Breton nation and awoke in them the temper of the Chouans who had resisted with such ferocity religious persecution, military conscription, and the policies of administrative centralization undertaken by monarchy and republic alike. Here, more clearly than anywhere else, were demonstrated the limits of centralization.

PART II
CLASS AND EDUCATION

5

The Education of the Élite

The system of secondary education established in the revolutionary–Napoleonic period, designed to train a civil, military,
and ecclesiastical élite, justified itself according to the principle
of the 'career open to the talents'. This principle served to
break the nobility's monopoly of positions of power and
made possible the consolidation of a governing class of both
aristocratic and middle-class elements. For the *idéologue*,
Destutt de Tracy, writing in 1800, the essential social distinction was that which divided the 'learned class', including
but superseding the privileged by birth, and the 'labouring
class'. The one required a summary practical education, the
other a lengthy and scholarly one, as befitted those destined
for the professions and public office, and never the twain
should meet.[1]

Such an analysis nevertheless glossed over the complexities
of French society. For there were not two but three main
social strata: the governing élite, the labouring masses, and,
between them, the *classes moyennes*. In the early nineteenth
century the primary-school system was ill-developed and
inadequate for the needs of the intermediate groups, who
would exercise pressure in order to gain access to the secondary schools. What interested them about secondary education
was not classical humanism and the education of a gentleman
but the qualifications that would equip them for the liberal
professions. How serious this pressure became depended very
much on the economic structure of the region. In prosperous
agricultural, commercial, and industrial areas the outlets for
the *classes moyennes* in the productive sectors of the economy
might tempt them away from a long, costly education towards prestigious but uncertain goals. But in more backward
areas where opportunities in trade and manufacture simply

[1] Detutt de Tracy, *Observations sur le système actuel d'instruction publique*
(Paris, An IX), pp. 2–5. This view has been reasserted by A. Prost, *Histoire de
l'enseignement en France, 1800–1967* (Paris, 1968), p. 10.

did not exist, secondary education would be used as an instrument of *déclassement*. This was a development that was viewed with anxiety by those already in positions of influence and gain. For as the middling groups turned their backs on their origins only to find the professions overcrowded and offices colonized, so they might become potential leaders of revolution which alone would hasten the 'circulation of élites'.

The main consideration of this chapter is the extent to which secondary schools were not so much confined to an oligarchy of landowners, bureaucrats, and professional men as open to fairly broad sections of the *classes moyennes*, tradesmen, artisans, *employés, petits fonctionnaires*, and even peasants. But such an assertion is meaningless unless we consider that secondary schooling was not homogeneous, but varied according to geography, syllabus, and ideology. There were *collèges royaux* (*lycées* before 1814 and after 1848) in departmental capitals that remained fairly inaccessible, yet the multitude of small towns with which rural France was studded each tended to have its own municipal college. The study of classics may have been a barrier or an irrelevance to many of the *classes moyennes*, but a good proportion of the small-town colleges taught more French than Latin. Lastly, there was a choice between colleges maintained by the state or municipalities and the secondary schools of the Catholic Church, which ranged from the most aristocratic establishments to the humblest seminaries.

The *collèges royaux* or *lycées* situated in large towns such as Rennes, Douai, and Nîmes not only charged high tuition fees but, beyond the daily reach of all but the inhabitants of the town and immediately surrounding area, also had a considerable population of boarders whose parents would have to pay again to have them lodged in the establishment. Distance was clearly socially selective. Of first importance then was the network of small-town colleges, in Ille-et-Vilaine at Dol, Fougères, and Saint-Servan, in the Gard at Bagnols, Alès, and Le Vigan, and all over the Nord from Cassel and Hazebrouck in Flanders to the industrial towns of Condé and Maubeuge in the south. The standards reached by these colleges have often been the subject of ridicule,[2] but they certainly served the purposes of the local clientele, providing a Latin education in their grammar classes up to the age of thirteen or fourteen, either to give a few years' varnish to the education of the local rural bourgeoisie or to provide an initial training in the classics, cheap and close to home, for a minority who hoped to complete their studies in the upper or humanities classes of the *collège royal*. The continuity between the municipal colleges of the early nineteenth century and those of the Ancien Régime was marked not only in their location and ecclesiastical direction but also in the modest social origins of their clientele.[3] Moreover, though the *internat* was part of the ideology of the secondary schools, an integral part of the virile education of the young man alongside his 'fortes études' in the classics, the proportion of boarders at the small-town colleges was very small. Many were day-boys, while those from further afield like the *poussoux* from Pleurtuit at the College of Saint-Malo, peasant sons who 'poussaient pour être prêtres', found it cheaper to lodge as *chambriers* with shopkeepers and working people in the town, receiving weekly provisions from their parents, than to pay the full boarding fees of the college.[4] The *chambriers*, far from disappearing with the Revolution, were a regular feature of life at small municipal colleges such as Dol and Fougères down to the 1870s. A survey undertaken in the Nord in 1865 revealed that 67 per cent of the pupils at the College of Le Quesnoy were boarders, as were 50 per cent at Avesnes and 43 per cent at Hazebrouck—all in the heart of rural areas— but that the proportion of boarders in the colleges of industrial towns was much smaller: 13 per cent at Armentières, 10 per cent at Saint-Amand, and one out of seventy pupils at the college of Condé.[5]

---

[2] P. Harrigan, *Mobility, Élites and Education in French Society of the Second Empire* (Waterloo, Ontario, 1980), p. 20.

[3] This point has been made by D. Julia and P. Pressly, 'La Population scolaire en 1789. Les extravagences statistiques du Ministère Villemain', *Annales E.S.C.*, 30e ann., No. 6, 1975, p. 1575. On the Ancien Régime colleges, the pioneering work by F. de Dainville, 'Effectifs des collèges et scolarité au XVIIe et XVIIIe siècles dans le nord-est de la France', *Population*, vol. x, 1955, 455-88 has been followed by W. Frijhoff and D. Julia, *École et société dans la France d'Ancien Régime* (Paris, 1975).

[4] E. Herpin, *et al.*, op. cit., pp. 224-5.

[5] A.D. Nord 1T 41/2, Statistic of 15 Nov. 1865.

In the second place, the secondary-school syllabus could be manipulated in order to open or close the flow of aspirants. Classical studies were the passport to the professions and to the élite in general, and were very much in demand in backward regions like Brittany. Attention was drawn in 1817 by the Comte d'Allonville, prefect of Ille-et-Vilaine, to the 'desire for instruction of all families elevated however slightly above the lowest classes of the population'.[6] Not all of them could aspire to the highest positions. In 1821, those attending the College of Saint-Méen in the clerical west of the department were 'for the most part sons of solid *laboureurs*' anxious to become clerks, *petits fonctionnaires*, and above all priests.[7] Jules Simon, a native of Vannes in the Morbihan, noted in his memoirs that 'before the Revolution, entering the priesthood was the only way for country-dwellers to move into the bourgeoisie'.[8] Little had changed in rural Brittany by the Restoration, and in 1823 the inhabitants of Saint-Méen converted their college into a *petit séminaire*, not least to serve as a powerful instrument of social promotion.[9] In commercial and industrial regions, on the other hand, the demand for Latin was much less. A school-inspector reporting from the booming linen town of Armentières, north of Lille, reflected that 'the soil of the mass of small towns will never support classical studies'. 'Everyone is involved in mercantile occupations', he wrote, so that the only studious pupils at the college were 'country children whose parents, perhaps because they have not the means, are foreign to the commercial mentality'.[10] So unsuccessful were the classics at Armentières that in 1831 the municipal council debated suppressing the posts of its Latin masters, though it contented itself with the gesture of cutting their salaries.[11] It was the French classes that generated what life the college had. Further north, in the Flemish colleges of Bergues, Cassel, and Hazebrouck, where teaching in the eighteenth

---

[6] A.N. F17 10213, prefect of Ille-et-Vilaine to Min. Int., 31 May 1817.

[7] A.D. I-et-V. 1T *Collège et Petit Séminaire de Saint-Méen*, sub-prefect of Montfort to prefect of Ille-et-Vilaine, 21 Dec. 1821.

[8] J. Simon, *Premières Annés* (Paris, 1901), p. 40.

[9] See above, p. 88.

[10] A.D. Nord 2T 1097, Inspector Agnant to Rector of Douai, 6 July 1822.

[11] A.D. Nord 2T 404, Insp. Acad. to Rector of Douai, 7 Aug. 1831.

century had been either in Latin or in Flemish, the teaching of French was an important innovation, although the *cours de français* were rarely put on a firm footing until after the July Revolution.[12]

There was after 1830 a movement in some commercial and industrial towns to break the monopoly of the classics and to make secondary education of more current appeal. At Lille, the campaign for change was led by a young doctor and public lecturer in botany, Thémistocle Lestiboudois (1797–1876), scion of the Lestiboudois dynasty which represented an enlightened, scientific tradition going back to Linnaeus and looked for a model to the *écoles centrales* rather than to that conservator of antiquity, the Napoleonic University.[13] His ambition was to overturn the tyranny of classics at the College of Lille, arguing that only one in ten of its pupils embraced a 'learned profession' requiring a classical education, the other nine going into commerce and industry, and that the lack of adequate training for commerce or the *grandes écoles* was driving pupils into commercial *pensions* or to Paris. A scheme was submitted to the municipal council of Lille in May 1832 proposing a 'bifurcation' of studies at the age of thirteen, with early leavers undertaking *études industrielles* which involved accounting, commercial law, and modern languages, a heavy dose of science for classical pupils in the rhetoric and philosophy classes, and a common trunk of classes in mathematics, drawing, French, and history.[14] It was adopted that autumn, but in the teeth of opposition from the Bureau d'administration of the college led by a conservative magistrate. Denounced as an 'insurrection against the University' by the Rector of Douai,[15] it was quickly disbanded. The main concern was that any dismantling of the classics in favour of a practical, French based course would provoke a surge of inferior

[12] E. Coornaert, 'Flamand et français dans l'enseignement en Flandre française des annexations au XXe siècle', *Revue du Nord*, liii, no. 209, 1971, 217-21.

[13] See above, p. 50.

[14] A.D. Nord 2T 404, Insp. Acad. to Rector of Douai, 4 July 1831; A.N. F17 8429, vice-president of Bureau d'administration of College of Lille to mayor of Lille, Oct. 1832; A.D. Nord 1T 56, commission du plan d'études au collège to municipal council of Lille, 25 Jan. 1837.

[15] A.D. Nord 1T 56, report to municipal council of Lille, 25 Jan. 1837.

elements into the secondary-school system. Such a course was acceptable only at the primary level, for there it would contain the *classes moyennes* within the primary-school system and maintain the colleges as a preserve of the élite.

The *écoles primaires supérieures* provided for in towns of over 6,000 inhabitants under the Guizot law were to serve exactly this purpose. Recognizing that existing elementary education was insufficient for the intermediary groups, they offered as an alternative to the Latin education of the colleges a few years' supplementary training in French grammar, calligraphy and drawing, arithmetic, geometry and book-keeping, geography and commercial law. This would divert the *classes moyennes* away from *fonctionnarisme* and the professions and provide generations of foremen and draughts-men, accountants and commercial clerks, for the French economy. Lille, which was not permitted to reform its college, set up an E.P.S. instead in 1837, attracting 'sons of artisans and small tradesmen' who were thus kept within the primary-school sector.[16] In general, the E.P.S. met with some success in seaports and industrial towns. Roubaix had an E.P.S. directed by the accomplished Van Eerdewegh, but no college before the abbé Lecomte founded Notre-Dame des Victoires in 1845,[17] while François Quiquet, director of the E.P.S. of Dunkerque, would also have a long and dis-tinguished career. In the Gard, the opening of the coal-basin of Alès by the railway linking it to the coast augured well for the E.P.S., founded in 1836. In Ille-et-Vilaine, the orientation of studies in the interests of a minority of rich 'omni-savans' at the College of Saint Servan had been criticized as early as 1806,[18] and was remedied by the establishment of an E.P.S. Rennes had only a *cours primaire supérieur* situated on a floor above the school of the Halle aux Toiles, but it

[16] A.D. Nord 1T 77/2, report of commission of municipal council of Lille, 12 Aug. 1837.
[17] See letter of Van Eerdewegh to municipal council of Roubaix, cited in Leuridan, op. cit., pp. 22–3.
[18] A.D. I-et-V. 1T *Collège de Saint-Servan*, two pamphlets of J.-M. Renou, signed 8 and 15 April 1806.
[19] A.M. Rennes R13, director of Cours supérieur, to mayor of Rennes, 30 July 1845.

attracted 'the children of industrialists (manufacturers or artisans), tradespeople and farmers, most of them being poor or comfortable, but rarely rich'.[19]

In two senses, however, the E.P.S. failed to establish a separate identity during the July Monarchy. Firstly, they never attained a level of study that clearly distinguished them from the elementary schools, and were starved of pupils by the Frères des Écoles Chrétiennes who, entrenched in the large towns, combined the overcrowding of classes at the lower end with a concentration of staff on the *classes d'honneur*, populated by sons of tradesmen and artisans who remained at school beyond the age of thirteen. This was clearly the case in the Nord,[20] in the Gard at Uzès and Bagnols, and in Ille-et-Vilaine at Vitré, where the rivals were the Frères de Ploërmel and the task of the E.P.S. was to provide 'an industrial and above all a sincerely national education' in order to fight them.[21] Secondly, many municipalities were too mean to finance an E.P.S. alongside the municipal college, and annexed one to the other long before the ordinance of 21 November 1841 permitted this officially. On the positive side, it was argued that now college regents were available to assist the E.P.S. director with parts of his syllabus. College principals, like the abbé Lecomte at Tourcoing, were glad to see a centre of rivalry with the 'French classes' of the college, where no Latin was taught, eliminated. But at a time when the director of the Cours primaire supérieur of Rennes, which was not annexed to the Collège royal, was having to defend himself against 'fantasmagorical rantings about *déclassement*', pointing out that two-thirds of his graduates went into apprenticeship or back to their farms, and that only a minority went on to the Collège or the École normale,[22] the original purpose of the E.P.S. was being vitiated. Instead of representing the summit of the education of the people, the *classes moyennes* slipped back on to the lower slopes of the education of the notables.

---

[20] P. Pierrard, *La Vie ouvrière à Lille sous le Second Empire* (Paris, 1965), pp. 356-7.

[21] A.N. F17 9826, Rector of Rennes to Min. Instr. Pub., 10 Aug. 1844.

[22] A.M. Rennes, R13, director of Course supérieur to mayor of Rennes, 30 July 1845.

In the face of this heavy admixture of popular elements in the municipal colleges, the social élite looked for ways, independent of government policy, of obtaining an education apart. One solution was the patronage of Catholic education. It would, of course, be rash to argue that social exclusiveness was the driving force behind Catholic education under the July Monarchy, or to claim that there was any social homogeneity in Catholic schools.[23] Political and religious explanations were powerful, and the Catholic Church was characterized by an immense flexibility, a readiness to adapt itself to every social need. That said, it is striking that over 45 per cent of the clientele of the Jesuit College of Brugelette in Belgium between 1835 and 1854 was composed of nobles, whether true or false.[24] The Assumptionist College of Nîmes grouped much of the Catholic élite of south-west France: 27 per cent of the pupils between 1845 and 1852 were of noble families, the sons of large landowners, army officers, magistrates, and barristers mingling with those of directors of ironworks at Alès and naval dockyards at Toulon, shipowners and *négociants* of Marseille and Cette.[25] At Assumption, the annual boarding fee was 800 francs; at Marcq-en-Baroeul, near Lille, the fee was 700 francs, but as one inspector pointed out, all the pupils were 'without the noble particle'. The foundation of Marcq in 1840 reflected 'a certain *amour propre* which, at Lille particularly, inclines the commercial notabilities to separate their children from those belonging to a lower class'.[26] It was patronized by an 'aristocratie d'argent', often of recent wealth, but which sought to distinguish itself not only from the modest clientele of the municipal college of Lille but also from the shopocracy so dominant in the ecclesiastical college of Tourcoing. A comparison of these two colleges with recruits to the Petit

---

[23] Similar reflections for a later period have been made by Anderson, 'The Conflict in Education. Catholic secondary schools, 1850-70: a reappraisal', in T. Zeldin, *Conflicts in French Society* (London, 1970), pp. 69-73.

[24] Archivium provinciae campaniae Societatis Jesu, M.S.E.F. 371, *Liste alphabétique des anciens élèves du Collège du Brugelette, 1835-1854* (Paris, 1875).

[25] A.A.A. Repertoire des élèves, 1845-52.

[26] A.D. Nord 2T 405, Inspector Le Bailly to Rector of Douai, 27 June 1842.

Séminaire of Cambrai illuminates the varieties of education offered by the Catholic Church.[27] The weight of landowners, liberal professions, and *fonctionnaires* was high at Marcq, without beginning to imitate Assumption. Industrialists and merchants represented about a half of the clientele at Marcq and Tourcoing, mill-owners, brewers, sugar-manufacturers, with *marchands-peigneurs* and other cloth merchants being in very heavy numbers at Tourcoing. There were more artisans and tradesmen at Tourcoing than at Marcq, but not as many as at the Petit Séminaire of Cambrai, which was thick with sons of cobblers, tailors, blacksmiths, joiners, masons, and, above all, weavers, especially the *mulquiniers* of the Cambrésis. Finally, a third of the seminarists were sons of *cultivateurs* and farmers, whereas they provided only 10 per cent of the recruits at Marcq and Tourcoing, and that figure does not include the 9 per cent of wage-earners, mainly agricultural labourers among the seminarists.[28]

After 1840 something of an 'aristocratic reaction' took place in the University.[29] It was characterized by a reinforcement of the monopoly of classical studies and by a bid to raise standards in the larger colleges to an excellence that could not be achieved by the small-town colleges, already corrupted by the annexation of the E.P.S. and their humble clients. The syllabus of 25 August 1840 announced by Victor Cousin, Minister of Public Instruction in Thiers's second ministry, ensured that between the sixth class and rhetoric, or between the ages of eleven and sixteen, college pupils would study nothing but the classics. Modern languages, which Salvandy had tried to promote in 1839, were confined to three classes between the ages of thirteen and

[27] Enrolment registers kept at Marcq-en-Barœul, at the Institution libre du Sacré-Cœur, Tourcoing, and, in the case of the Petit Séminaire of Cambrai, at the diocesan archives of Lille. The number of cases is respectively 307, 538, and 1108. The sample from the Assumptionist College, only 86, is clearly unreliable.

[28] The popular and rural nature of recruitment to seminaries in the early nineteenth century is discussed for the diocese of Rennes by Lagrée, *Mentalités*, pp. 178–85, and for that of Bourges by C. Dumoulin, *Un séminaire français au XIXe siècle: le recrutement, la formation, la vie des clercs à Bourges* (Paris, 1978), pp. 49, 149–50.

[29] D. Julia and P. Pressly, art. cit., p. 1547, advance a similar thesis.

Fig. 3. Recruitment to Catholic Secondary Schools, 1828–1852

Profession of Father	Collège de l'Assomption, Nîmes, 1845–52	Marcq-en -Barœul, 1840–52	Collège Communal, Tourcoing, 1845–51	Petit Séminaire, Cambrai, 1828–51
landowners, rentiers	13.9	22.5	6.5	2.1
liberal professions	29.1	9.8	4.3	3.1
management	9.3	0.7	0.5	0.1
public officials	16.3	6.5	3.3	1.3
industrialists, entrepreneurs	2.3	25.4	21.6	3.2
bankers, merchants	26.7	21.8	31.6	10.2
tradesmen, shopkeepers	–	1.3	11.0	7.8
artisans	1.2	1.3	5.4	21.1
employés, petits fonctionnaires	–	0.3	2.0	7.9
peasants	1.2	10.4	10.8	30.3
wage-earners	–	–	2.8	8.6
servants	–	–	0.2	1.2
orphans	–	–	–	3.1

fifteen. Most significantly, the study of sciences, not only physics, chemistry, and natural history, but also mathematics, was postponed until the two years of philosophy, taken at the age of seventeen and eighteen. As far as the structure of the secondary-school system was concerned, a bill of the education minister Villemain in 1844 divided colleges into first and second orders. The first-order colleges were required to employ eight University masters, at an annual cost of 20,400 francs; the second-order colleges must have at least four masters and a budget of 5,800 francs. The latter would be qualified to teach Latin only in the grammar classes, after which pupils would transfer to study their humanities at a first-order college, usually a *collège royal*. This was conceived as a massive blow dealt at the small-town colleges, operating on tiny budgets and seeking to retain their Latinists for as long as possible. The inhabitants of Le Quesnoy in the Nord petitioned the Chamber of Peers, protesting that the 'immediate result of these arrangements would be the ruin of municipal colleges in small towns like our own'.[30]

[30] A.N. CC 468/513, Petition of sixty-five inhabitants of Le Quesnoy to Chambre des Pairs, 15 Mar. 1844.

And from the other end of France, Jean-Marie de la Mennais replied, 'You are killing all or almost all the colleges of small and middling towns, you are killing all the *pensionnats*, you are killing the private institutions.'[31] It was a measure of centralization, encouraged by the development of the railways, that threatened the local colleges which were the traditional stepping-stone of the *classes moyennes* towards the élite. On the positive side, the measure meant the elevation of the municipal college of Lille to the status of *collège royal* in 1845, in a bid for the custom of the local bourgeoisie.

Two factors lay behind this reform of the University. In the first place, it was a response to the development of Catholic colleges like Marcq and Brugelette, a bid to stop the flight of the better-heeled clientele. But secondly, it represented a retraction of the élite, a concern with the 'excess of educated men' and an overcrowding of the professions that was the inevitable consequence of the expansion of a system of secondary education intended to produce only a limited number of ruling cadres. The defensive reflex of the University to check *déclassement* was paralleled by attempts by the professions themselves to raise standards and limit recruitment to their ranks. 'La France des notables' was consolidating itself.[32]

Very little has been written about private education outside the domain of the Catholic Church, no doubt because *universitaires* regard it with scorn. Yet private education only flourished in order to meet needs that were felt, and the reorganization of the University in the 1840s in many ways sanctioned an education that was impractical. The syllabus of 1840 required eight long, arduous, and costly years of study after the age of eleven before the *baccalauréat* could be obtained. But the anxiety of pupils and families was that this qualification should be acquired as quickly, easily, and cheaply as possible. The conflict of interests gave rise, under

---

[31] J.-M. de la Mennais, *De l'avenir réservé aux collèges communaux par la loi Villemain* (Paris, 1844). Copy of pamphlet in A.F.P. 101.

[32] On this, see L. O'Boyle, 'The Problem of an excess of educated men in western Europe, 1800-50', *J.M.H.*, 4, 1970, 471-95; G. Weisz, 'The Politics of medical professionalization in France, 1845-48', *Journal of Social History*, 12, 1978, 3-30; J. Léonard, op. cit., pp. 82-5, 92-5, 213-19.

the July Monarchy, to a whole industry of crammers, together with cribs, primers, and interlinear translations, a veritable 'counterfeit of science'. Those who saw that the true purpose of classical education was being perverted by such practices did not mince their words in denouncing them. In 1849, d'Alzon attacked the 'kind of rage or monomania' that led families to make use of 'hothouse studies' outside the college, driving their children in front of the examiners at times before they had entered the rhetoric class.[33] The Rector of Douai, pillorying such 'deserters' in 1855, declared that 'the essential task of University education is not to manufacture *bacheliers* in a hurry but to train men with a serious intellectual discipline and intellects that are highly cultivated'.[34] It was the eternal battle in the world of education between the missionaries and those who wanted only to get on.

Bachotage of this nature might legitimately be attacked when the pupils were aiming at the *baccalauréat* and the liberal professions. But cramming became normal, even necessary, under the regime of 1840 when admission to the *grandes écoles* was at stake. The postponement of mathematics and sciences to the philosophy classes resulted in a *surménage* that could leave only confused and superficial ideas, while practical questions such as the fact that the maximum age of admission to 'la Borda' (the Naval School at Brest) was sixteen turned candidates for such schools away from the *collèges royaux* into the arms of the preparatory courses and crammers.[35] In 1843 a number of notables of Rennes petitioned Villemain to point out that while classical education was imperative for theologians, lawyers, doctors, and men of letters, it was not necessary in order to train engineers, army and navy officers, industrialists and merchants. They proposed the annexation of a five-year science-based course to the college, the examination for the military and naval schools to be taken after four years, and that for the

---

[33] *Maison de l'Assomption, Rapport de M. l'abbé d'Alzon, directeur*, 1849.

[34] A.D. Nord 1T 32/4, Rector of Douai to Insp. Acad. Lille, 29 Oct. 1855.

[35] This is a point omitted by G. H. Weiss, 'Origins of a Technological Élite: engineers, education and social structure in nineteenth-century France' (Harvard D.Phil. thesis, 1977), in his treatment of private preparatory institutions, pp. 283–5.

Polytechnique at the end of the full term.[36] Equally, the heavily classical course of Cousin was of no interest to the *classes moyennes*, who would not in any case be inclined to complete the whole course. As Ferdinand Roux, principal of the college of Alès, argued, viewing the growth of the mining, metallurgical, and silk industries of the Cévennes, some provision had to be made for those destined for industry and commerce, and the E.P.S., still part of the primary-school system, was not adequate for the task.[37] Curiously enough, it was another native of Alès, Jean-Baptiste Dumas (1800–84), who had trained as a pharmacy student in Geneva after 1816, become a chemist, and was now Dean of the Science Faculty of Paris, who was campaigning in higher places against the syllabus of 1840. Denouncing the complete 'black-out' of scientific training before the age of seventeen, he proposed a system of 'scientific colleges' where those who were going to leave school at sixteen would do three years' classical training between the ages of eleven and thirteen, then three years of mathematics, physics, chemistry, accounting, and drawing, while still keeping up with history and geography, modern languages and Latin authors.[38] It was this plan that Salvandy attempted to launch in 1847, under the title of 'special education', though it met with very little success.

It was in the wake of 1848 that the University began to lose its clientele at a rapid rate on all sides. The conservative interpretation of the Revolution was that it had been caused by *déclassés* and intellectuals, educated above their station, ambitious to succeed and finding their paths blocked by the notables of the July Monarchy. In order to prevent a similar recurrence, the Falloux law abolished the E.P.S. which over-educated the people, drawing a firm line between primary and secondary education, and emancipated the Catholic colleges. The value of filtering some of the population of the secondary schools away from the liberal professions and towards scientific and engineering careers was now appreciated.

[36] A.N. F17 6890, Petition of notables of Rennes to Min. Instr. Pub., 7 June 1843.
[37] A.D. Gard 1T 71, Ferdinand Roux to Rector of Nîmes, 9 May 1846.
[38] J.-B. Dumas, *Rapport sur l'enseignement scientifique dans les collèges, les écoles intermédiaires et les écoles primaires* (Paris, 1847).

The project of Hippolyte Fortoul, Napoleon III's Minister of Public Instruction, launched under the statute of 30 August 1852 as 'bifurcation', accordingly divided pupils at the third class, at the age of fourteen, between the literary section, leading to the *baccalauréat ès lettres*, and a scientific section, crowned by the *baccalauréat ès sciences*.[39]

These changes proved of very little benefit to the University. The attempt to separate secondary education, that of the notables, from elementary education, that of the people, failed to reckon with the *classes moyennes* and their specific needs. These needs must somehow be met, and if it were not to be by the cumbersome machinery of the University other ways would be found. At the College of Dunkerque, the E.P.S. run by François Quiquet since 1838 was reformed as a *cours spécial* or *école commerciale* in 1850.[40] At Roubaix, the E.P.S. directed by Van Eerdewegh, most of whose sixty-five pupils in 1849 were 'sons of foremen employed in the mills and factories of the town and surrounding area',[41] continued after 1850 as an independent *école de commerce* training draughtsmen and accountants for the town's commercial houses.[42] A similar commercial school, advertising its courses of preparation for the Arts et Métiers, the mines' school of Saint-Étienne, and the École Centrale, flourished at Valenciennes.[43] In the countryside, private *pensionnats* hovering in the uncertain regions between primary and secondary education catered for a rural bourgeoisie of prosperous peasants, millers, brewers, and innkeepers. The Cambrésis was thickly studded with them: at Bantigny the *instituteur communal*, Bourgeois, who had a successful *pensionnat* attached to his school since 1851, opened a private boarding-school in 1861, drawing attention to the 'numerous young men trained by myself and employed

[39] On this, see R. Anderson, *Education in France*, pp. 66–72, and R. Raphaël and M. Gontard, *Un Ministre de l'Instruction Publique sous l'Empire autoritaire: Hippolyte Fortoul, 1851–1856* (Paris, 1975), 113–47.

[40] A.D. Nord 1T 107/6, primary-school-inspector Bourgeois to Rector of Douai, 30 June 1850.

[41] A.D. Nord 1T 74, report of primary-school-inspector Marre, 1 Nov. 1849.

[42] A.D. Nord 125/10, letter of Van Eerdewegh to Insp. Acad. Douai, Sept. 1863.

[43] A.D. Nord 2T 1094[1], Charles Gruson, director of *pension* at Valenciennes to Insp. Acad. Douai, 25 April 1856.

today either in the administration, or in medical schools, or in education' (he claimed to have instructed fifty *instituteurs*).[44]

Similarly during the 1850s the Catholic Church perfected a whole battery of institutions designed to fulfil the needs of a petite bourgeoisie anxious to separate its own children from the lower classes of the primary schools by the payment of fees, yet sceptical about the value of a long college education. New enterprises came from so many quarters within the Church, moreover, that internal tensions occasionally resulted. In the first place, the teaching congregations developed *pensionnats* and *demi-pensionnats*, charging 350 or 400 francs for boarders and between two and five francs a month for day-boys. The Frères de Ploërmel, on the invitation of Brossays Saint-Marc, opened the Pensionnat Notre-Dame-du Thabor at Rennes in September 1848 and reorganized one at Saint-Servan in 1855, while across the Rance estuary, first at Dinard and then at Saint-Briac, Frère Bénigne trained scores of young sailors in navigation between 1850 and 1870 for their masters' ticket.[45] The Frères des Écoles Chrétiennes were hampered by their dedication to free education, but the mayor of Saint-Malo overcame this issue in 1852 by promising that the Frères could run the École primaire supérieure in the town without accepting payment, while fees would be paid only to English, drawing, and singing masters brought in from the outside.[46] Boarding fees were not a sensitive issue, and the Frères des Écoles Chrétiennes founded the Pensionnat Saint-Louis de Gonzague at Alès in 1856, which included preparation for the Arts et Métiers and the school of master-miners set up in 1843, and in the Nord *demi-pensionnats* at Tourcoing in December 1848 and at Lille in 1861. These *pensionnats* posed a threat to the clientele of the municipal colleges, drawn from the same *classes moyennes*. At Alès, the *pensionnat* of the Frères was partly a response to the fact that since the Falloux law, the college had become a rump dominated by

[44] A.D. Nord 1T 117/1, prospectus of Bourgeois, 24 Sept. 1861.

[45] A.F.P. Dinard St. Énogat and Saint-Briac.

[46] A.F.E.C., NC 818, mayor of Saint-Malo to Frère directeur, Saint-Malo, 19 Oct. 1852; superior-general to mayor, 15 Nov. 1852.

the Protestants.[47] So successful were they, that ecclesiastical as well as lay colleges suffered from their competition. In 1860, the principal of the ecclesiastical college of Saint-Malo protested to Brossays Saint-Marc that such competition should be left to 'the University and the enemies of the clergy', not invited from the Frères. Brossays Saint-Marc, with characteristic pugnaciousness, bullied Frère Philippe into reducing the number of staff at his school from four to two.[48]

Secondly, however, the ecclesiastical colleges were responding more sensitively in their own right to the demands of the *classes moyennes*. The College of Immaculée Conception, established by secular priests in the Ursuline convent of Sommières, near Nîmes, in 1845, was developed after 1848 as a professional school, channelling its pupils towards the Arts et Métiers, veterinary schools, the Post Office and customs service, commerce and wine-growing.[49] At Roubaix the municipality reached an agreement with the abbé Lecomte for the addition of an *école supérieure* to the college of Notre-Dame des Victoires, aimed at 'the small shopkeeper and honest artisan', who was not tempted by the classical course of the college.[50]

There was a possibility that bifurcation might have revived the fortunes of the University. It certainly replied to a demand for scientific education. The years 1852–6 witnessed a heavy depopulation of the literary classes of *lycées* in favour of the sciences, not only in industrial towns like Lille but also in military, legal, and administrative centres like Rennes.[51] Bifurcation, brandishing the intellectual superiority of the University, was a riposte to the advantage given to Catholic institutions by the Falloux law. But in two respects the measure was not enough. Firstly, Fortoul's scheme was not accessible to the large numbers of artisans' and shop-

[47] A.F.E.C., NC 377/1, Frère Lédard to Frère assistant, 27 Feb. 1865.

[48] A.F.E.C., NC 818, principal of College of Saint-Malo to Brossays Saint-Marc, 12 Dec. 1860; Brossays Saint-Marc to superior-general, 29 Dec. 1860.

[49] E. Capelle, *Un moine: le père Jean, abbé de Fontfroide, 1815–95* (Paris-Toulouse, 1903), p. 16.

[50] A.D. Nord 1T 77/2, report of commission of municipal council of Roubaix, 6 Aug. 1852. Cf. above, p. 11.

[51] Anderson, *Education in France*, p. 99.

keepers' sons who might profit from a scientific *lycée* education but had never studied Latin. These *classes moyennes*
remained in the 'commercial' and 'professional' classes of the
municipal colleges, in the commercial *pensions*, or in the
pensionnats of the Frères, which continued where the E.P.S.
had left off. Moreover, those who did undertake the scientific
course frequently came to reject the world of trade as too
prosaic for their ambitions, and were lost to the productive
sector. The *proviseur* of the Lycée of Douai reflected that
too often the sons of industrialists and large-scale farmers
who initially envisaged following their fathers were stimulated
by obtaining the *baccalauréat ès sciences* to attempt the
grandes écoles.[52] If bifurcation had any clear effect, it was
to multiply the number of candidates for Saint-Cyr, Centrale,
and Polytechnique. Secondly, those traditionalists who stood
by classical studies as a virile education that strengthened
the character, as an aesthetic education that elevated the
soul, or merely as the necessary gateway to the professions
denounced bifurcation as a disruptive factor in secondary
education. It never really caught on in Catholic colleges.[53]
Indeed, Catholic principals tended to see themselves as
defenders of the old system. The superior of the College of
Saint-Joseph at Lille declared that 'we are resisting with all
our strength the forces that are precipitating the University
in the direction of the almost exclusive study of the natural
sciences, and we are faithful, as far as is possible, to the old
traditions of the Schools'.[54] The abbé Lecomte, principal of
the College of Tourcoing, had lectured his prize-day audience
of mill-owners, cloth-merchants, and their families in 1839
on the need to 'fight the materialistic and almost mechanical
tendencies of our society by an intellectual culture'.[55] Now
he steadfastly refused to admit bifurcation at Tourcoing.
There was no contradiction in this: the captains of industry

[52] A.N. F17 6845, report of *proviseur* of Lycée of Douai, 1864.

[53] In 1858, 49 per cent of the upper classes of the Lycée of Rennes were studying science, whereas the figure for Saint-Vincent was 22 per cent and Saint-Sauveur,
Redon, 27 per cent. A.N. F17 8899², Insp. Acad. to Rector of Rennes, 29 Mar. 1858.

[54] A.D. Nord 1T 125/8, report of superior of Collège Saint-Joseph, Lille,
8 Feb. 1856.

[55] Arch. Institution libre du Sacré-Cœur, Tourcoing, prize-day speech of
abbé Lecomte, 17 Aug. 1839.

looked to the abbé Lecomte not to teach them about textiles but to legitimize their new wealth and power by conferring on their sons the traditional culture of the élite.

The Catholic defence of classical humanism—for d'Alzon's campaign against pagan classics in favour of patriotic texts was an aberration—was symptomatic of its growing appeal to the richest classes of society, who wanted to mingle with the aristocracy and receive education more than knowledge. The popular profile of many of the *lycées* and colleges can be gauged from a survey undertaken in the Nord in December 1855.[56] At Douai, once the seat of the Parlement of Flanders and now of a Cour impérial, the headquarters of an Academy and an important garrison town to boot, 45 per cent of the clientele were sons of rentiers, magistrates, civil servants, army officers, and members of the liberal professions. On the other hand, this proportion was equalled by sons of industrialists and tradespeople (the inquiry does not distinguish merchants from smaller tradesmen). The college of the seaport of Dunkerque was much more heavily colonized by trade, and its profile differed little from that of the Lycée of Lille. Among the smaller colleges, Saint-Amand and Hazebrouck drew over half their clientele from *classes moyennes*—*commerçants*, artisans, and *cultivateurs*—fortified at the *arrondissement* capital of Hazebrouck by a good dose of sons of *fonctionnaires*. At Armentières, the rapidly growing textile town on the Lys, 85 per cent of the school population was drawn from the *classes moyennes*, mainly tradespeople, and a petition circulating in 1865, calling for Catholic education in the town, naturally favoured not a high-flown college but 'an intermediary institution entrusted to the Marists of Beaucamps'.[57] At Condé, on the Escaut, 11 per cent of the school population were sons of working men, while at Bailleul, in Flanders, a third were drawn from the ranks of the peasantry. Small-town colleges attended to a large extent by the sons of local farmers were not a feature peculiar to the Nord. In the Gard, it was the case not only at Bagnols but also, if we are to believe

[56] A.D. Nord 2T 263/4, Statistic of December 1855; the sample is of 1,039 cases.
[57] A.D. Nord 1T 42/1, principal of Armentières to Insp. Acad. Douai, 12 June 1865.

Fig. 4. Recruitment to Lycées and Colleges of the Nord, 1855

	Dunkerque	Bailleul	Hazebrouck	Armentières	Douai	St. Amand	Condé
rentiers	12.6	12.1	13.3	—	12.2	—	10.3
liberal professions	6.6	3.0	2.4	1.8	9.4	5.1	8.4
public officials	18.5	8.1	14.8	3.6	23.0	8.5	4.7
industrialists	8.4	4.0	12.8	8.2	38.0	20.3	10.3
commerçants	33.5	12.1	26.7	55.5	7.3	16.9	23.4
artisans	13.2	27.3	3.8	5.5	—	18.6	20.5
cultivateurs	7.2	32.3	25.7	23.6	10.1	23.7	11.2
ouvriers	—	1.0	0.5	1.8	—	6.8	11.2

Alphonse Daudet, at Alès, at which he was a *maître d'études* in 1857. The middle school of Sarlande, as Alès became in his fiction, was composed of 'fifty-odd rascals, chubby mountain-people of twelve to fourteen years old, sons of enriched *métayers* whose parents had sent them to college to have them made into *petits bourgeois* for 120 francs a term'. They were 'coarse, insolent, arrogant, and amongst themselves spoke a vulgar Cévenol patois that was completely incomprehensible to me'.[58]

By contrast, the bourgeoisie proper tended to gravitate towards the Catholic colleges. This view, contested by one recent historian,[59] was the cause of much embarrassment to University *fonctionnaires* at the time. In 1864 the anti-clerical Rector of the Academy of Rennes, Alfred Magin, reported that 'the preference of the children of rich families is for the ecclesiastical houses, and . . . their presence tends to attract the sons of public servants. It is *bien porté*, as we say, and by following the current favoured in all the confessionals one can acquire an aristocratic perfume or varnish.'[60] The principal of Saint-Amand in the Nord described that there were, *extra muros*, 'a large number of tradesmen, artisans, shopkeepers of all sorts, it is they, along with the peasants, who provide me with most of my pupils'. For the rich and educated minority, however, living *intra muros* and strongly influenced by the clergy, 'the College' was Notre-Dame des Anges, patronized because it was 'more aristocratic'.[61]

Responsibility for setting the University back on its feet, in this as in so many other things, must be attributed to Victor Duruy. What was required when he became Minister of Public Instruction in 1863 was a system of non-classical secondary education that would recover the *classes moyennes* from private institutions, without in any way degrading the

[58] Alphonse Daudet, *Le Petit Chose* (first edn., 1868; Paris, 1947), p. 160.

[59] P. Harrigan argues by contrast that during the Second Empire, Catholic schools drew on 'the aristocracy, the peasantry and the urban lower middle-class, but not the bourgeoisie', 'Catholic Secondary Education in France, 1851–1882' (Univ. Michigan, Ph.D. thesis, 1970), p. 254.

[60] A.N. F17 6848, report of Rector of Rennes, 1864. The Rector of Douai made similar observations in A.N. F17 6845.

[61] A.D. Nord 1T 59, principal of Saint-Amand to Insp. Acad. Douai, 21 May 1864.

classical studies which alone could guarantee *lycées* and municipal colleges against defection to Catholic schools. Special education, a four-year programme of non-classical studies, based on modern languages, history and geography, and the natural sciences, formalized by the law of 21 June 1865, was Duruy's response to this challenge. Like the E.P.S., it caught on in industrial towns and centres of trade, not least because in most cases they were only refashioned versions of the 'commercial' or 'professional' courses that had survived at those colleges through the 1850s. In the Nord, special education flourished at Dunkerque and Armentières, Valenciennes and Saint-Amand, Condé and Maubeuge. In the Gard, Ferdinand Roux, former principal of Alès and now director of the École de Cluny, set up to train masters for special education, returned to Alès in September 1867 to set the college on its new footing, and numbers there rose rapidly from ninety to two hundred. In Ille-et-Vilaine, the port of Saint-Servan and Fougères, where the woollen industry was giving way to the thriving manufacture of shoes, special education was similarly welcomed. But the new course was adopted not only for economic reasons. In towns such as Hazebrouck and Saint-Amand, it was a stick with which to beat rival Catholic institutions. In the case of small-town colleges whose Latinists now left at a very early age along the newly constructed roads and railways in order to complete their studies at larger centres, a successful special-education course could mean the alternative to closure. This was clear in the Nord, at such colleges as Cassel and Le Cateau, in the Gard at Bagnols and Le Vigan, where the municipal council was on the verge, between 1865 and 1867, of replacing the college by an E.P.S., and in Ille-et-Vilaine at Dol, which was scarcely more than a glorified primary school.

No study of special and classical education in the 1860s can ignore the work of Patrick Harrigan, based on Victor Duruy's inquiry into secondary education of 1864, presented both statistically and analytically.[62] The regional perspective

[62] P. Harrigan with V. Neglia, *Lycéens et collégiens sous le Second Empire, Étude statistique sur les fonctions sociales de l'enseignement secondaire publique d'après l'enquête de Victor Duruy, 1864–65* (Paris, 1979); P. Harrigan, *Mobility, Élites and Education in French Society of the Second Empire* (Waterloo, Ontario, 1980).

Fig. 5. Distribution of Pupils between Classical and Special Education, 15 November 1865[63]

	Total Number	Classical	Special	Primary
Dunkerque	199	57.8	42.2	—
Armentières	219	16.4	48.4	35.2
Hazebrouck	61	34.4	55.7	9.8
Valenciennes	355	49.8	43.7	6.5
Saint-Amand	113	25.7	38.0	36.3
Condé	70	18.6	62.8	18.6
Maubeuge	171	29.2	48.0	22.8
Alès	126	46.8	38.9	14.3
Bagnols	62	22.6	67.7	9.7
Le Vigan	47	36.2	48.9	14.9
Fougères	85	20.0	80.0	—
Saint-Servan	170	25.3	42.4	32.3
Dol	77	15.6	28.6	55.8

is nevertheless still useful, both because other sources such as the enrolment registers of schools offer statistically signifi-cant evidence for individual establishments which the Duruy survey does not provide,[64] and because the regional dimen-sion in Harrigan's work is not fully developed.

Special education was clearly more accessible to the *classes moyennes* than classical education. There were of course clear limits to that accessibility. The attraction of pupils from elementary, especially village, schools was difficult, and such pupils arriving at college were often so coarse and slow that one year in a preparatory class was a 'precious noviciate'.[65] Moreover, it was beyond the means of many modest families to stay the whole five years of the course, even supposing that the college offered it. Many left for farms, workshops, or offices in order to earn an honest living after one, two, or three years. The Rector of Douai pointed out in 1865 that 'the many youths belonging to families of medium or small property' required no education more exalted than that offered by the old E.P.S., and regretted

[63] *Statistique de l'enseignement secondaire*, 1868.

[64] Harrigan provides firm evidence of the social origins of pupils in classical and special education for 4,103 cases in the whole of France in 1864. The enrol-ment registers of the Lycée of Rennes for 1863–70 in A.D. I-et-V. 10Ta 32–4 give 566 cases.

[65] A.D. Nord 2T 63/5, Insp. Acad. to Rector of Douai, June 1866.

their passing. Besides, he could not but be aware of the correctness of some of the adversaries who argued that 'the system of education now being organized has nothing special, nothing professional' about it, and would not produce the material sought by hard-headed employers.[66]

According to Harrigan's analysis, classical education was preferred by the élite of landowners, professional men, and 'businessmen' together with lower professionals, petty civil servants, salaried employees, and workers interested in social mobility; the traditional *petite bourgeoisie*, consisting of shopkeepers, artisans, and peasants tended to opt for special education.[67] The comparison with the records of the Lycée of Rennes, the only *lycée* in our three departments for which enrolment registers of the period survive, to a large extent confirms this pattern. The sons of landowners, public officials, and professional men accounted for 62 per cent of those reading classics in 1863–70, but only 18 per cent of those on the special course. In the *lycée* of a large provincial town such as Rennes the *classes moyennes* would be less represented than in a small-town college, and the sons of tradesmen, peasants, and artisans who attended accounted

Fig. 6. Social origin of pupils at Lycée of Rennes, 1863–1870[68]

Profession of Father	Classical	Special
landowners, rentiers	19.4	6.8
liberal professions	16.4	3.8
management	4.1	6.0
public officials	26.3	7.5
industrialists, entrepreneurs	2.8	8.3
bankers, merchants	17.3	27.0
tradesmen, shopkeepers	3.0	7.5
artisans	3.2	9.0
employés, petits fonctionnaires	6.2	18.0
peasants	0.5	1.5
wage-earners	0.7	3.8
servants	—	0.7

[66] A.N. F17 8708, Rector of Douai to Min. Instr. Pub., 28 Feb. 1865.
[67] Harrigan, *Mobility*, pp. 25, 28–9.
[68] A.D. I-et-V. 10T, Lycée of Rennes, registres d'inscription, 1863–70.

for 6.7 per cent of those on the classical course, 18 per cent of those on the special course. But the sons of *employés* and *petits fonctionnaires* were also thick on the special side, and the sons of merchants and entrepreneurs too seemed to gravitate in that direction. Rennes may of course be an exception. The commercial and industrial bourgeoisie was notoriously isolated there. At Tourcoing, where the textile *patronat* dominated town life, Lecomte's successor as principal of the municipal college, the abbé Leblanc, maintained a healthy disdain for special education, arguing that it was only a revamped *cours de français* and that twenty years' experience had shown that 'a liberal education, far from being an obstacle to success in industry, is usually a condition of superiority. I could cite a large number of examples among our young men established at Lille, Roubaix and Tourcoing.'[69]

A further point in Harrigan's analysis concerns the ambitions of secondary-school pupils. Special education limited expectations, classical education excited them. Only 10 per cent of the sons of the commercial bourgeoisie on the classical course intended to enter commerce or industry, while 73 per cent of those on the special course had this as their aim. Two-thirds of shopkeepers' sons and half the sons of peasants on this side intended to return to the shop or farm.[70] This was in no way surprising, as special education, short, non-classical, and sanctioned only by a diploma, excluded those who pursued it from entering the higher professions. It was not for all that without its uses. A college education, seen as quite different from the elementary school, was made available to prosperous families anxious that their sons should 'se dégrossir un peu'.[71] The principal of the College of Cambrai reported that the 'enlightened classes' saw in special education only 'the old *enseignement primaire supérieur*, neither more nor less' and studiously avoided it, but that 'the special courses have credit above all among the middling and merchant class and among the rural populations, who look to them to make their children fit

[69] A.D. Nord 1T 60, abbé Leblanc to Insp. Acad. Lille, 21 May 1864.
[70] Harrigan, *Mobility*, pp. 50–1.
[71] A.N. F17 8702, Rector of Douai to Min. Instr. Pub., 17 April 1865.

for employment in the shortest possible time'.[72] Beyond that, special education offered an opportunity to rise from the world of the farm and *boutique* to the Arts et Métiers, the veterinary schools, employment in the post or tax offices, customs, or elementary education, or to a clerkship with a notary, broker, or merchant. But to enter the liberal professions, classical studies capped by the *baccalauréat* provided the only avenue. If we trace pupils who were at the Lycée of Rennes between 1863 and 1870 through to the year-books of the Association d'anciens élèves it becomes apparent that the son of a peasant, cattle-merchant, bookseller, or *instituteur* could rise to the rank of pharmacist or veterinary surgeon with a special education, but to become a doctor he required the classical *baccalauréat*. The taste of the classics in itself seemed enough to stimulate ambition. At the end of the century, the son of a shopkeeper who became a doctor at Dunkerque, wrote that 'until the age of fifteen I had a great deal of enthusiasm for commerce but the study of classics made me blush to help my parents in the shop. It seemed to me that working with one's hands was unworthy of a young man who was sharpening his wits by translating Cicero and Livy.'[73]

The scope and limit of ambition was nevertheless decided in the end not by the vagaries of a particular course of education but by the pattern of opportunity set by the economic and social structure of the region. Harrigan makes the point that while industrialization tended to raise social expectations, it tended to place a drag on achievement, so that social mobility was higher in poor, agricultural areas. The explanation seems to be that in prosperous industrial (usually coinciding with prosperous agricultural) areas, sons of artisans, shopkeepers, and peasants were encouraged to return to the farm or trade, while in backward regions *déclassement* took place 'less because of expanded opportunity than because of economic forebodings'.[74] Harrigan proves his point by comparing northern and eastern France with southern and western

---

[72] A.D. Nord 2T 397, principal of Cambrai to Insp. Acad. Douai, 6 April 1865.
[73] A.N. F17 13940, Dr Lancry of Dunkerque to abbé Lemire, deputy of Hazebrouck, 19 Feb. 1899.
[74] Harrigan, *Mobility*, p. 93.

France, taking the Saint-Malo–Geneva line so familiar to historians of education as the divider. A more precise comparison, taking the Nord with six other departments of the north, the Gard with eighteen other departments of the Midi, and Ille-et-Vilaine with eleven other departments of the west, is nevertheless instructive.[75] In the north, the liberal professions represented a particularly stable élite, three-quarters of sons following their fathers into the professions. The tendency to move away from mercantile and industrial occupations was much stronger, although 22 per cent of industrialists' sons entered the *grandes écoles*, Polytechnique, and Centrale, and may well have returned to reinforce both the family firm and the local *patronat*. Most interesting, however, is the fact that the tradesmen, peasants, and workers of the north were the most stable of the three areas, no doubt because of the prosperity of trade and agriculture in the north. In the west of France, the most backward region, society was simple and stratified. The isolation of the *négociants* and *entrepreneurs* is again clear: movement out of trade into the traditional élite was more difficult here than elsewhere. On the other hand, small tradesmen's sons had a tendency to become *fonctionnaires* and the sons of *employés* and *petits fonctionnaires* were well placed to enter the professions. As far as the sons of artisans, workers, and peasants were concerned, there was one guaranteed means of escape from their condition: as *instituteur* or priest. It was nevertheless in the Midi that social mobility was greatest, a phenomenon as well known[76] as it is difficult to explain. Rates of *déclassement* among sons of tradesmen, artisans, workers, and especially peasants, only 18 per cent of whom returned to the soil as against 43 per cent in the west and 65 per cent in the north, were uniformly higher here than in the other two regions. There was a certain stability among *petits fonctionnaires*, if only because *fonctionnarisme* had already achieved its goal. Agricultural and industrial decline, especially in the silk industry, relative to

[75] Harrigan, *Lycées et collégiens*, Tables 116, 118, and 121 provide 1,298 cases.
[76] A. Girard, *La Réussite sociale en France: ses caractères, ses lois, ses effets* (Paris, 1961), pp. 71–87, 171–88.

the expanding north, was no doubt the major factor, but such a decline was common to the west as well. Whether political and religious radicalism played a part in the Midi, it is impossible to say. But the role of the schools as levers of social mobility was closely defined by circumstances.

It is clear then that the *classes moyennes* might be motivated by two forms of ambition, either to succeed in the world of trade or industry which required no Latin education, or to obtain a Latin education in order to move into the priesthood, professions, or public office. The University ultimately catered very little for their needs. Despite a local initiative at Lille, there was no question of annexing 'commercial' courses to colleges in the 1830s, for fear of provoking an invasion of the middling sorts, and the E.P.S. that were offered as an alternative were of too low a standard and abolished in 1850. Those looking to trade or industry were therefore forced towards the rash of commercial schools and private *pensionnats* that sprang up in the 1850s to meet their needs, and it was not until the mid-1860s that the University developed special education in order to recapture them. The element of the *classes moyennes* that sought a Latin education and crowded into the colleges early in the century met with little joy either. Fearful of *déclassement*, the notables of the July Monarchy tried to guide them into E.P.S. and then (when these were merged back into the colleges) castrated the municipal colleges and stepped up standards of classical education in the élite *collèges royaux*. As a result, these middling sorts moved towards the crammers or whatever the Catholic education system had to offer.

For it was the Catholic Church that designed its education system with flexibility and sensitivity to the needs of all social groups. To the aristocracy and bourgeoisie who wanted a pure classical education far from contact with grocers' sons, it offered *pensions* and colleges not unlike English public schools. For those of modest means, including peasants and artisans, who wanted a cheap Latin education, it offered seminaries, not all the clientele of which went on to become priests. And for the *classes moyennes* turning towards trade and industry it developed the higher classes of the large urban schools of the Frères, which became *pensionnats* in the 1850s,

Fig. 7. Career paths of Pupils at Lycées and Colleges in Three Regions of France, 1864–1865

		Reproduction	Tradesmen Artisans Peasants Workers	Landowners Merchants Industrialists	Officers Fonctionnaires	Legal and Medical Profession	Teachers Artists Priests	Grandes écoles
Landowners	W	8.3	11.1	2.8	16.7	44.4	2.8	13.9
	N	5.9	23.5	—	20.0	35.3	3.5	11.8
	S	35.7	13.3	1.0	11.2	23.5	8.2	7.1
Liberal Professions	W	57.9	5.3	5.3	21.0	—	5.3	5.3
	N	76.1	6.5	4.3	13.0	—	—	—
	S	53.3	16.7	16.7	3.3	—	6.7	3.3
Public Servants Officers Professeurs	W	53.8	23.1	—	—	7.7	—	15.4
	N	45.8	4.2	4.2	—	41.7	—	4.2
	S	43.5	15.2	17.4	—	19.6	—	4.3
Industrialists	W	30.8	38.5	—	7.7	7.7	—	15.4
	N	22.2	16.7	—	16.7	22.2	—	22.2
	S	14.3	9.5	23.8	19.0	23.8	9.5	—

Négociants	W	41.7	25.0	—	16.7	16.7	—	—
	N	25.0	43.7	—	18.7	8.3	4.2	—
	S	34.9	18.6	1.2	16.3	18.6	2.3	8.1
Tradesmen	W	47.1	11.4	1.4	24.3	4.3	8.6	2.9
	N	55.4	8.1	—	18.9	2.7	9.5	5.4
	S	47.0	6.1	13.6	7.6	6.1	4.5	15.1
Artisans	W	35.0	17.5	—	15.0	—	25.0	7.5
	N	28.1	6.2	6.2	25.0	3.1	9.4	21.9
	S	24.3	13.5	5.4	16.2	2.7	5.4	32.4
Employés	W	34.5	10.3	6.9	—	31.0	10.3	6.9
Petits	N	40.0	23.3	23.3	—	20.0	6.7	6.7
Fonction-	S	45.5	13.6	13.6	—	15.9	9.1	11.4
naires								
Peasants	W	42.6	8.2	—	4.9	4.9	37.7	1.6
	N	64.6	3.0	—	6.1	9.1	13.1	4.0
	S	17.6	5.9	—	17.6	26.5	17.6	14.7
Workers	W	16.7	8.3	—	11.1	—	63.9	—
	N	52.2	—	—	30.4	—	8.7	8.7
	S	8.3	8.3	8.3	16.7	8.3	33.3	16.7

and also added commercial courses to selected colleges in industrial areas, long before special education was invented.

The dangers of ambition would in fact sort themselves out. *Déclassement* was to a large extent a spectre conjured up by the professional, propertied classes. In backward areas, where agriculture, trade, and industry offered no openings, then of course young men would be stimulated to seek a Latin education in order to escape from their origins. But where economic life was flourishing, they would look for careers in productive occupations. A growing capitalist economy was a precondition of the fulfilment of Guizot's advice: 'Enrichissez-vous!'

6

The School of the People

Whereas the main function of secondary education was to control the movement of intermediary groups into the élite, the principal concern of elementary education was to draw the mass of the population into the schools to be civilized. Before 1881-2 elementary education was neither free nor compulsory, and yet the notion that it was conjured out of nothing by the founding fathers of the Third Republic is a myth of their own creation. Popular education was expanding long before that, under the July Monarchy and especially under the Second Empire. The cost of schooling was of course one factor that determined the level of attendance, but other factors, such as the lie of the land and the rhythm and demands of the world of work—given the prevalence of child labour among those classes living on or near the level of subsistence—also came into play, and these can be properly examined only as part of the pattern of local life.

Secondary education claimed to produce individuals whose character had been tempered by the rigours of the *internat* and by the intellectual gymnastics of a classical training, their sensitivity and generosity refined by the contemplation of the beautiful; in a word, its model was the *honnête homme*. The ambition of most of these individuals, on the other hand, was merely to obtain the qualifications that would equip them for entry into the faculties and *grandes écoles*, and hence for the liberal professions or government office. The purpose of elementary education, similar though pitched at a lower level, was to bring the labouring classes within the bounds of acceptable social behaviour, to make civilized men out of savages. What amounted to internal colonization, a *mission civilisatrice*, took several forms: the propagation of the French language amongst those for whom it was a foreign tongue, the enlightenment of populations that remained illiterate, and the imposition of some sort of social order on the idle and rebellious, a work of 'moralization' that used religion, the fear of God, as its necessary sanction. There could be no instruction without education. Given the tension between backward areas and those which were economically advanced, the main question that must therefore be confronted is whether industrialization in nineteenth-century France was a stimulus to school attendance and the improvement of literacy, or whether on the other hand it was a retarding factor.

Just as the presence of a small college in the local *arrondissement* capital was vital to the chances of the *classes moyennes* obtaining a secondary education, so a complete framework of *écoles communales* was a precondition of the people's access to primary schooling. The Guizot law required every commune in France to maintain an elementary school, but clearly some communes were larger than others, and in areas where the population was scattered in isolated homesteads or hamlets rather than concentrated in the commune's *bourg*, travel to the village school would be difficult. This was the case in the Flemish polderland, behind Dunkerque and Gravelines. It posed problems in the Causses, the high limestone plateau cut by deep ravines which includes the canton of Trèves in the Gard. And in Ille-et-Vilaine, especially south-west of a line between Saint-Méen and Retiers, the communes based on the ancient Celtic parish or *plou* were unusually large, while communications in the Breton *bocage*, along sunken lanes winding between steep banks, were arduous and almost impossible in winter. In the first years after the Guizot law many communes in cantons such as Janzé were devoid of schools, so that the school situated in the *chef-lieu*, duly equipped with a *pensionnat*, served a wide area.[1] Moreover, the authorization accorded to some pairs or even trios of communes in Brittany, too poor to maintain a school separately, to maintain one together, did nothing to ease the isolation of many inhabitants of moorland and heath.

Where the fabric of elementary schools was inadequate, clandestine classes, held by mercenary itinerant schoolteachers, frequently made shadowy appearances before

---

[1] A.N. F17 12371, Rector of Rennes to Min. Inst. Pub., 22 April 1835.

vanishing again. An inspector described these 'strolling teachers' as 'ignorant, uncouth men, without title or guarantee, unknown to the authorities who should be supervising them', serving only to perpetuate the prejudice of equally ignorant parents that a regularly maintained school was unnecessary.[2] In the mid-1850s, following the decree of 31 December 1853 on supply teachers, schemes were devised to delegate *sous-maîtres* to hold winter schools in isolated parts of Ille-et-Vilaine and the Causses, but little was achieved.[3] Improvement came more rapidly after the law of 10 April 1867, which included provisions for the maintenance of hamlet schools. The development of these schools was of especial importance in Ille-et-Vilaine, but needs multiplied as fast as provisions to meet them. The development of the coalfields around Bessèges and Saint-Ambroix created a new demand for teaching congregations in the Gard in the 1870s.[4]

A second factor determining school attendance was, very obviously, its cost. School pence averaging a franc a month was no mean sum to demand of a poor peasant or artisan earning three francs a day, especially if he had a large family. But although free elementary education was not universalized until the law of 16 June 1881, this law was only the final term of a process that was almost as old as the century. The Frères des Écoles Chrétiennes made the provision of free education a condition of their establishing a school and resolutely defended this principle when municipalities sought to go back on their engagement. Large towns, thriving on the revenues of *octroi*, could afford to institute free education, and the Guizot law provided for the free education of indigents, as a charge on the parish. By 1843, 49 per cent of children at *écoles communales* in the Nord were educated free. The question of free education was highly contentious, especially in the wake of the law of 30 December 1853, which empowered communes to reduce expenditure by imposing

---

[2] A.N. F17 12474, report of *Inspecteurs généraux des études*, 10 Dec. 1837.
[3] A.N. F17 9373, Min. Instr. Pub. to prefect of Ille-et-Vilaine, 21 Aug. 1855; A.N. F17 9325, primary-school-inspector of Le Vigan to Insp. Acad. Nîmes, April 1856.
[4] See above, p. 140.

school pence in schools that had hitherto been free, and after that of 10 April 1867, proferring state and departmental subsidies to communes prepared to stump up most of the cost of free education. The opponents of free education argued that only what was paid for was appreciated: paupers whose chidren's education was financed by the commune were always less assiduous in their school attendance than paying pupils, and there were movements from the 1850s to deprive such families of poor relief if they did not send their children regularly to school. Indigents, along with factory children, would have become the only category constrained to compulsory school attendance before 1882. Arguments were also advanced that under free education the poor ended up paying for the rich, and that the parental responsibility or duty to provide education would only be felt if they paid school fees themselves. Supporters of free education, like Frère Philippe, superior-general of the Frères des Écoles Chrétiennes, pointed out that the distinction between indigents on public relief and poor workers and tradesmen was minimal: the *classes laborieuses* would let their children roam the streets rather than pay a fee to school them.[5] Others noted out that the imposition of a monthly fee induced poor parents to keep their children away for the whole month if they had missed a few days at the beginning, while graduating school fees by age provoked early leaving. The provision under the Falloux law that school pence would henceforth be collected not by the *instituteur* himself but by the municipal receiver or *percepteur* ended the flexibility available to *instituteurs* of tailoring their fees to families' means or accepting payment in kind, while it was argued in Brittany that 'the country-dweller is reluctant to make an agreement which might set him at odds with an agent of the Revenue', and would prefer to patronize private or cheap clandestine schools instead.[6] Lastly, advocates of free education claimed that fee-paying established an invidious distinction between paying and pauper children, and relegated the latter to 'classes de charité' or 'classes des pauvres', especially in nuns' schools,

---

[5] A.F.E.C., NC 3771/1, Frère Philippe to mayor of Alès, 30 Mar. 1854.
[6] A.D. I-et-V. 1N, report of Insp. Acad. Rennes to Conseil général, 16 Aug. 1851, pp. 205-6.

described by the primary-school inspector of Redon as 'a hothouse where a mass of pitiful children languish in immobility, boredom and silence'.[7] The problem was that it was this very distinction that kept the *classes moyennes* in the primary schools, especially in the case of girls. The Comité d'arrondissement of Avesnes warned in 1841 that if such segregation were abolished in the local girls' schools, 'the parents of fee-paying pupils would withdraw their children from these establishments at once'.[8]

The only way of determining the impact of free education on school attendance is to examine the question, canton by canton, in each department.[9] If we compare the proportion of elementary-school children educated free of charge and the rates of attendance of children between seven and thirteen in 1860, it becomes apparent that there was a correlation between rates of free education over 70 per cent and school attendance of over 80 per cent in highly industrial or populous cantons like Roubaix in the Nord and Beaucaire and La Grand'Combe in the Gard. Conversely, there was a link between free education enjoyed by under 30 per cent of the school-going population and attendance of less than 70 per cent in the Gard at Alès-West and the Catholic Cévenol canton of Sumène, in Ille-et-Vilaine in the poor fishing canton of Cancale, and on the woodland and heath of Pipriac and Plélan in the west and of Liffré to the north of Rennes. In the Nord, this was clearly the case in the declining weaving cantons of Carnières and Clary in the Cambrésis. When one municipal council there attempted to cut down the list of those educated free in 1869, the local school-inspector reported that 'if we close the school to the children of weavers who cannot afford the fees or the school equipment, we risk seeing such schools fall into desuetude'.[10]

There were, nevertheless, exceptions to these patterns. On the one hand, there were cantons where, although the proportion of schoolchildren educated free was very low—

---

[7] A.N. F17 9319, report of primary-school-inspector of Redon, 31 Oct. 1855.

[8] A.D. Nord 1T 90/3, Comité d'arrondissement of Avesnes, 19 Feb. 1841.

[9] États de situation, 1860, in A.N. F17 10411, 10412, 10416.

[10] A.D. Nord 1T 67/8, primary-school-inspector of Cambrai to Insp. Acad. Douai, 25 Feb. 1869.

under 30 per cent—rates of school attendance were in excess of 80 per cent. It might be supposed that the key factor here was wealth; in fact, the most significant force was a 'blue' mentality, anticlerical or Protestant, avid for instruction almost as a political commitment. In Ille-et-Vilaine, the two cantons in this category were Rennes North-East and Louvigné-du-Desert, a veritable nursery of *instituteurs* in the nineteenth century. In the Nord, there were eight such cantons: Cambrai-West, and the others concentrated in the *arrondissement* of Avesnes. This was an *arrondissement* that was not poor, but it was no richer than *Flandre flamingante* or parts of the Douaisis. What characterized it above all was its anticlericalism. In the Gard, free education was concentrated in the Catholic north and east of the department. Of the thirteen cantons with low free education and high attendance rates, nine had Protestant majorities, whether they were prosperous, wine-growing cantons like Vauvert or austere, mountainous bastions like Saint-Jean-du-Gard. On the other hand, there were cantons where over 60 per cent of the school population was educated free of charge, yet school attendance was below 80 per cent. This could be observed at the centre of large cities such as Rennes, Lille, Tourcoing, Douai, and Saint-Amand. If paupers were lax in their attendance at school it was not, as critics of free education argued, because they did not appreciate what they did not pay for, but because there were limits to what free education could achieve. Even if schooling were free, poor working-class families that sent a child to school sacrificed an instrument of labour, the income from which—however meagre—could make the difference between surviving and succumbing.

In the last resort, then, attendance at elementary school was determined by the pattern of rural and urban economies, and specifically by the use of child labour. Three aspects should be discussed: the seasonal rhythm of labour, the age at which productive labour started, and the cyclical movement of the economy. In both country and town, but especially in the former, there was a *morte saison* during which work slackened off, and this was the period of maximum attendance at school. As a rule, schools began to empty with the coming of the fine weather at Easter-time, and would start to fill again after the harvest was in at All Saints. In the Avesnois, where *vaine pâture* survived, children disappeared to mind cattle in May. July and August was the high season for gleaning in the fields, while in October in the Cambrésis hops were being harvested and sugar-beet dug up. One *instituteur* referred to the 'three winter months' during which poor peasant children attended school.[11] In the Gard, spring was announced in the *mas* or farmhouse by the emergence of the silkworms that had to be fed vast quantities of mulberry leaves. Cévenol peasants went down to the plain for the grain harvest followed by the *vendanges* in September, returning to the mountains in October for the grapes and, much more important, the gathering and drying of chestnuts. Many schools would not be at full strength until Christmas, after the harvest of olives.[12] In Brittany, the annual exodus took place between the haymaking in May and the apple harvest at the end of November. But because communications were so difficult in winter, spring brought the infants back to school along the sunken roads, just as the older ones were departing for the fields.[13]

If one problem was the number of months a year that were spent at school, another was the age at which children first appeared in the class-room, and that at which they left. Few spent more than two or three years at school, between the ages of eight and eleven. The traditional rite of passage that marked the end of study was the First Communion, taken at eleven in most regions, often at ten in Brittany. An *instituteur* of Rennes who had long taught in the Morbihan told of children who came to school only a few months before their Communion, in order to learn their catechism.[14]

[11] A.D. Nord 1T 67/25, report of *instituteur* of Ferrière-la-Grande to primary-school-inspector of Avesnes, 3 Dec. 1864.

[12] On this, see the brilliant survey by J.-N. Pelen, *La Vallée longue en Cévenne: vie, traditions et proverbes du temps passé, Causses et Cévennes*, no spécial, 1975.

[13] A.N. F17 9317, report of primary-school-inspector of Fougères, 6 Mar. 1855. This 'diachronic' phenomenon is noted for the west in general by G. Desert, 'Alphabétisation et Scolarisation dans le Grand-Ouest au XIXe siècle', in D. Baker and P. Harrigan, ed., *The Making of Frenchmen*, p. 162.

[14] A.N. F17 10792, submission of Vallée, assistant teacher of Rennes, 2 Feb. 1861.

Motte-Bossut, the cotton-spinner of Roubaix, vowed that the children working in his mills were released to school for three weeks before their Communion.[15] First Communion often marked the beginning of adult, working life, but this was not a hard and fast rule. In the Gard, a child could expect to tend to the farmyard, and stone and weed fields from the age of four or five.[16] Small girls were employed in damp cellars in the Cévennes to open silk cocoons, while in the Nord girls were set to lace-making in Flanders at the age of seven and boys in the Cambrésis of the same age were employed in the domestic weaving industry.[17] In the towns, the structure of the working class was reflected in the school system. Those who were to be apprenticed to a trade might stay on at school to the age of twelve, thirteen, or beyond. One of the criticisms levelled at the Frères des Écoles Chrétiennes was that they concentrated their facilities on the education of the artisan élite, leaving hundreds of paupers who needed their free schooling to roam and beg in the streets.[18]

In between the labour aristocracy, which acquired a full elementary education, and the children of casual labourers and the chronic poor, whose instruction, if they received any, was the fruit of municipal relief, was the industrial proletariat, the work-force of the textile mills, mines, and iron-works, the martyrs of the industrial revolution. It was not unusual for children to be taken on as *galibots* or pit-boys, in the coal-mines at the age of ten, or to assist the adult weaver as *rattacheurs*, piecers mending the broken threads, or *bâcleurs*, replacing the used bobbins in the cotton-mills, at the age of six. No legislation was passed to regulate this state of affairs until the factory act of 22 March 1841 which laid down the conditions of child labour in factories and workshops employing more than twenty persons. This

[15] A.D. Nord M 613/1, Motte-Bossut to prefect of Nord, 9 Jan. 1853.

[16] *Journal des Écoles primaires du Gard*, 1838, p. 130.

[17] A.N. F17 4717, report of Inspecteur des mines of Alès, 15 Dec. 1869; A.D. Nord 1T 124/8, mayor of Méteren (Bailleul S.W.) to Rector of Douai, 18 Oct. 1851, repudiating the need for a girls' school in the commune; 1T 107/4, report of Hilaire, primary-school-inspector of Cambrai, 4 April 1851.

[18] On the situation at Tourcoing, A.D. Nord M 613/5, Rector of Douai to mayor of Tourcoing, 6 Sept. 1844; 1T 107/2, report of sous-inspecteur Ernotte, 10 July 1847.

law prohibited child labour under the age of six and limited the workday to eight hours for children between eight and twelve and to twelve hours for those between twelve and sixteen years of age. Night-work was outlawed for children under thirteen and Sunday work for those under sixteen. Moreover, provisions were made for the education of factory children: those under twelve must attend a public or private school in the locality; those over twelve would also have to unless they could supply a certificate from their mayor attesting that they had received an elementary education.

The department of the Nord is a highly suitable one for a study of the application—or non-application—of this law. It was clearly against the interests of the textile *patronat* ruled by the hard-headed, entrepreneurial mentality of emergent capitalism. Whereas the reformers attributed the source of urban degeneration and brutalization to excess labour, the Chamber of Commerce of Lille was of the opinion that insalubrious and overcrowded housing and poor diet were to blame, and argued that any banning of child labour would only increase vagabondage and prostitution.[19] The presidents of the Chambres consultatives des Manufactures of Roubaix and Tourcoing reported more frankly that the working day of children could not be reduced: the working day was geared to the market, twelve to fourteen hours on average, and including night-work when demand was high. The division of labour around the machine meant that spinners, *rattacheurs*, and *bâcleurs* worked as a unit: if one stopped, the others would also be obliged to. Moreover, children had to start work at an early age in order to acquire the necessary dexterity.[20]

However bitter the class conflict, employers and workers were agreed that children should be set to work as early as possible: there was, in that sense, as much collusion as conflict. The reasons are not difficult to find. A survey by Dr Villermé published in 1840 calculated that the income of

---

[19] A.D. Nord M 611/3, Chamber of Commerce of Lille to Min. Agric. Comm., 27 April, 21 August 1840.

[20] A.D. Nord M 611/1, Nadaud and Cordonnier, presidents of Chambres consultatives des Manufacturers of Roubaix and Tourcoing to Min. Agric. Comm., respectively 17 and 16 Aug. 1837.

a cotton mill-hand, his wife, and child amounted to 915 francs a year, of which 798 francs went on the rent for an attic or cellar and food. The remainder, 117 francs a year, had to serve for linen, clothes, heating, lighting, furniture, tools, and so on, and that assumed that the family was spared from illness, unemployment, or alcoholism.[21] The verdict of school-inspector Bernot, in 1851, was more cynical. He attacked working-class parents for bearing children only to reap the meagre fruits of their labour, for they 'spend in a single day at the cabaret more than the child earns in a week'. Poverty or vice, the result was the same: the child did not go to school.[22]

A third obstacle arose over the enforcement of the law of 1841. Disciplining industrialists for the exploitation of child labour was difficult in a region where municipal councils and other powerful bodies were in the hands of those same industrialists, impossible when the commissions of inspection set up under the law were also dominated by them. The Commission d'inspection du travail des enfants dans les manufactures, set up at Lille in October 1843, was presided over by the Catholic and royalist sugar-manufacturer, Kolb-Bernard, who stated that the purpose of the law of 1841 was 'not to deal severely with the masters but to moralize the workers'. Co-operation between the administration and industrial chiefs was of the essence, and temporary mitigations of the law were therefore in order.[23] The subcommission of Roubaix went even further. Faced by a wall of mill-owners who threatened to dismiss all children under twelve, and even all those under sixteen, if they were obliged to bow to the law, it resigned in November 1844.[24] No enforcement of the law could properly be envisaged until a salaried inspector had been appointed, but this was not achieved until 1852. How much changed then would be difficult to estimate: Dupont, the Inspecteur général du travail des enfants

---

[21] L.-R. Villermé, *Tableau de l'état physique et moral des ouvriers employés dans les manufactures de coton, de laine et de soie* (Paris, 1840), i, 99–101.

[22] A.D. Nord 1T 107/4, school-inspector Bernot to Rector of Douai, 1851.

[23] A.D. Nord M 611/8, report of Kolb-Bernard, 5 Feb. 1844.

[24] A.D. Nord M 613/1, Commission d'inspection du travail des enfants dans les manufactures, Lille, 6 Nov. 1844.

dans les manufactures prosecuted Motte-Bossut of Roubaix on four counts of infraction in 1853, but the police court imposed a fine of only thirty-seven francs.[25]

The form of schooling that was developed for factory children was the *école de midi*, one hour's instruction every day before lunch in the nearest public or private school. It was a device that was totally unsatisfactory. The prefect of the Nord, the Vicomte de Saint-Aignan, tried in 1844 to tighten up the process by requiring a foreman or adult worker to escort the children to school, but a 'committee of spinners' protested to the Minister of Commerce that this meant laying off workers not covered by the law and the enforced shutting-down of machinery for several hours.[26] In 1853 the prefect was struggling to prohibit the use of evening classes as an alternative to midday classes: the advantage to the manufacturer was that the former impinged not on the working day but on the workers' free time, though by nightfall the children were exhausted and the streets dangerous.[27] Meanwhile, the midday classes were shown to be largely ineffective. Employers reduced the allotted time to half an hour or took the hour from the lunch-break. The child-workers, the first victims of compulsory education, employed a hundred dodges to avoid the school-room. When they did arrive, the found them already populated by apprentices, while in boom periods, when thousands of children switched from day-school to the workshops, overcrowding in the midday schools was intolerable. Adolphe Morel, *instituteur* at Moulins-Lille, judged that while the apprentices might be taught the rudiments of literacy, the children from the linen-mills were of an ignorance and coarseness such that a minimum of religious instruction and 'moralization' was all that could be achieved.[28] By 1860, the uselessness of the midday classes had become plain, and officials like Inspector Dupont and the Prefect Vallon, with increasing support in the municipal council of Lille and

[25] A.D. Nord M 613/1, Dupont to prefect of Nord, 4 Feb. 1853.
[26] A.D. Nord M 611/8, petition of Comité des filateurs to prefect of Nord, 20 Aug. 1845.
[27] A.D. Nord M 611/11, prefect of Nord to mayor of Lille 6 Jan. 1854.
[28] A.D. Nord M 613/16, Adolphe Morel, *instituteur* of Moulins-Lille, to Dupont, Mar. 1853.

the Conseil général of the Nord, started to campaign for the prohibition of all child labour under the age of twelve, which alone could ensure a decent intellectual and physical development before a lifetime in the workshops began.[29]

Not only were the clear provisions of the law of 1841 ignored; there were also several loopholes in the law that were keenly exploited by industrialists—and, indeed, by working-class families. The main one concerned the limitation of the law to workshops employing more than twenty people. Coal-mines were not affected, and were regulated only by the law of 1813 that prohibited children under ten from working in the pits. There was an urgent need for revision, and though some companies like the Société anonyme des hauts fourneaux, forges et laminoirs of Anzin set up *écoles de fabrique*, in this case run by the Sœurs de Saint-Vincent de Paul, the provision of schooling was usually inadequate.[30] Besides this, workshops employing less than twenty workers were shielded from the law: at Comines, in the canton of Quesnoy-sur-Deûle, there was a mass of small workshops making ribbons and employing large numbers of Belgian children who could not be subjected to the law, while similar exploitation took place in the nail-making shops of Saint-Amand.[31] At Fougères, in Ille-et-Vilaine, one shoe-factory owner employed six hundred workers in 1869, including numerous children, but at any attempt to enforce the law he could require them to work at home and operate a 'put-out' system.[32] The whole rural textile industry, whether in Flanders or the Cambrésis, the Cévennes or the heathlands of Brittany, was beyond the reach of the law of 1841.

Lastly, it should be made clear that economic cycles had a considerable influence on school attendance. The proportion of children attending school rose steadily in the

---

[29] A.D. Nord M 613/14, Dupont to prefect of Nord, 2 Dec. 1860; A.N. F12 4717, prefect of Nord to Min. Agric. Comm., 31 Mar. 1861.

[30] A.D. Nord M 613/14, prefect of Nord to Min. Agric. Comm., 19 Dec. 1860; Dupont to prefect of Nord, 24 June 1860.

[31] A.D. Nord M 613/14, Dupont to prefect of Nord, 19 April 1861; M 611/15, Dupont to prefect, 24 July 1861.

[32] A.N. F12 4717, Massieu, Ingénieur des mines of Rennes to Min. Agric. Comm., 16 Oct. 1869.

nineteenth century, increasing prosperity being a more important factor than legislation. In 1880, 806 of every thousand children between the ages of five and fourteen in the Nord attended school. But this proportion had first been attained in 1863, and in 1882, after the introduction of free, compulsory elementary education, the proportion was only 847 in a thousand.[33] This gradual improvement had nevertheless been interrupted by periodic hiccups, which underlie the fragility of the conquest of the working population by the primary school. In a country that was still largely agricultural, the state of the harvest was the main factor determining the buoyancy of the economy until the last third of the nineteenth century. A bad harvest or harvests, since they often tended to follow in pairs, precipitated a *crise de subsistances* which swallowed up most of the resources of poor peasants and urban populations in the purchase of bread. Education then became a luxury, and the major episodes of *disette* and *cherté* in the nineteenth century —1846-7, 1853-6, 1860-2, 1866-8, 1873-4—were accompanied by slumps or slower growth rates of the school population.[34] The price of wheat is only a guide to *crises de subsistances*: in Ille-et-Vilaine in 1853-4, the school population continued to rise, despite the exorbitant price of wheat, because the subsistence crop of buckwheat was excellent; in 1854-5 the buckwheat crop also failed, and the school population declined. In the Gard, where the pattern of economic life was very different, the sluggish growth of the school population in the middle years of the century must be attributed to other factors. The population scarcely increased between 1861 and 1876, and declined thereafter, but two massive economic problems must bear some responsibility: the pebrine disease that undermined silk-raising after 1850, and the phylloxera epidemic, that struck just before 1870. In the more industrial Nord, the harvest cycle had some bearing on industrial activity, since

[33] M. Leblond, 'La Scolarisation dans le département du Nord au XIXe siècle', *Revue du Nord*, lii, no. 206, 1970, 387-98.

[34] *Crises de subsistances* also affected the population of colleges and *pensions*, as in the Gard in 1828 and 1838. On the Ancien Régime situation, see F. de Dainville, art. cit., p. 463.

a decline in demand from the countryside inevitably provoked wage-cuts, short time, and unemployment in the
textile industry. But foreign markets had an influence that
was just as important. The American Civil War caused a
cotton famine that undermined the northern cotton industry,
but linen manufacturers at Lille and Armentières and woollen
manufacturers at Roubaix and Tourcoing were able to take
on the unemployed labour force and experienced a boom in
the mid-1860s. It was not until American cotton began to
flow again, creating a glut, and the harvest of 1867-8 was
ruined that a 'price scissors' occurred and the growth of the
school population suffered. Such was the interweaving of
school attendance with the patterns of the French economy.

Patterns of life and labour made full school attendance
impossible to achieve. But even when children did attend
school, there could be no guarantee that they would be
turned out into the world enlightened, moral, and speaking
French. The French Revolution made it clear that national
unity would be chimerical unless the whole population living
within the borders of France both spoke and wrote the lingua
franca, French. All regimes, imperial, royalist, or republican,
subscribed to this essential truth. Yet by the mid-nineteenth
century, little had been done to eradicate not only languages
like Breton, Flemish, and German from French soil, but also
the degenerated forms of old French, patois, that was spoken
in most departments of the Midi.[35] Breton had shrunk further
west than Ille-et-Vilaine, but the peasantry spoke a *patois
gallot* that became all but incomprehensible in communes
bordering the Morbihan.[36] In the Restoration period, Flemish
was spoken in most communes of the *arrondissement* of
Hazebrouck and much of that of Dunkerque, while in the
region of Lille the popular tongue was *rouchi*, based on old
French but with 'expressions of Celtic, Walloon, German,
Latin and Spanish origin'.[37] In the Gard, the population spoke
a Languedocian dialect, especially strong in the mountainous

[35] E. Weber, *Peasants into Frenchmen. The Modernization of Rural France,
1870–1914* (London, 1977), pp. 67–70.
[36] A.N. F17 9326, report of primary-school-inspector of Redon, 27 Mar. 1856.
[37] A.D. Nord 2T 1097, report of Bernot, primary-school-inspector of Lille,
1 April 1856.

districts.[38] The resilience of these languages and patois was based on the fact that they expressed the intimacy of the family, the weight of tradition, the local community against officialdom, the province against centralization. Concessions had to be made by the authorities, notably in Flanders, where the Flemish clergy were the most vehement defenders of their language. The teaching of the catechism in Flemish was authorized under the July Monarchy, and the most the Second Empire could do was to oblige Mgr Regnier to bring out a catechism in both Flemish and French in 1863.[39] Further south, in the *arrondissement* of Douai, the school-inspector reported that 'patois is in such general use in the region that it would be impossible to eradicate it; besides, for most children, French is like a foreign language'.[40] If that was the experience of the mass of the population, it was scarcely consoling to learn that of fifty-four *élèves-maîtres* recruited from the five departments of Brittany in 1845, to the École normale of Rennes, twenty-four were Breton-speaking while 'the thirty others come from the depths of the countryside with an unintelligible patois and extremely bad pronunciation'.[41]

The fact that teaching French to children in primary schools could be as difficult as teaching Latin to the pupils of secondary schools was only the first obstacle to the spread of literacy. The concept of the 'three Rs' was ill-developed in the nineteenth century: the skills of reading on the one hand and writing and arithmetic on the other belonged to different traditions, and it was possible to be versed in one while being totally unacquainted with the other. As Philippe Ariès has said, one was 'associated with literary and religious culture, the other with manual arts and commercial practice'.[42] Reading meant access to the culture of the Churches,

---

[38] A. Audiganne, op. cit., ii, 152.

[39] See G. Cholvy, 'Enseignement religieux et langues maternelles', p. 37.

[40] A.D. Nord, 1T 107/3, report of primary-school-inspector of Douai, 15 Sept. 1865.

[41] A.N. F17 9651, report of director of École normale of Rennes to its Commission de surveillance. 30 Nov. 1845.

[42] P. Ariès, *Centuries of Childhood*, p. 297. See also J. Meyer, 'Alphabétisation, Lecture et Écriture. Essai sur l'instruction populaire en Bretagne du XVIe au XIXe siècles', *Congrès national des sociétés savantes*, Reims, 1970 (Paris, 1974), 336–43.

and even then it was imperative only for Protestantism, founded on the private meditation of the vernacular Bible. In the Catholic Church, it was possible to learn the catechism, prayers, psalms, and litanies by rote, in order to chant or sing them in unison, and learning to read Church Latin (albeit without understanding it) remained as important a skill as learning to read French. Writing was a skill of a different order, practised by the scribe or writing-master, entering the schoolroom as calligraphy—the inclined, cursive, and round scripts required by the Guizot law, with Gothic added in 1851. The art of writing, moreover, was closely intertwined with reckoning, bookkeeping, and the verification of deeds.

The dissociation of these two traditions was evident in the schools. The revolutionary achievement of the mutual schools was to tackle reading, writing, and arithmetic at the same time, dividing pupils into a score of circles, finely graded by ability and exercised by monitors. The lay brothers, by contrast, followed the precepts of de la Salle, whose *Conduite des écoles chrétiennes* ordained that 'it is essential not to start children on writing until they can read passably, otherwise there is a danger that they will never learn to read'.[43] The risk in practice was that the overcrowded lower forms, entrusted to a novice without the *brevet*, were paralysed by the boredom of catechism, religious instruction, and reading, while the minority who reached the top form were 'crammed' not only with drawing, history, geography, and singing but also with the writing, grammar, and arithmetic at which they were still imperfect.[44] At Nîmes, the pedagogy of the Frères des Écoles Chrétiennes was outstanding in the more mechanical skills taught to an élite of their pupils: writing, arithmetic, algebra, geometry and accounting, drawing and even architecture. On the other hand, the reading method was said by the Comité local to be 'still extremely defective, and the tone [of recitation] generally monotone and shrill'.[45]

---

[43] J.-B. de la Salle, *Conduite des écoles chrétiennes* (Paris, 1837 edn.), p. 42. See also H.-C. Rulon and P. Friot, *Un siècle de pédagogie dans les écoles primaires, 1820-1940* (Paris, 1962), pp. 103-14.

[44] A.M. Rennes R11, report of Comité d'instruction primaire of Rennes, 1836.

[45] *Journal des Écoles Primaires du Gard*, report of Comité communal, 30 May 1836, p. 87.

The mutual schools and the schools of the Frères des Écoles Chrétiennes were, of course, confined to the towns. In the countryside, the backwardness of teaching methods was even more marked. Writing could not be taught alongside reading for the simple reason that in poor communes desks and even benches were in short supply. An *instituteur* near Fougères reported in 1861 that at his school a shift system operated, with pupils coming in turns to the tables to write, then going back to stand in enforced idleness along the walls. It was a recipe for boredom, frustration, indiscipline, and disgust with school life.[46] As far as reading was concerned, La Mennais noted in 1831 that peasants were only interested in having their sons decipher leases and other manuscripts, and 'if what we call "contracts" were abolished entirely, the number of school attenders would fall by a half'.[47] The main concern in rural schools, especially in Catholic regions, was the teaching of catechism that was the passport to First Communion. In Flanders, a school-inspector reported in 1838 that three-quarters of the pupils' time was taken up by the study of catechism.[48] Nearly twenty years later, it was said in the Breton *arrondissement* of Redon that 'if the catechism were not a religious obligation, most schools would be empty in this part of the countryside'.[49] For girls, even more than for boys, schools were for learning the catechism. A religious congregation like that of the Sœurs de Saint-Vincent de Paul, charged with poor relief as well as the education of pauper girls, was obliged by its statutes only to teach catechism, sewing, and spinning in *petites écoles*.[50] But there were also school-inspectors like Dalimier in Ille-et-Vilaine, a man not lacking in shrewdness, who stated in the manual on the education of girls he published in 1843 that there was no need to perfect girls in the art of writing, 'especially in the countryside'.[51]

[46] A.N. F17 10792, submission of *instituteur* of Saint-Hilaire des Landes, 25 Jan. 1861.
[47] A.F.P. 80, La Mennais to Min. Instr. Pub., 22 Mar. 1831.
[48] A.D. Nord 1T 107/1, inspector Carlier to prefect, 1838.
[49] A.N. F17 9317, report of primary-school-inspector of Redon, 9 Feb. 1855.
[50] A.N. F17 12470, Rector of Rennes to Grand Maître, 30 April 1847.
[51] M.-J.-M. Dalimier, *La Pédagogie des écoles rurales, principalement à l'usage des institutrices* (Rennes, 1843), p. 127.

Lastly, writing was in some sense a luxury, beyond the reach of the poor. Until about 1850, parents regularly paid pro rata for the education received by their children. At Pleurtuit on the Rance in 1835, for example, reading lessons cost 75 centimes a month, reading and writing one franc, reading, writing, grammar, arithmetic, and the metric system 1 franc 25.[52] Hard-pressed families of peasant-fishermen could economize by opting for a limited instruction. Apart from that, pens and paper, which were rarely furnished by the commune, were an additional expense. An *instituteur* of Ille-et-Vilaine asked in 1861 how he could 'have the books, paper, pens, etc., so indispensable for instruction procured for the children of poor parents who want for clothes and often even for bread'.[53] Moreover, the indigent children who were relegated to 'charity classes' in the nuns' schools would rarely receive an education which included the art of writing. It is not surprising, then, that the records show in the mid-nineteenth century a penumbra of semi-literates who could read but not write. The conscription registers that are the principal source for this analysis reveal that the proportion of conscripts that could only read was between 15 and 25 per cent in parts of Brittany and the Cévennes. Among women, rates of partial literacy were undoubtedly higher.[54]

In a general sense, it may be said that before 1880 teaching methods in elementary schools were poor. The process of learning was formal, mechanical, even ritual, very much tied to memorization and recitation. Once a child could read passably, he was given a psalter, de la Salle's *Devoirs du chrétien*, described by Pauline Kergomard, the reformer of infant education, as an 'absolutely unintelligble theological work',[55] or, as the school-inspector of Cambrai, Hilaire, complained, arid little history textbooks packed with place-names, and 'baroque and ill-sounding words'.[56] In the

---

[52] A.N. F17 11192, municipal council of Pleurtuit, 8 Mar. 1835.

[53] A.N. F17 10792, submission of *instituteur* of Sens-de-Bretagne, 26 Jan. 1861.

[54] The census of 1872 for Ille-et-Vilaine suggests that partial literacy rates for those over twenty were 18.3 per cent for men, 24.3 per cent for women.

[55] A.N. F17 10847, report of Pauline Kergomard, 18 Sept. 1879.

[56] A.D. Nord 1T 107/4, report of primary-school-inspector of Cambrai, 4 Apr. 1851.

opinion of school-inspector Bernot, the child found his way around his reader by a combination of memory and guesswork, but it was 'the only one from which he was able to read'.[57] At the same time, spelling was learned under endless dictation, grammar by learning rules by rote, and writing by copying pages of models. No thought was given to composition, concern being for the form not the content of the language. Gradually, there arose a ground swell demanding a pedagogic revolution: the submissions of some *instituteurs* to the Emperor in 1861 called for teaching that was less abstract and more concrete, using weights and scales, maps and engravings, while others requested the reintroduction of *conférences cantonales* in which *instituteurs* could discuss methods and strengthen the bonds of fraternity.[58]

The impact of schooling on literacy may best be analysed under three heads: the difference between the generations, the difference between one region and another, and the difference between occupational groups. The *Enquête sur le travail* of 1848 noted of Ille-et-Vilaine, that 'in the generation that has grown up in the last twenty years education is fairly general amongst boys, but girls' education still leaves much to be desired'.[59] Those born after 1825 clearly had the advantage, and the achievements of the Guizot law were palpable. Even so, too much must not be made of the generation gap. Contemporaries were ever critical of the illiteracy of conscripts who came up before the military commission at the age of twenty: many who had been away from school for eight or ten years had forgotten most of what they had learned. Moreover, it would be wrong to imagine that rates of literacy improved steadily with the decades. The Maggiolo inquiry into marriage signatures undertaken at the beginning of the Third Republic reveals that in the Nord, between 1786-90 and 1816-20, the proportion of men who signed their marriage acts declined from 51.4 per cent to 48.5 per

---

[57] A.D. Nord 1T 107/4, Bernot, primary-school-inspector of Lille to Rector of Douai, 1851.

[58] A.N. F17 10776, submissions of *instituteurs* of Thun St. Martin, Fourmies, and Annappes (Nord), 27, 29, 31 Jan. 1861; F17 10781, submissions of *instituteurs* of Congenies and Mialet (Gard), 28, 29 Jan. 1861.

[59] A.N. C954, *Enquête sur le travil agricole et industriel, 1848, Ille-et-Vilaine.*

cent, that of women from 29.3 per cent to 23.6 per cent.[60] The first century of the Nord's integration within the Kingdom of France was one of massive progress in literacy; it climbed from sixty-fourth to twenty-sixth position among the future departments of France. But by 1872-6, it was the only department north of the Saint-Malo–Geneva line that had a literacy rate of less than 70 per cent.[61]

The imaginary line dividing the north and east of France, that of enlightenment, from the south and west, that of backwardness, has become engrained in the analyses of most French historians. For our purposes it is not particularly useful, not only because of the peculiar position of the Nord but also because some Mediterranean departments such as the Gard were well in the vanguard of literacy. In 1866, 79.1 per cent of men signed their marriage signatures in the Gard, as against 69.6 per cent in the Nord and 57.8 per cent in Ille-et-Vilaine, truly part of the backward west. Curiously enough, however, women in the Nord were more literate (55.7 per cent of signatures) than women in the Gard (53.6 per cent). Reverting to the conscription registers of 1860,[62] it seems that the correlation between school attendance and literacy is fairly close. High rates of school attendance by boys and high literacy rates (over 80 per cent) were evident in the Protestant Cévennes, the Avesnois, and at Saint-Malo. At the other end of the scale, attendance rates of under 70 per cent dragged literacy rates below 50 per cent in much of Ille-et-Vilaine, especially the *arrondissement* of Vitré and Redon, and in the Nord, in the depressed weaving cantons of Clary and Carnières. There is no instance of high literacy coexisting with low school attendance; on the other hand, in highly industrialized cantons like Roubaix and La Grand'Combe, literacy was very low despite high, subsidized school attendance.[63]

---

[60] B.N. 4° Lf[242] 196 *Statistique retrospective, état récapitulatif et comparatif indiquant, par département, le nombre des conjoints, qui ont signé leur acte de mariage au XVIIIe et XIXe siècle* (Paris, n.d.).

[61] M. Fleury and P. Valmary, 'Les Progrès de l'Instruction élémentaire de Louis XIV à Napoléon III', *Population*, no. 12, 1957, 84.

[62] A.D. Gard, Ille-et-Vilaine, Nord, series R. For the Nord, the more complete registers of 1858 have been used.

[63] Similar conclusions are reached in F. Furet and J. Ozouf, *Lire et écrire. L'alphabétisation des Français de Calvin à Jules Ferry* (Paris, 1977), i, 273-4.

Various explanations, cast in the mould of the studies of André Siegfried, have been advanced for the geography of literacy. The dichotomy between illiterate *bocage* and literate openfield[64] may be clear in Brittany, but in the Nord the situation is more confused. While between 1750-90 and 1866 the proportion of men signing their marriage registers in the openfield of the Cambrésis rose from 55.2 per cent to 68.3 per cent, in the Flemish interior it leaped from 51.2 per cent to 70.6 per cent.[65] It has been said that towns were more literate than the countryside, and this is clear in Ille-et-Vilaine, where Rennes, Saint-Malo, Fougères, and Redon stand out as citadels of culture in a backward desert. On the other hand, a distinction has been drawn between Ancien Régime towns, dominated by the Church, magistracy, and universities, and mushroom towns, the product of rapid industrialization.[66] In the Nord, the *arrondissement* of Lille, the most urban and industrial, was the most illiterate between 1790 and 1880, while that of Avesnes, the most rural and agricultural, was the most enlightened.[67] In the industrial region of Lille, where there was no slack season for schooling, where child labour in the factories began at eight, and periods of unemployment only increased mendicancy, little else could be expected. Far from promoting literacy, early industrialization actually depressed it.[68]

Such an assertion may be upheld with confidence only in the light of a breakdown of literacy rates by occupational groups. This is possible from the conscription registers, but it is a task that has been attempted only too rarely.[69]

---

[64] F. Furet and J. Ozouf, *Lire et écrire.* i, 180-9.

[65] Ibid. 233, 247. The figures for 1750-90 were taken from Fontaine de Resbecq, op. cit.

[66] Furet and Ozouf, *Lire et écrire*, i, 240-2. See also Furet and Ozouf, 'Literacy and Industrialization: the case of the *Département du Nord* in France', *Journal of European Economic History*, vol. 5, no. 1., 1976, 24-5.

[67] Furet and Ozouf, *Lire et écrire*, i, 233: 'Literacy and Industrialization', 22-3.

[68] Compare with the conclusions of M. Sanderson, 'Literacy and Social Mobility in the Industrial Revolution in England', *P. & P.*, no. 56, 1972, 75-104.

[69] One exception is A. Corbin, 'Pour une étude sociologique de la croissance de l'alphabétisation au XIXe siècle. L'instruction des conscrits de l'Eure-et-Loir, 1833-1883', *R.H.E.S.*, no. 53, 1975, 99-120. The registers of the *conseils de révision* in A.D. Gard, Nord, and I-et-V (series 1R) form the basis of Fig. 8.

Fig. 8. Literacy by Occupational Groups, 1858 and 1860

	Gard 1860	Nord 1858	Ille-et-Vilaine 1860
Shopkeepers, Innkeepers, *Cafetiers*	80.2	92.7	68.1
Woodworkers	92.1	80.6	71.5
Cobblers	86.9	85.4	66.7
Tailors	90.3	87.8	53.1
Metalworkers (farriers, blacksmiths etc.)	84.5	82.6	54.7
Stoneworkers	76.9	69.0	50.3
Sailors, Fishermen	81.2	68.7	63.4
Textile Workers	69.8	42.7	64.3
Miners	41.1	36.6	50.0
Cultivateurs	66.4	87.9	39.2
Day-labourers	28.6	43.2	25.5
Domestiques (Farm-hands, Servants)	52.3	37.9	22.2

Shopkeepers and café-owners, important channels of communication and often political leaders, were among the most literate of the sample, especially in the Nord. The clear distinction was between the artisans and the industrial workers. Woodworkers, that is to say, sawyers, carpenters, joiners, coopers, cabinet-makers, and wood-turners, artisan metalworkers—farriers, blacksmiths, and wheelwrights—together with tailors and cobblers had literacy rates of over 80 per cent in the Gard and the Nord, over 53 per cent in Ille-et-Vilaine, the leading role of woodworkers and cobblers being especially significant in the last department. Stoneworkers—quarriers, masons, and stone-cutters—were slightly less literate, as were sailors and fishermen, but again on the Breton coast, their relative literacy was important. The question of textile workers is more weighty. In Ille-et-Vilaine, the hemp-weavers concentrated in the south-east of the department had literacy rates that set them apart from the mass: we are reminded of the weavers of Paul Bois's Sarthe. On the other hand, the *taffetassiers, faiseurs de bas*, and carders of the Gard's silk industry were among the less literate, while the huge textile population of the Nord reveals the true impact of industrialization, whether factory labour or depressed hand-loom weaving. The overall literacy figure was 42.7 per cent, within which the combers were an élite with 50 per cent literacy, followed by weavers with 43.1 per cent, threadmakers with 42.9 per cent, spinners 40.6 per cent, and the *rattacheurs*, tied to the mules and jennies from the earliest age, 28.8 per cent. The difference between traditional crafts and large-scale industry dominated by the division of labour is also clear in the Nord. Whereas the literacy rate of artisan metal-workers was 82.6 per cent, that of the labour force in the large metallurgical plants was 63.5 per cent and nail-makers rated only 30.8 per cent. Among miners, literacy was similarly low: 41.1 per cent in the Gard, 36.6 per cent in the Nord; the figures for the lead-mines of Pont-Péan in Ille-et-Vilaine are of little value. Finally, the myth of the uniform ignorance of the peasantry must be challenged: in Ille-et-Vilaine, peasants were the least literate of our ten groups, and their performance in the Gard was not good, but peasants in the Nord, from the farmers of Flanders to the *cultivateurs* of the Avenois, were second only to tradesmen in literacy, another telling reflection on the results of industrialization. That said, class differentiation existed as much in the countryside as in the town: the literacy of agricultural labourers and farm-hands was far lower than that of peasant-proprietors and farmers, with less than a quarter of their number being literate in Ille-et-Vilaine.

The campaign for literacy, the results of which were very uneven, was only one aspect of the *mission civilisatrice*, and not necessarily the most important. Ignorance was only one scourge of the mass of the population in nineteenth-century France: vagrancy and begging, drunkenness and debauchery, dirtiness and 'hatred of the rich'[70] scandalized the notables and seemed to threaten the very fabric of society. It may be that all these problems could be traced to the birth-pangs of industrialization within a basically poor and backward economy. Archaic agricultural systems and decaying artisanal industries gave rise to thick wedges of under- and unemployed, numbers that were swollen by *crises de subsistances* and cyclical depression. But contemporaries saw the root of the

[70] A.N. F17* 105, report of inspector delegate Verrier on canton of Alès, 8 Nov. 1833. See also L. Mazoyer, 'La Jeunesse villageoise du Bas-Languedoc et des Cévennes en 1830', *Annales H.E.S.*, x, 1938, 502-7.

evil in the vice of idleness, and as late as 1882 a school-inspectress touring Brittany warned not only of the physical disorder occasioned by mendicancy but also of the 'moral disorder resulting from an intractable penchant for the nomadic, irregular, idle life'.[71] One observer of manners in the Gard, putting aside the poor, hard-working, and sober Protestant population of the Cévennes, spared no criticism for the 'opulent and lazy' inhabitants of the Vaunage or the workers of Nîmes who 'are rarely to be found at home: they are either at work or at the café', where 'even children are to be seen smoking, swearing, drinking, gaming and indulging in every vice'.[72] But immorality could also be the result of overwork. For just as the factory system left many artisans idle, so it riveted a new proletariat to machines for long periods, brutalizing and degrading them. Bernot, the school-inspector of Lille, reflected that work taken to the degree of wage-slavery 'loses its moralizing character and becomes only a means of providing for material needs and of satisfying nothing but appetites and gross instincts'.[73] Similarly, attacks could be levelled against the dirty, disease-ridden, and uncouth populace as if it were the fruit of their own vice, and only a minority recognized that they were the effects of chronic poverty in the countryside or of overcrowding, slum-dwelling, lack of sanitation, and malnutrition in the growing towns.[74]

Nineteenth-century liberal society could not be expected to tackle these problems at the root, which was poverty, nor to interfere with a capitalism the profitability of which depended on exploitation. In 1849, Kolb-Bernard declared that poverty was 'a normal occurrence, the sign of a well-ordered society'. The key notion was that of patronage: 'moralize the masses and you will have done more to fight poverty and misery than all the innovators'.[75] The task

[71] A.N. F17 10864, report of Mme Dillon, Inspectrice générale des écoles maternelles, Mar. 1882.

[72] E. Frossard, *Tableau pittoresque et moral de Nîmes* (third edn. Toulouse, 1854), pp. 86–9, 297–8.

[73] A.N. F17 9330, report of Bernot, 1 Apr. 1856.

[74] See the report of the Conseil de salubrité of the Nord, 1 Apr. 1832, in L.-R. Villermé, op. cit., pp. 13–14.

[75] Cit. P. Pierrard, 'Un grand bourgeois', 401.

was not to change society but to stabilize it. The most important means of achieving order and stability was education, and the core of education was religion. It is important to distinguish, as contemporaries did, between education and instruction. For many conservatives, instruction—the propagation of literate skills—was a social evil, provoking pride, ambition, *déclassement*, and, ultimately, anarchy. If instruction were necessary, only coupling it with education could neutralize the dangers inherent in it. Writing in 1836 to the mayor of Maubeuge about the foundation of a school of the Frères des Écoles Chrétiennes, Sculfort, an industrialist and administrator of the hospice civil, pointed out that 'if children are taught only reading, writing, the letter of the catechism, etc. etc., that is doubtless a great good, but it is not enough. They must be taught to love religion, inspired with the taste for work, order and subordination, so that having trained obedient and respectful children, we shall then have industrious workers, faithful servants, honest citizens and good fathers.'[76] After the events of 1848, the fear of instruction separated from education grew even more acute. The reaction of de Tarteron, *conseiller-général* of Sumène, was not isolated.[77] Mgr Regnier urged in 1849 that 'it is not enough for the salvation of France that the people know how to read. The assassins and incendiaries of Paris knew how to read.'[78] The reaction of the Protestant notables was more restrained. Gustave Fornier de Clausonne observed in 1852 that discontent and rivalry resulted from 'the smallest dose of the least knowledge' in a sea of 'pre-existing ignorance'. The general diffusion of enlightenment would in fact have a civilizing effect. On the other hand, he said, primary education should inculcate 'habits of respect, obedience, submissiveness, regularity, order, and one might add correctness, cleanliness, industry, attention and thoughtfulness'.[79]

The burden of educating the people fell, of course, on the elementary schoolteachers. They were to be apostles of

[76] A.D. Nord 1T 124/7, Sculfort to mayor of Maubeuge, 28 Apr. 1836.
[77] See above, pp. 79.
[78] A.D. Nord, 2V 114, Mgr Regnier, letter of Mar. 1849, quoted in his *Mandement de Carême*, 14 Jan. 1872.
[79] A.D. Gard, Archives du Château de Clausonne Meynes, 90, G. Fornier de Clausonne, 'Quelques directions pour les institutrices', M.S. 1852.

civilization, a new sort of priesthood. Protestant *institutrices*
for Fornier de Clausonne were 'a Providence for the children',
while for Job Renault the Frères de Ploërmel were 'the visible
guardian angel of childhood'.[80] Yet the role of the schoolteacher was nothing if not ambiguous. Charged with the
instruction of French children, they were the epitome of
the *déclassé* and *demi-savant* held in horror by the conservative notables. Supposed to stand apart from society,
since their role in the class-room was *in loco parentis*,
they were in fact *hommes du peuple*, inclined to drink
and smoke in the local cabaret along with everyone else.[81] The
memoranda of the *instituteurs* in 1861 reveal much about
their embarrassment. That of Gallargues (Vauvert) reflected
that 'country-dwellers are far from believing that teaching is
a priesthood' and that they see in the *instituteur* 'little more
than a mercenary or day-labourer like any other'.[82] The
parents who paid school pence felt entitled to call the tune.
The *instituteur* of Vieux-Vy, north of Rennes, said that in
the commune the teacher was 'everyone's subordinate; from
the mayor to the last navvy, all have the right to criticize
him, to chide him and to check his behaviour'.[83] The modesty
of the *instituteur*'s stipend obliged him to a pluralism of
employments that diminished his independence. An *instituteur* near Cambrai asked whether it was possible to be secret-
ary to the mayor without being 'the mayor's man',[84] while
another in the canton of Trélon protested that as lay clerk
required to ring the mass and Angelus, sing, prepare the
candles, look after the vestments, and sweep the church, he
was no better than 'the very humble servant of the parish
priest', regarded by the local population as 'their old *clerc* or
maître d école' from before the Revolution.[85] The ambition
of every *instituteur* was that he should be regarded as a *fonctionnaire*, paid by the state, answerable only to the Academy

[80] J. Renault, *Guide de l'Instituteur Chrétien* (Ploërmel, 1868), p. 11.

[81] A.D. Nord 1T 107/4, reports of primary-school-inspectors Bernot and
Hilaire, 1851.

[82] A.N. F17 10781, submission of *instituteur* of Gallargues, 30 Jan. 1861.

[83] A.N. F17 10792, submission of *instituteur* of Vieux-Vy-sur-Couesnon,
31 Jan. 1861.

[84] A.N. F17 10776, submission of *instituteur* of Paillencourt, 28 Jan. 1861.

[85] Ibid., submission of *instituteur* of Féron, 26 Jan. 1861.

and the prefecture, freed from the tyranny of ignorant peasants and 'all parochial suzerainty'.[86]

There were three principal aspects of the civilizing mission, incumbent on *institutrices* as well as *instituteurs*, religious congregations as well as lay teachers. The first aim was to save the poor from idleness and vice and to enforce work-discipline; the second, as the mayor of Vitré put it in 1861, was 'the moralization of the labouring classes';[87] the third, to achieve minimum standards of cleanliness, which was next to godliness. Education, in fact, amounted to little else than the imposition of a bourgeois ethic on the mass of the population in the interests of the established order.

The enforcement of work-discipline, sometimes known as 'le goût du travail' was at one extreme a crusade for social redemption, rehabilitating girls who had fallen from virtue and taking care of orphans and foundlings. In garrison towns like Rennes, where prostitution was rife, the Sœurs de Saint-Vincent de Paul were active from the First Empire founding *ateliers de charité* or *ouvroirs* where girls would be 'snatched from vice', set to work spinning wool and linen, and brought up to be 'excellent mothers'.[88] At Cassel, in Flanders, the Filles de l'Enfant-Jésus set up a similar charity workshop to 1840 to cure the 'coarse and idle youth' of its 'libertine morals and thieving habits', supervising lace-making with a few hours a day given over to reading and catechism.[89] Such establishments were largely financed by the product of the work of their inmates, something that could provoke accusations of exploitation or unfair competition. Orphanages were often of similarly early origins, run for the most part by religious congregations, though a Maison des orphelines protestantes du Gard with lay teachers was founded in 1822.

---

[86] A.N. F17 10781, submission of *instituteur* of Milhaud (Gard), 30 Jan. 1861. See also P. V. Meyers, 'Professionalization and societal change: rural teachers in nineteenth-century France, *Social Hist.*, Vol. 9, no. 4, 1976, 542–58, and J. Ruffet, 'La Liquidation des instituteurs-artisans', *Les Révoltes logiques*, no. 3, 1976, 61–76.

[87] A.D. I-et-V 30 e 56, mayor of Vitré to prefect of Ille-et-Vilaine, 1861.

[88] A.M. Rennes R13, petition of Bureau d'administration of École gratuite des pauvres jeunes filles of Parish of Saint-Aubin to municipal council of Rennes, 2 May 1807.

[89] A.D. Nord 1T 90/3, Comité local of Cassel to sub-prefect, president of Comité d'arrondissement of Hazebrouck, 11 Nov. 1840.

Its task was to ensure that 'young orphan girls receiving a Christian education should be preserved for ever from the sufferings of destitution, the shame of mendicity, the opprobrium of vice and the infamy of crime'.[90] Orphanages for boys were rare before the July Monarchy, and were then often set up in the country in order to train farmhands, countering the effects of the 'manque de bras' in rural areas at the same time as they inculcated virtue. Establishments of this sort were set up at Launay, a hamlet near the Petit Séminaire of Saint-Méen, the superior of which was put in charge in 1841;[91] and Nîmes in the same year, founded by the Conférence de Saint-Vincent de Paul, and moved to Courbessac as a 'colonie agricole' in 1861; and at Mettray in the Nord in 1840, serving as a 'penitentiary colony' for young delinquents as much as an orphanage.[92]

The principle of the education of poor children differed very little from that of the education of orphans: both came under the rubric of charitable works. The curate of Saint-Étienne, Lille, submitting a project in 1817 for a school for poor girls of Lille to be run by the Sœurs de Sainte-Thérèse, underlined that 'schoolmistresses will never forget that after the reading and catechism lessons, the work-class is one of the capital features of the establishment', and no effort should be spared in inspiring 'le goût du travail'.[93] In the mid-century, there existed girls' schools that were only glorified sweat-shops, whether that of Carnières, in the Nord, where a score of girls were employed all day at the spinning-wheel,[94] or the lace-making shops of the Hazebrouck–Bailleul area where the girls were lucky to receive an hour's

---

[90] B.M. Nîmes, Fonds légal 517, Letter of provisional Conseil d'administration of Maison des orphelines protestantes du Gard 'aux Chrétiens du département du Gard', 8 Sept. 1822.

[91] A.N. F17 9326, report of Insp. Acad. Rennes, 3 May 1856.

[92] A.D. Nord 1N93, report of A. de Melun to Conseil général of Nord, 27 Aug. 1851, pp. 202–3. On Mettray, see M. Foucault, *Surveillir et punir: naissance de la prison* (Paris, 1975), pp. 300–3.

[93] A.D. Nord 1T 115/9, statutes submitted by abbé Desnoyer to prefect of Nord, 21 June 1817.

[94] A.D. Nord 1T 107/4, report of Hilaire, primary-school-inspector of Cambrai, 4 Apr. 1851.

catechism a day.[95] The difference between schools which were also *ouvroirs* and workshops that were also schools was only one of degree, and the system was perpetuated rather than relieved by the ministerial circular of 13 August 1852, which provided for *asiles-ouvroirs* in which three hours a day were set aside for schooling. Bernot, primary-school-inspector of Lille, made the point that the issue was not the necessity to drill lazy children in 'the habits of order and industry', since they 'already work far too much'.[96] The scourge was not idleness but child labour. Moreover, lace-makers and mill-hands, deft in their specific skill, were quite ignorant of the arts of needlework and dress-making. A movement of opinion started during the Second Empire demanding that the working-class family that was being destroyed by factory labour be reconstituted by teaching girls the skills of the housewife, above all how to mend clothes. Only such economies could help to balance the family budget and restore order and harmony to the household.[97]

It was nevertheless schooling that had to bend to the demands of labour, not the other way around. Another example of this is the role of the infant schools or *salles d'asile*. It was clearly in the interests of employers that female labour should be released for the fields and the workshops, and the function of *salles d'asiles* was to look after children between the ages of two and seven during the working day. Labouring families saw the value of these institutions, even where they did not appreaciate the value of primary schools proper. Women were liberated to earn, whether in the mills of Tourcoing, or the silk-spinning of the Cévennes, or for collecting oysters in the bay of Cancale. The nuns who ran them had to synchronize their classes to the rhythms of labour, staying open in August for the harvest and gleaning—80 per cent of the infants in the

---

[95] A.D. Nord 1T 107/7, report of Flament, primary-school-inspector of Dunkerque and Hazebrouck, 24 Oct. 1853.

[96] A.D. Nord 1T 116/1, Bernot to Rector of Douai, 27 Oct. 1852.

[97] A.N. F12 4717, circular of prefect of Nord to local officials and notables, 16 Nov. 1861. See also letter of Charles Lecluse of Conseil des prudhommes of Roubaix, published in *Réforme social* of Brussels, 27 May 1870, cited in L. Trénard, ed., *Histoire des Pay-Bas Français. Documents* (Toulouse, 1974), pp. 326-8.

salle d'asile of Marchiennes, near Douai, in 1868, were children of day-labours[98]—and from 7 a.m. to 7 p.m. at Lille, to 'relieve the parents' during the twelve-hour day, especially when pressure was put on them by the president and vice-president of the Comité des salles d'asile, respectively the wives of industrialists Wallaert and Scrive.[99] To a large extent, children who went to work at eight in the textile mills of Lille or the sugar-beet factories of the Douaisis, obtained all the education they would ever receive in the *salles d'asile.*[100] Moreover, during hard winters, notably after a poor harvest and rising grain prices, as in 1853-4, the *salles d'asile* provided shelter, food, and clothing. Not only were they institutions of education, they were also part of the apparatus of poor relief. The only problem was that there were simply not enough of them. Largely confined to the towns, they could not even there cope with the needs of the working population. And so a rash of *garderies* sprang up, uncontrolled, in the hands of incompetent and illiterate women, tolerated only because the official provision of facilities was insufficient. In 1851, the town of Tourcoing alone had forty-five of them.[101]

The inculcation of work-discipline was only one aspect of the campaign to 'moralize' the masses. Another was to impose some sort of 'order' on a population that was corrupt, rebellious and violent. The role of the school was central, but just as at one end of the spectrum it merged into the business of social redemption and poor relief, so at the other it was only one instrument of the disciplining and regimentation of society, and may be coupled with prisons and the army. Much has been said about the birth of the disciplined society in the nineteenth century, the transition from repression by an often ritualized force to the constraining of the individual within a rigid framework of behaviour. The

[98] A.D. Nord 1T 116/12, Registre nominatif des élèves, *arrondissement* of Douai, 1868.
[99] A.N. F17 10869, Madame Monternault, deléguée spéciale, to Rector of Douai, 30 April 1875.
[100] A.D. Nord 1T 116/6, reports of school-inspectors Bernot 14 Jan. 1854, and Hilaire, 10 Dec. 1855.
[101] A.D. Nord 1T 66/3, Commissaire de police, Tourcoing, to Procureur de la République, Lille, 31 Aug. 1851.

introduction of mutual education is almost a caricature of this development, based as it was on the ranking of individuals and circles of pupils, the elaboration of a precise timetable, the multiplication of surveillance by the monitorial system, manœuvring between exercises at the sound of a bell, whistle, or wooden clapper, and a scale of rewards and punishments that stimulated emulation and eliminated corporal violence.[102] It is significant that Benjamin Appert, the missionary of mutual education in the Nord, not only founded mutual schools for poor children but also set up regimental schools, first at Douai and then on a national basis under the War Minister, Gouvion Saint-Cyr, together with a school of prisoners at the military prison of Montaigu, until he fell foul of the authorities in 1822 for trying to spring a fellow Bonapartist.[103] Constantly defending mutual education against accusations that it was irreligious, instruction without education, he affirmed that 'the experience of six years' application proves that children and soldiers become docile and religious in these institutions'.[104] Religious congregations were themselves no strangers to prisons, not only in the case of nuns, but also in that of the Frères des Écoles Chrétiennes, who undertook surveillance and elementary education at the Maison Centrale de Force et de Correction at Nîmes between 1841 and 1848. After they left, a prisoner noted that 'this gaol that armed and brute force had difficulty in containing has been calmed and regenerated by poor men of religion with nothing but their rosaries to defend them'.[105]

In a more general sense, elementary schools sought to implant notions of duty, social subordination, and respect for authority. At the summit of the hierarchy was God, and religion was brought in as a powerful force to support the social order. A report of 1829 praised the achievements of the Frères des Écoles Chrétiennes in the coastal towns of

---

[102] R. Tronchot, op. cit., i, 170–86.

[103] B. Appert, *Dix Ans*, i, 98–102, 165.

[104] B. Appert, *Traité d'éducation élémentaire d'enseignement mutuel pour les orphelins et les adultes des deux sexes* (Paris, 1822), p. 78. His later work included *Bagnes, prisons et criminels* (Paris, 1836).

[105] A.F.E.C. NC 716/2, 'Mémoire sur la Maison Centrale de Nismes par un prisonnier', n.d.

Brittany, including Saint-Malo, approving the manner in which children were taught their 'duties towards God and Man'.[106] One of the basic texts of the teaching congregations, the *Civilité chrétienne*, which can be traced back to the *Civilité puérile et honnête* concocted by Erasmus, albeit a manual of etiquette essentially for a bourgeois audience, nevertheless urged respect for God, magistrates, churchmen, and one's parents. Guizot therefore said nothing new in his letter to *instituteurs* of 18 July 1833 when he informed them that they were 'one of the guarantees of order and social stability'.[107] Unfortunately this did not mean, in fact, as inspector Dalimier counselled, that every school-room was 'the image of a properly civilized society'.[108] For many *instituteurs*, the banning of corporal punishment was a decisive disadvantage. That of Dol, near Saint-Malo, warned that a child who was accustomed to obey only 'the brutality of a father', would settle for nothing less from a schoolteacher, and that problems with discipline accounted for one in every four desertions from the teaching profession.[109] The brutality of the Frères was doubtless exaggerated by anti-clericals, but the evidence goes to show that corporal punishment remained one of the weapons of their armoury.

Given these difficulties, the transmission of moral and religious sentiments was seen by educators as very much the responsibility of women. The school-inspector of Valenciennes stated in 1850, 'to educate girls is to open a school in the heart of every family'.[110] And in Bernot's opinion, 'it is by means of girls rather than boys that the dissolute morals of our manufacturing populations will be reformed'.[111] For most people, in the middle of the nineteenth century, morality was inseparable from religion,

---

[106] A.N. F17 10214, inspector Jégou to Rector of Douai, 15 July 1829.

[107] F. Guizot, *Rapport au Roi par le Ministre Secrétaire d'État au département de l'Instruction publique sur l'exécution de la loi du 28 Juin 1833* (Paris 1834), pp. 76–81.

[108] Dalimier, op. cit., p. 79.

[109] A.N. F17 10792, submission of *instituteur* of Dol, 24 Jan. 1861.

[110] A.D. Nord 1T 107, report of school-inspector of Valenciennes, 10 July 1850.

[111] A.D. Nord 1T 67/6, report of Bernot, school-inspector of Lille, 12 July 1853.

and powerless without it. Religion not only laid down the canons of virtue, it also provided the sanction of salvation or damnation. For inspector Dalimier, 'education is made by religion'; it was the 'powerful and superhuman reason which will correct vices and reform habits'.[112] And Mère Saint-Félix, superior of the Sœurs de l'Immaculée Conception of Saint-Méen, advised her nuns: 'give greatest care to the religious instruction of your pupils. Form them in the ways of piety.'[113]

The links between morality, religion, and social order were equally strong in the *salles d'asile*. Public funds were available in much smaller quantities for the *salles d'asile* than for elementary schools, and they were very much the products of private initiative. The social élite was active in the propagation of infant schools, not only for charitable reasons, but also in order to moralize the *classes laborieuses* at the earliest age. The Protestant Consistory of Nîmes founded a *salle d'asile* soon after 1830, the first outside Paris. The royal ordinance of 22 December 1837 governing *salles d'asile* triggered off a spate of foundations, notably under the auspices of Conférences de Saint-Vincent de Paul. These were often dominated by clerico-legitimist notables, like that of Rennes which founded a *salle-d'asile* in 1839 and interlocked with the directors of the *Journal de Rennes*, launched in 1844.[114] The contribution of pupils of the Lycée and colleges of the department to the funds of the *salles d'asile* was praised in 1858 by the Rector of Rennes, declaring that they would render 'an immense service to the suffering classes and bind by one more link the needy to those whose situation permits them the happy practice of charity'.[115] The large industrial companies were clearly interested in the *salles d'asile* not only to liberate a female work-force but also to discipline new generations of workers. The Compagnie houillère of la Grand'Combe and the Compagnie des Forges et Fonderies of Bessèges founded *salles* in the 1850s, and in

---

[112] Dalimier, op. cit., p. 4.
[113] P. A. Fouqueray, op. cit., p. 211.
[114] A.N. Rennes, R27, *Œuvre des salles d'asile*, statutes of 1840. Lagrée, *Mentalités*, pp. 338–42.
[115] A.N. F17 10868, Rector of Rennes to school-inspectors, 19 Jan. 1858.

the Nord the Anzin company was only one important patron. Most characteristic of the *salles d'asile*, however, was the role of ladies of the nobility and *haute bourgeoisie*, the wives of prominent notables in the provinces, first as *dames inspectrices* under the ordinance of 1837, then, when inspection was bureaucratized under the decree of 21 March 1855, as *dames patronnesses*, raising funds by means of bazaars, lotteries, and charity concerts, and exercising an informal supervision of standards.

The moralizing function of the *salles d'asile* was outlined by Guizot in a circular of 4 July 1833, stipulating that they should inculcate 'habits of order, discipline and regular occupation that are a beginning of morality'.[116] The least that can be said about the education offered by the *salles d'asile* is that such counsels were taken to unwarranted extremes. A well-furnished *salle d'asile* was dominated by the *gradin*, or tiered rows of seats, and it was perched on these that the infants spent most of the day. Marie Loizillon, an inspectress appointed under the decree of 1855 and later denounced by Brossays Saint-Marc as 'that freethinker', was scathingly critical of the Filles de la Sagesse at Lille in 1858, noting that 'they sacrifice everything to posture, order, cleanliness and silence'. Such excess 'produces parrots and machines, but one thing is certain, it does not develop moral life in these poor little creatures'.[117] The overwhelming amount of time given to religious education and the force-feeding of encyclopaedic facts according to a pedagogy that took no account of curiosity or imagination were also criticized. Pauline Kergomard, visiting Ille-et-Vilaine in 1879, reported that 'biblical history and above all the catechism are taught without pause or respite', while the children's 'heads are crammed with definitions and words that are meaningless to them'.[118] It was such an archaic pedagogy that she set out to reform.

Cleanliness was a final aspect of the moralizing mission. For most educationists, it was synonymous with order and discipline. Dalimier urged that the cleanliness of girls not

---

[116] Copy in A.D. Gard 1T 99.
[117] A.D. Nord 2T 61/7, Marie Loizillon to Rector of Douai, 28 April 1858.
[118] A.N. F17 10847, report of Pauline Kergomard, 18 Sept. 1879.

only inspired sentiments of moderation and restraint but also 'offers the sensible image of inner purity and innocence'.[119] The issue was of no small importance. In 1853, the school-inspector of Rennes reported that dirtiness and drunkenness were the twin sores of the Breton countryside.[120] Instruction after instruction pressed home the necessity for the war on dirt, but conditions were unfavourable. To begin with, the schoolhouses reflected the budget of communes and the life-style of the surrounding area. At Rosult, near Saint-Amand, in 1859, the old schoolhouse had grown so overcrowded that the class had to be moved to an 'old stable, a low, dark and dank building'.[121] In the *arrondissement* of Hazebrouck the schoolhouses were built of clay, without windows, but then so were most of the other habitations.[122] The environment made matters no easier. The *instituteur* of La Guerche in Ille-et-Vilaine reported in 1861 that children often came to school from great distances, along bad roads, and arrived 'their clothes wet and covered in mud'. In Brittany, he added, children spent much of the year dressed only in goatskins anyway.[123] At La Vernarède, in the coal-basin of Alès, the school-inspector noted in resignation that 'people say that the children are dirty because at La Vernarède there is very little water and lots of coal'.[124]

Even if instruction were balanced by education at school, the success of moralization might still be jeopardized by other factors. Early school-leaving removed the adolescent from the tutelage of the lay teacher or religious congregation and exposed him to pernicious influences, not least that of *mauvais livres* and *mauvais journaux*. The story of the education of the popular classes involves not only what went on in the class-room, but also what happened after they left school. Two possibilities were open to the notables if they were not to lose their influence over the people. Firstly,

[119] Dalimier, op. cit., p. 17.
[120] A.N. F17 9314, report of primary-school-inspector of Rennes, 3 Sept. 1853.
[121] A.D. Nord 1T 110/4, report of Délégation cantonale of Saint Amand R.G., 30 Jan. 1859.
[122] A.N. F17 9330, report of Insp. Acad., Douai, 1 May 1856.
[123] A.N. F17 10792, submission of *instituteur* of La Guerche, 25 Jan. 1861.
[124] A.D. Gard 1T 718, primary-school-inspector of Alès to Insp. Acad. Nîmes, 4 June 1874.

they could seek to control the literature to which the lower classes had access and, secondly, they could group apprentices and young workers into organizations that were variously evening classes, *catéchismes de persévérence*, youth clubs, or a combination of all three. The notion of patronage was central, indeed the common word for such institutions was *patronages*.

One of the earliest such schemes was promoted in 1818 at the École du Béguinage of the Frères des Écoles Chrétiennes at Douai by its founder, Édouard Deforest de Lewarde. In order to maintain a supervision of pupils after they left, between the ages of thirteen and fifteen, they remained *agrégés* of the school. In return for being apprenticed to a trade, receiving additional lessons in drawing and architecture, and being entitled to a dowry when they got married, the *agrégés* undertook to avoid cabarets, gambling-dens, and public entertainments, and to cultivate 'devotion to their duties, love of work, sobriety and the company of persons fearing God'.[125] Besides such enterprises, there was a concern that ordinary people should be insulated as much as possible from dangerous writings. There was, clearly, in the nineteenth century a gulf between the cultivated élite and the illiterate mass. The educated classes had access to municipal libraries, the stock of which was largely theological and legal works inherited from dissolved monasteries, with little or no books on science, history, or the arts.[126] They also gathered in *chambres littéraires*, rather like clubs, going back to the end of the Ancien Régime, in the case of that of Rennes (1775) or to the Revolution, such as the salon des *négociants* of Lille.[127] On the other hand, newspapers were becoming more available through the *chambres de lecture* attached to many bookshops, while in the countryside *colporteurs* hawking almanacs, saints' lives, chivalrous and

[125] F. Capelle, op. cit., pp. 47-52.

[126] On municipal libraries after the Revolution, see A.D. I-et-V. 14Ta; E. le Glay, *Mémoire sur les Bibliothèques publiques et les principales bibliothèques particulières du département du Nord* (Lille, 1841); H. Rivoire, *Statistique du département du Gard* (Nîmes, 1842), pp. 380-1, and A.D. Gard, 4T 26.

[127] A.D. Nord 1T 217/1, Commissaire central de police, Lille, to Prefect of Nord, 18 May 1820; C. Barré, 'La Chambre de Lecture de Rennes, 1775-1875' (*mémoire de maîtrise*, Rennes, 1972).

picaresque novels, useful advice on the arts of writing, craftsmanship, and love, treatises on magic like the *Petit Albert*, jokes, pornography, engravings, and trinkets, periodically became the victims of official fears for social order.[128] Such influences could not be countenanced by the notables, either Catholic or Protestant. The Consistory of Nîmes founded a Bibliothèque populaire protestante in 1827, under the motto 'civiliser sans pervertir'.[129] The year before, at Lille, Édouard Lefort, the publisher and active promoter of social Catholicism, was the force behind the creation of a Société Catholique des Bons Livres, which lent improving books free of charge. The Comte de Muyssart, Mayor of Lille, reported that the aim of the society was to 'fight the pernicious effect produced by the multiplication of so many sinful works under the press' and to contribute to the propagation of 'those salutary doctrines that are the only foundation of virtue and the pledge of the stability of all social institutions'.[130]

The Revolution of 1830 and the social discontent that followed it at Lyon and elsewhere disconcerted not only the displaced legitimist élite but also the new constitutional and Protestant leadership.[131] The *question sociale* was announced in no uncertain terms, and the necessity of moralizing the *classes laborieuses* both in school and outside it could not be neglected. A new wave of enterprises sought to wean the agitated masses away from politics and show them the way towards self-improvement. Édouard Le Glay, archivist of the Nord, helped to found a Bibliothèque charitable at Cambrai in 1830, appealing to others to 'substitute for the vain and delusive theories of politics the taste and love for letters'.[132] The Rector of Nîmes pressed the

---

[128] On the question of *colportage*, see P. Brochon, *Le Livre de colportage en France depuis le XVIe siècle* (Paris, 1954), G. Bollême, *La Bibliothèque bleue. La littérature populaire en France du XVIe au XIXe siècles* (Paris, 1971); and J.-J. Darmon, *Le Colportage de librairie en France sous le Second Empire* (Paris, 1972).

[129] *Journal des écoles primaires du Gard*, 1837.

[130] A.D. Nord M222/799, statutes of 15 Mar. 1826.

[131] L. Mazoyer, 'Rénovation intellectuelle et problèmes sociaux: la bourgeoisie du Gard et l'instruction au début de la Monarchie de Juillet', *Annales H.E.S.*, vi, 1934, 20–39.

[132] E. Le Glay, op. cit., p. 227.

Protestant mayor, Girard, to add to the existing free public lectures in chemistry and applied geometry, aimed at the labour aristocracy, other lectures in history, natural history, French language and literature. Such a programme would serve to direct the ideas of youth 'towards a useful goal, and to distract them from political discussions'.[133] One important development was the multiplication of evening classes for apprentices and young workers, largely under the supervision of the Frères des Écoles Chrétiennes. There were foundations at Valenciennes in 1832, Armentières in 1833, Nîmes in 1834, notably for silk-weavers, and Alès in 1836. At Lille, Édouard Lefort broadened the scope of the Bibliothèque des Bons Livres by founding a *patronage*, the Société Saint-Joseph in 1836. Meeting on Sundays and Mondays in order to tempt youths away from the cabarets, it offered 'honest amusement' in the form of cards, dominoes, chess, and billiards in the winter, archery, boules, skittles, and gymnastics in the nearby country at Esquermes in summer. The main problem was that less than a third of the clientele was made up of workers, and that an artisan élite of mechanics, locksmiths, printers and bookbinders, tailors and cobblers. The rest was composed of clerks, tradesmen, liberal professions, and even army officers.[134] In 1844, Kolb-Bernard and his colleagues were campaigning for the establishment of an evening class in the poor working-class parish of Saint-Sauveur, but with little success.[135] The apprentices of Rennes were perhaps more fortunate. There, the Société de Saint-Vincent de Paul founded an Œuvre des apprentis in 1843, entrusted to the Frères des Écoles Chrétiennes. Sundays were taken by up catechism, reading and writing lessons, mass, games, and religious instruction at the Archiconfrérie de la Sainte Vierge in the evening, and classes were run both on weekday evenings and at midday. The *œuvre* catered for 116 apprentices in 1846 and, combined in 1847 with the classes set up by Brossays Saint-Marc for

---

[133] A.N. Nîmes R272, Rector to Mayor of Nîmes, 12 July 1831; A.D. Gard 10T 31, Rector to Conseil académique, 13 June 1831.

[134] E. Huard, *Édouard Lefort, Président de la Société de Saint-Joseph de Lille* (Lille, 1893).

[135] A.F.E.C., NC 654/1, Comité d'administration des Écoles Chrétiennes, Lille, to superior-general, July 1844.

workers of between sixteen and twenty-five years of age, it formed a highly satisfactory *Œuvre du Patronage*.[136] The Catholics had clearly made their mark, although in the Gard the Protestant consistories were active founding evening classes, not only at Nîmes—attended by eighty people in 1836 as against two hundred at the Frères[137]—but also in small towns like Anduze where the *instituteur* underlined the need to find the means 'to make the people content with its position and to make that position better'.[138]

The classes and libraries destined to improve the lower classes received very little support from municipalities, let alone from higher authorities. They were overwhelmingly the fruit of private patronage. No firm initiative from the government came until the Second Republic, a circular of Hippolyte Carnot, dated 8 June 1848, urging officials and municipalities to organize public lectures by *lycée* or college *professeurs*, in order to popularize the masterpeices of French literature and history. Little happened before the following spring, and the experience served only to illuminate the cultural chasm that existed between the educated minority and the scarcely literate populace. The masters of the Lycée of Douai, discovering that their readings from Thiers and Lamartine drove the working class away, reflected that the tone was 'serious and elevated . . . and perhaps the customs of the region do not lend themselves to the gravity of such pastimes'.[139]

The events of 1848-9 confused the thinking of notables about instruction and education in this as in other fields. The first response of the authorities was repression. A law of 27 July 1849 obliged *colporteurs* to seek authorization from the prefects, on pain of imprisonment, and in 1852 hawked material was subjected to stringent censorship.

---

[136] A.F.E.C., NC 777, Dufretay, *notaire* and president of Société de Saint-Vincent de Paul, Rennes, to superior-general, 17 Feb. 1845, 23 Nov. 1846, and to Frère directeur. 25 Jan. 1847.

[137] *Journal des écoles primaires du Gard*, report of school-inspector and Comité communal, Nîmes, 30 May 1836, p. 127.

[138] A.D. Gard 1T 571, private *instituteur* of Anduze to Comité local, 7 Aug. 1837.

[139] A.D. Nord 1T1/4, masters of Lycée of Douai to sub-prefect of Douai, 16 June 1849.

A decree of 29 November 1851 gave prefects the right to close bars after a single misdemeanour, though the opposition of the brewing interests rendered it largely ineffective. In 1851 there were 1,600 bars in Lille and its suburban communes, one for every seventy inhabitants, and bars were where the working man not only drank but also read the newspapers. They were the popular *bourse aux nouvelles*.[140] In the general revulsion against instruction, evening classes petered out, and *instituteurs* who thought of reopening them were closely scrutinized. The Protestant *instituteur* of Crèvecoeur, reported by Hilaire, the Catholic primary-school-inspector of Cambrai, as having 'advanced political opinions' and anxious to use the class only to 'undertake proselytism in favour of the Protestant religion', could expect no authorization from the Rector of Douai.[141] In the 1850s there were virtually no evening classes except those set up by industrialists for their young workers and those run by the Frères des Écoles Chrétiennes that survived in a clerical climate.

Repression was only one side of the coin. The Catholic Church in the era of the Falloux law was as active in the domain of post-school *œuvres* as it was in those of primary and secondary education. At Rennes in 1849 a group of magistrates and ecclesiastics, together with the printer, Hippolyte Vatar, launched an Œuvre des Bon Livres in order to combat 'this overflowing of incendiary and corrupting writings which, hawked from one end of France to the other, infect even the cottages of the poor with their savage doctrines'.[142] In the South of France, an Association des bibliothèques paroissiales, of which Emmanuel d'Alzon was a leading light, was set up in May 1850, and by 1855 parish libraries stocked with pious literature had been established in all but thirteen of the 249 parishes of the diocese of Nîmes.[143] Catholic *patronages* also enjoyed a new prosperity, notably the Œuvre de la Jeunesse set up in 1858 in the vast building erected at Lille at the expense of the

---

[140] P. Pierrard, *Vie ouvrière*, p. 288.
[141] A.D. Nord 1T 114/3, school-inspector of Cambrai to Rector of Douai, 22 Oct. 1851.
[142] *Journal de Rennes*, 10 Aug. 1849.
[143] S. Vailhé, op. cit., ii, 183–4.

Comtesse de Grandville on the site of the old Hôtel de la Monnaie. Originally designed to be a Jesuit College, it became a *patronage* because of the opposition of Mgr Regnier to the scheme, though it remained in Jesuit hands. Calling on the Frères des Écoles Chrétiennes to assist with the work, the Jesuit Père Cœurdacier declared that nineteen out of every twenty pupils of the Frères leaving school 'lose their Christian habits. This contagion is appalling. Our *patronage* is their last chance of salvation.'[144] Unfortunately, Jesuits and Frères disagreed about the clientele to be attracted. Père Cœurdacier increased the subscription from five to ten centimes a day in 1860 and noted subsequently that 'the social standing of the *patronage* has improved notably: our youths are of the better class of workers. Many of them are employed as clerks in offices; others are sculptors, printers or upholsterers.' They had 'nothing in common with the vulgar thread-makers and spinners of Saint-Sauveur or Wazemmes'.[145] Eventually, in 1865, the Jesuits were eased out of *La Monnaie*, and the Frères took full control.

The Catholics did not have a monopoly of the patronage of schemes to moralize the worker, at least after the Empire took an anticlerical turn in 1860. Another series of private initiatives originated in the democratic, even republican tradition that threw up the Ligue de l'Enseignement. The ethic of this movement was not to reinforce the traditional social hierarchy and the influence of the Church, but it was in no way socialistic. The key notion was that of self-help, the possibility for the working man to improve his position within bourgeois society by assimilating its values of instruction, thrift, and temperance. At Nîmes, Ali Margarot, the Protestant banker and future mayor of the city, who had visited the doyen of the movement, Jean Macé, in Alsace, founded a Bibliothèque populaire in 1864. Aimed at the 'industrious worker', and opening in the evenings and on Sundays in order to suit him (as municipal libraries did not), it made available books on history, biography, travel and geography, science, political economy and legislation, French and foreign novels, for an annual subscription of two

[144] A.F.E.C., NC 654/1, Cœurdacier to superior-general, 5 Jan. 1858.
[145] P. Pierrard, *Vie ouvrière*, p. 410.

francs.[146] In the Nord, a Société d'instruction populaire was formed at Lille by a mechanic in 1865 and dominated by metal-workers,[147] but at Roubaix the initiative came from the *patronat*. Charles Junker, manager of a textile mill and himself an Alsacian, convinced that the class struggle was the result of the ignorance of the working class, launched a Cercle des Travailleurs in 1869 in order to 'instruct and moralize the worker, to inculcate gradually the sentiments of duty'.[148] Authorized in February 1870, its goal of class peace was very difficult to achieve in the period of the Commune, and under the inspiration of the *ingénieur* Émile Moreau it became very much a force for radicalism and anti-clericalism.[149] Enlightenment took longer to reach Ille-et-Vilaine, but the Société Franklin founded a library at Redon in 1872 and the republican leadership of Rennes, headed by Waldeck-Rousseau and Le Bastard, set up a Bibliothèque populaire there in 1878.[150]

It was not until the latter part of the Second Empire, and notably under the auspices of Victor Duruy, that the government became an active promoter of evening classes and of popular libraries annexed to *écoles communales*. In the year they took off, 1865-6, organized overwhelmingly by lay *instituteurs* rather than teaching congregations, 5,700 people or about 1.4 per cent of the population attended in the Gard, 7,100 or 1.2 per cent in Ille-et-Vilaine, and 13,100 or 1.0 per cent in the Nord.[151] The proportion of illiterates who attended was not high—about a sixth of the total, drawn essentially from the class of conscripts who had been made brutally aware of their ignorance, but the courses were popular because of their obvious practicality. At Marchiennes, in the Nord, artisans requested lessons in arithmetic,

---

[146] A.D. Gard 4T 30, statutes of 1 Dec. 1864.

[147] A.D. Nord M222/722 *bis*, statutes of 14 Sept. 1865.

[148] A.D. Nord, CC729, Charles Junker, 'Avant-projet de cours, bibliothèque et cercles populaires' addressed to 'Messieurs les Fabricants de Roubaix', 1 Oct. 1869.

[149] R. Talmy, 'Tendences anticléricales', pp. 42-5.

[150] A.D. I-et-V 14 Ta5, statutes approved by prefect of Ille-et-Vilaine, 9 Jan. 1872, 1 June 1878. On the Société Franklin, see N. Richter, *Les Bibliothèques populaires* (Paris, 1978), pp. 56-8.

[151] A.N. F17 12324, replies of prefects to ministerial circular of 10 Oct. 1866.

the metric system, and drawing, necessary to their trades.[152] At Grand-Fougeray in Ille-et-Vilaine 'a certain number of young *laboureurs'* anxious to acquire agricultural expertise turned up, but the bulk of the clientele was composed of masons, joiners, carpenters, and shoemakers eager to improve their skill in drawing, measurement, accounting, and letter-writing.[153] Unfortunately, the initial blaze of evening classes did not continue. The year 1865–6 remained the high point in the Gard and Ille-et-Vilaine; only in the Nord did numbers increase until 1867–8. Neither the lack of public funds nor the economic depression of the late 1860s contributed to the success of the enterprise.

A circular of Duruy to Rectors dated 8 October 1867 stated that the foundation of school libraries would serve as an invaluable compliment to the evening classes, and he noted that the propagation of improving literature had 'preserved some counties of England from more than one riot'. The idea was not original. Rouland, a previous Minister of Public Instruction, had been urging the foundation of school libraries to which local inhabitants would have access since 1859, envisaging that they would be stocked by useful volumes of the *Bibliothèque des campagnes* such as Eugène Loudun's *Victoires de l'Empire.*[154] There was enthusiasm for school libraries from the *instituteurs* who submitted memoranda in 1861. In the Nord, the *instituteur* of Brunémont (Arleux) suggested that books on ethics, history, agriculture, and science would be the most conducive to self-help. That of Illies (Seclin) noted that 'a choice of good books in the countryside would be the best way to prevent the infiltration of novels', while one from Valenciennes urged that 'anything that resembles a novel in any way should be ruthlessly eliminated': the taste for 'frivolous readings' was the slippery slope to addiction to pornography.[155]

The number of school libraries increased slowly during the

[152] A.D. Nord 2T 1092, Insp. Acad. to Rector of Douai, 23 June 1866.
[153] A.N. F17 9262, *instituteur* of Fougeray to primary-school-inspector of Redon, 16 Mar. 1866.
[154] A.D. Nord 1T 67/19, circular of Rouland, 20 Aug. 1859. The prefect of Nord replied 31 Oct. 1859 that all departmental funds had already been allocated.
[155] A.N. F17 10776, submissions of *instituteurs* of Valenciennes, Illies, and Brunémont, 31 Jan., 1 and 2 Feb. 1861.

1860s. By 1870 the Gard had fifty-six of them, dense around Alès and in the Protestant communes near Le Vigan and in the Vaunage, but absent in vast tracts of the Cévennes, the northern part of the department, and the cantons of Beaucaire, Aramon, and Marguerittes, all heavily Catholic.[156] There were other obstacles to the success of this campaign for moral enlightenment. The 'taste for reading' was undeveloped among the peasantry, and, as a school-inspector of Le Vigan noted in 1877, this was not surprising, given the way that reading was taught in the schools, the monotonous intoning of ecclesiastical texts.[157] Moreover, the rural populations who were attracted to reading would not necessarily be drawn by the works of popular science, political economy, agriculture, and hygiene that figured prominently in the official catalogue and arrived in bundles at the local railway stations. These dry volumes remained unread on the library shelf, while it was the histories of the campaigns of Napoleon I, stories of voyages round the world, the popular novels of Mayne-Reid and Canon Schmidt, and the illustrated series such as the *Musée des familles* and the *Magasin pittoresque* that attracted readership. Finally, there was the advent of the popular press, reaching wider audiences by serializing novels, reducing costs—the *Petit Journal*, launched in 1863, was the first newspaper to retail at one *sou*—and adopting the tactics of the 'hard sell'. Brossays Saint-Marc could declaim in 1867 against the *mauvais journaux*, 'the diabolical invention of the *romans feuilletons*', the pernicious influence of which substituted for 'respect for authority the disdain and hatred of any social superiority',[158] but his anger was only a measure of the failure of the crusade of moralization.

A study of the 'school of the people' again reveals the importance of private and local initiatives as against governmental schemes of national significance. Itinerant teachers fulfilled the needs of isolated or upland communities where the 'one commune, one school' provision of the Guizot law broke down. And even where *écoles communales* were set

---

[156] A.D. Gard, 1T 885, *Bibliothèques scolaires, 1862–80.*
[157] Ibid., primary-school-inspector of Le Vigan to Insp. Acad. Nîmes, 29 Jan. 1877.
[158] A.D. I-et-V. IV 28, Brossays Saint-Marc, *Mandement de Carême*, 1867.

up, private and sometimes clandestine teachers could often offer more competitive rates. Free education did not have to wait until 1881, but was inaugurated by the Frères in many large towns and was developed in some parts of the country-side long before then. Private *garderies* and *salles d'asile* to look after infants and to liberate female labour flourished long before a state system of nursery schools, as did evening classes, *patronages* or youth clubs, and popular libraries to combat the influence of 'mauvais livres'.

The Churches, both Catholic and Protestant, were at the origin of many of these initiatives. The utilitarian strain, seeking to train a new race of 'economic men' who calculated the benefits of hard work, sobriety, and thrift for themselves and for society, was much weaker in France than in England, clear only in mutual education and the Ligue de l'Enseigne-ment. Far stronger was the religious tradition, that insisted that instruction was dangerous without education and that moralization was impossible without the sanction of religion. Both approaches nevertheless had certain assumptions in common. In paticular, they believed that problems of crime and poverty could be solved not by changing the environ-ment but by reforming men, by converting them to a new bourgeois code of values. What both failed to understand was that the labouring classes had no interest in being reformed; if they attended school it was in order to get on, to obtain material advantages.

In the end, of course, the campaign to bring the popular classes within school walls and to moralize them depended on social and economic conditions. Even if education were made free there existed a poverty threshold beneath which school attendance was impossible, not least because of the contribution of child labour to the family economy. Growing prosperity tended to push up attendance but even then it is clear that the early phase of industrialization was of no benefit to the wage-earning working class, who rarely attended school and were the least literate of the labouring population.

7

The Republican Church

Few have indulged in the rhetoric of education more than the politicians and notables of the Third Republic. Ceaselessly they broadcast its merits as a universal panacea. New generations, schooled and vigorous, would spring up prepared for *revanche* against Germany. Whereas in 1848 a Republic had been founded among people lacking the virtue to sustain it, and their ignorance had played into the hand of feudalism, despotism, and the government of priests, now a Jacobin alliance of democratic bourgeoisie and people would be constructed to defeat the champions of the Ancien Régime and educated in their rights and duties as citizens of the Republic. Finally, despite the Paris Commune, they argued that there was no social question, no class conflict, no classes even, but only *couches sociales*. Society was composed of 'individuals in a process of rising', and popular enlightenment was crucial to the encouragement of this capillary action. The inculcation of ideas of self-help, hard work, and thrift would permit these individuals to acquire property, the condition of moral and material emancipation. Moreover, spreading science in waves would reduce the gulf between the educated élite and the illiterate mass, enabling the people to participate in the common patrimony of humanity, and promote the 'fusion of classes'. As the sub-prefect of Vitré told pupils of the municipal school there in 1878, 'our government of peace and work, of order and liberty, can only gain from this diffusion of science. To enlighten intelligences, is that not to pacify and unite them?'[1]

The republicans are usually credited with the foundation of universal, popular education, enshrined in the laws of 16 June 1881 making education in the *écoles communales* entirely free, and that of 28 March 1882, making it lay and compulsory between the ages of six and thirteen. The precise extent of their contribution must nevertheless be calculated

[1] *Avenir de Rennes*, 2 Aug. 1878, prize-day speech by sub-prefect of Vitré.

more exactly. It is true that between the school year beginning in 1880 and that beginning in 1881, when universal free
education came into force, the primary-school population
rose by 5.6 per cent in the Nord, 6.4 per cent in the Gard,
and 10.4 per cent in Ille-et-Vilaine, whereas the difference
the previous year had been increases of only 1.5 per cent,
0.7 per cent, and 1.0 per cent respectively. Two qualifications must nevertheless be made. Firstly, municipal councils,
on whose ordinary revenues and special rates the financing
of free education principally fell, had been supporting
primary schools to a greater or lesser extent since the Guizot
law. On the eve of completely free education in 1879-80,
the proportion of children educated free in *écoles communales* was 73 per cent in the Gard and the Nord and 60 per
cent in Ille-et-Vilaine. The greater leap in school attendance
in Ille-et-Vilaine under the 16 June 1881 law may be accounted
for by the fact that the communes of this backward, agricultural region could not match the funds that other departments allocated to free education. The role of the state
was correspondingly greater in what was in some sense an
administrative colony, a target for economic imperialism.
High rates of free education were not confined to lay schools.
On the contrary, free education was more widespread in
schools run by religious congregations than in lay schools in
the Nord and the Gard, though not in Ille-et-Vilaine, where
the proportion of girls educated free in nuns' schools was
markedly lower than that in the minority of lay schools.
Secondly, the growth of the school population, relative to
the growth of the population as a whole, was no greater
in the ten years after 1881 than during the years 1861-72,
dominated by the Liberal Empire, when gigantic strides were
made on many fronts in the propagation of education. In the
Nord, the school population grew at just over twice the rate
of the total population in both periods, although marginally
slower in the 1880s. In Ille-et-Vilaine, where the overall
population remained fairly stable, the school population
increased five times as fast under the Empire as under the
Republic. In the Gard, finally, the school population grew
in 1861-72 while the overall population fell, but fell in the
period 1881-91, while the population of the department

increased, Once free compulsory education is placed in
context, it becomes apparent that the groundwork for universal education had already been laid, notably during the
Second Empire.

It was, of course, one thing to decree that school attendance was compulsory between the ages of six and thirteen,
and quite another to enforce it. Under the law of 1882
commissions scolaires were organized to draw up lists of
those of school age in the commune and to ensure that they
attended school. But the commissions were nothing other
than committees of municipal councils, and elected representatives would not risk alienating voters by adopting too
draconian a policy. As we have seen, schools were sometimes boycotted for political reasons, especially in the
time-lag between a laicization and the foundation of a
private school run by a religious congregation, and the
Waldeck-Rousseau ministry strove to reinforce the *commissions scolaires* at the time of the passage of the Associations
Law in order to fight such absenteeism. More important,
however, were the demands of labour on the population of
school age, as pressing at the end of the nineteenth century
as at the beginning. The *commissions scolaires*, often composed of employers of labour, frequently found it extremely
difficult to reconcile public and private interest.

Free education broke down the barrier of school pence,
but most if not all of those children whose contribution
to the family economy was indispensable would have been
exempted from payment under the old system as indigents.
Besides, the hidden cost of loss of labour or earnings by
attendance at school had to be placed in the balance. For this
reason the impact on school attendance of seasonal rhythms
of work, early leaving for gainful employment, and the
cyclical movement of the economy was still considerable.[2]
In December 1884, the *délégué cantonal* of Marguerittes,
near Nîmes, complained that since June the schools had been
'almost deserted, the children staying away as if the law of
28 March did not exist'. The reason, a local mayor explained,
was that, as ever, the olive harvest followed the *vendanges*
and the children did not return to school until it was

[2] See above, pp. 214–22.

over.[3] More precise evidence is available to show that the effect of the law of 1882 on seasonal school attendance was minimal. The matriculation register of Bazouges-sous-Hédé in Ille-et-Vilaine demonstrates that in the 1870s indigent pupils attended school for an average of ten months of the year, fee-paying children for six months; between 1882 and 1890, the average stay of all pupils was seven months and three weeks.[4]

The problem of school-leaving at an early age should have been countered by the institution of free, compulsory education. This is certainly what was predicted by the municipal council of Visseiche (canton of La Guerche) in 1875.[5] However, the persistent problem of children starting work at an early age had been thrown into relief by the paternalism of the Moral Order regime, revising the law of 1841 on child labour by one of 19 May 1874. Under this, the minimum age at which a child could start work was twelve, although a subsequent decree reduced this to ten in specified industries, notably textiles. In addition, children between the ages of twelve and fifteen who did not have the *certificat d'instruction primaire élémentaire* were limited to six hours' work per day instead of eight, and were obliged to attend school for the period remaining. As always, there was many a slip between legislation and enforcement. There was one Inspecteur divisionnaire du travail des enfants dans les manufactures for six departments of the west of France, including Ille-et-Vilaine, who might closely supervise the large-scale shoe factories of Fougères but was unable to check every small workshop for infractions.[6] In the Gard, the tradition of taking girls away from school at the age of eleven and setting them to work in the silk mills did not change, and trumpetings of the inspector's imminent arrival gave employers time to hide their illegal labour force. As the primary-school-inspector of Le Vigan declared, 'it is true that there is a law on child labour, but the law is not enforced'.[7] Evasion was more

---

[3] A.D. Gard 1T 475, Délégué cantonal of Marguerittes to prefect of Gard, 8 Dec. 1884; mayor of Bezouce to prefect, 14 Dec. 1884.

[4] A.D. I-et-V, registre matricule of Bazouges-sous-Hédé.

[5] A.N. F17 12221, municipal council of Visseiche, 4 July 1875.

[6] A.D. I-et-V. 1N, report of Inspecteur divisionnaire, 1890, p. 507.

[7] A.D. Gard 1T 718, primary-school-inspector of Le Vigan to Insp. Acad. Nîmes, 30 April 1876.

difficult in the textile region of the Nord, but the mill-owners were able to bend the law in their favour. The idea of half-time schools for teenage workers was anathema to them, and the Inspector of the Academy was himself opposed to troops of factory children invading municipal schools for several hours a day. There was simply not enough room in the schools, the confusion in the class-room would be reminiscent of the midday schools of the July Monarchy, and the *institu-teur* teaching without respite would soon be driven to exhaustion.[8] The only alternatives were factory schools or evening classes, such as that set up by the Marists at Halluin, near Lille, in 1876.[9] But the former were private schools, controlled by the employers, who could whittle away teaching time to insignificant amounts, and the latter postponed schooling to the end of the working day, something that the authorities of the Nord had tried to avoid in the 1850s.[10] In the event, children under fifteen did work more than six hours a day, and when tolerance of this irregularity was withdrawn in 1879, it was not only the industrialists but also the Inspecteur divisionnaire at Lille, fearing that the children would only roam the streets, who protested.[11]

It is instructive that the notables of the Opportunist Republic, so closely interlocked with the economic life of the country, did nothing to reform the child-labour laws. The law of 28 March 1882 on compulsory education up to the age of thirteen should have made a difference, but in the Nord the prefect, Jules Cambon, circulated sub-prefects and mayors to underline the exemptions to the minimum age of twelve still holding under the law of 1874, and to clarify the fact that parents with children under thirteen might apply to the *commissions scolaires* for exemption from either the morning or the afternoon session.[12] Ardouin-Dumazet

---

[8] A.D. Nord 1N 120, report of Insp. Acad. Lille, 1876, pp. 559-60; A.N. F12 4765, Inspecteur divisionnaire de Lille to Min. Agric. Comm., 28 Dec. 1880.

[9] A.F.M., BEA 660, Halluin, Mayor of Halluin to director of Marist School, 10 April, 1876.

[10] See above, p. 219.

[11] A.N. F12 4765, Inspecteur divisionnaire of Lille to Min. Agric. Comm., 3 April 1880.

[12] A.D.N. F12 4765, Prefect of Nord to sub-prefects and mayors of Nord, 10 Nov. 1882.

later claimed that compulsory education had ruined the lace-making schools of Bailleul, more workshops than schools, but lace-making was already declining in the face of the mechanized tulle industry of Calais and Caudry.[13] It was nevertheless imperative that child-labour legislation be dovetailed to the official school-leaving age of thirteen. Nothing was achieved until the law of 2 November 1892 banned child labour in industry unless the child in question was equipped with the *certificat d'études primaires*, established under the law of 1882. No provision was made for agriculture, and indeed nothing could be done about preventing the peasant family working as a unit. In the Causses around Le Vigan, the children of peasant families with insufficient land to provide occupation or sustenance were hired out as shepherds for the summer months, from the age of ten or less, for the paltry sum of fifteen or twenty francs.[14] Moreover, setting the *certificat d'études* as a hurdle was itself counter-productive. Once this modest certificate had been obtained, the constraints of compulsory school attendance fell away. As with the *baccalauréat*, families grew anxious that it should be attained as soon as possible, which was frequently at the end of the *cours moyen* of the elementary school, at the age of eleven. The *certificat d'études* replaced, or at least reinforced, First Communion as the rite of passage which marked the end of childhood and schooling. Inspectors of Academies never ceased to complain that too few pupils continued their studies in the *cours supérieur*, up to the age of thirteen, with the result that by the time they came up before the military commissions at the age of twenty many had subsided into complete ignorance.[15]

Finally, school attendance was affected by the cyclical movements of the economy, notably the agricultural depression that set in during the last two decades of the nineteenth century. Between 1896 and 1906, the population of Ille-et-Vilaine fell by 1.6 per cent, evidence of the rural depopulation

---

[13] Ardouin-Dumazet, *Voyages en France*, 18 (Paris–Lille, 1899), pp. 185–6.

[14] A.D. Gard 1T 877, primary-school-inspector of Le Vigan to Insp. Acad. Nîmes, 31 July 1902.

[15] A.D. Gard, Conseil général, report of Insp. Acad. Nîmes, 26 June 1888; A.N. F17 11926, Insp. Acad. Rennes, to rector of Rennes, 5 Sept. 1894; A.D. Nord 1N 152, Conseil général, report of Insp. Acad. Lille, 22 Aug. 1908.

towards Paris. The impact on school attendance was, nevertheless, much greater, as it fell during the same period by 8.3 per cent. The Inspector of the Academy of Rennes was in no doubt about the cause in 1913, as he described 'low attendance, explained above all by the labour crisis'.[16] The population was declining, the exodus was above all rural, and of young people. As the *manque des bras* grew more severe in the countryside, so the premium on child labour in the fields increased, and the class-rooms of the village schools grew emptier.

Besides the loss of child labour, basic wretchedness was an obstacle to full attendance. Children who came ragged and shoeless to school were often treated as outsiders by their better-dressed comrades, and subsequently they might not come at all. Since their parents were unable to afford books, pens, and paper, they sat uselessly at the back of the class, wasting their own time and disturbing the concentration of others. And their poverty was not helped by the fact that their parents, by sending them to school, had forfeited the few *sous* they might have earned in the fields or in the streets. It is significant that the campaign to promote school attendance among this residue of the poor was started by private initiative. From the accession of the republicans to power there spang up, here and there, *sociétés du sou des écoles laïques*, organizations of professional men, merchants, entrepreneurs, journalists, commercial travellers, and *employés*, shopkeepers, and café-owners, dedicated to raising funds from private subscriptions, charity fêtes, concerts, lotteries, and street processions, in order to increase attendance at lay schools. Distributing clothes, school materials, and savings books in reward for assiduity and even compensation to families for the loss of their children's labour, their efforts were aimed above all at the indigent classes. Their activities, fitting into the patterns of sociability of the Midi or the northern industrial towns, were usually inspired by political motives. The Société du Sou of Nîmes, founded in 1878, claimed to have 'prepared the way for the laicization' of the municipal schools, and after 1881 to have promoted 'attendance of the newly regenerated schools'.[17]

---

[16] A.D. I-et-V. 1N, Conseil général, report of Insp. Acad. Rennes, 1913.

[17] A.M. Nîmes, R 265, Council of Société du Sou des Écoles laïques to mayor of Nîmes, 24 Oct. 1882.

Another cluster of *sociétés du sou*, set up in the coal-basin of the Cévennes in the Combes era, was founded to combat the private schools 'supported by the rich and clerical families of region, and' more powerful still, by the mining companies'.[18]

Such societies were established in only a small proportion of communes, usually in the larger centres. The official instrument of the promotion of school attendance, theoretically required to be established in every commune under the law of 28 March 1882, were the *caisses des écoles*. In some areas, like the poor Breton *arrondissement* of Redon, they performed an invaluable function. Unfortunately, such subsidies to pauper children had the effect, as the Inspector of the Academy of Rennes put it, of restoring 'the division into indigents and non-indigents that the law of 1882 had expressly intended to abolish'.[19] This division was again apparent when *cantines scolaires* or free school dinners were introduced in the larger towns for the poor. The idea was not new. There is evidence that even before the Guizot law, the Frères de Ploërmel instituted a *trempage*, or soup, into which the school children, travelling from the extremities of the large Breton communes to school, would be able to dip the bread they had brought with them, in order to save the time and trouble of returning home.[20] This was nevertheless a service that had to be paid for. In the *salles d'asile*, soup-kitchens were a familiar feature, especially during hard winters, though it was only the poorest families who were excused payment. Free school dinners in the *salles d'asile* were not provided at Rennes until the radical municipality of Edgar Le Bastard in 1882.[21] An Œuvre des cantines scolaires des écoles laïques was set up at Lille in 1893, but its benefits were confined to those in receipt of poor relief. The socialist mayor of Roubaix and his deputies who introduced a motion into the Conseil général of the Nord, of which they were all members, that 50,000 francs be

[18] A.D. Gard 1T 692, sub-prefect of Alès to prefect of Gard, 11 Mar. 1909.
[19] A.D. I-et-V. 1N, Conseil général, report of Insp. Acad. Rennes, 1903, p. 632.
[20] A.F.P. 80, La Mennais to Min. Instr. Pub., 22 Mar. 1831.
[21] A.M. Rennes R 27, report of Le Bastard to municipal council of Rennes, 27 May 1882.

made available for the *cantines scolaires* of their town, met with no response.[22]

The subsidizing of *caisses des écoles* and *cantines scolaires* from public funds was itself a political issue, because these funds were directed uniquely to municipal schools. In Ille-et-Vilaine, a running battle took place between those who believed that all the needy should receive relief, irrespective of whether they attended a lay or Church school, and those who followed the principle that public monies should be devoted only to public schools, a principal embedded in republican thinking since 1886. A petition gathered by the conservative notables of the Comité de l'Union national in 1896, calling for the sharing of aid between public and private schools alike, was turned down by the municipality of Rennes early in 1897. But it was a conservative alliance of moderate republicans and *ralliés* that won the municipal elections of 1900, and mayor Pinault opened the flow of aid towards the private schools. Yet when the conservatives were replaced in 1908 by a radical council under Jean Janvier, the grant was once again confined to public schools, on the ostensible ground that the issue was one of education not of poor relief, but in fact the purpose was to end subsidies to private schools rivalling the *écoles laïques*.[23]

The republicans saw it as their mission to propagate enlightenment all the way down the social scale. Before the Third Republic, instruction in elementary schools was handicapped by the archaic nature of pedagogy. There was none the less a growing awareness of the need for new teaching methods, an awareness that filtered through to school-inspectors and to a large section of the teaching body by the end of the nineteenth century. In the first place, it was felt that schoolchildrem had become mystified by words, whereas they should be put in touch with things. Learning should be intuitive, through the senses, based on observation and experiment. Geography textbooks listing the rivers of the world should be thrown away and replaced by

[22] A.D.Nord 1N 136, Conseil général, motion of socialist councillors of Roubaix, 22 Aug. 1892.

[23] A.M. Rennes R 12, dossier of debate on relief for poor schoolchildren, 1896-1911.

wall-maps. The *leçon des choses* was introduced, starting with palpable sticks, stones, and shells in order to unveil the secrets of the world, and *musées scolaires* were built up to provide the apparatus of the new learning. School walks and visits to farms and factories became in vogue, although the *instituteur* of Navacelles (Saint-Ambroix) who planned to spend most summer afternoons in the open air, teaching agriculture, science, geography, and history from the sur-roundings was restrained by the primary-school-inspector, given the need to maintain discipline and a rigid timetable.[24] Secondly, emphasis had to be placed, right from the infant school, less on exploiting the memory, to permit the force-feeding of facts, more on the development of the mental faculties, awakening the curiosity and (for the first time) the imagination of the child.

There can be no doubt that pedagogic methods did im-prove towards the end of the nineteenth century. On the other hand, there were always limits. The persistence of patois in both country and town meant that French was still an alien language, the mechanics of which had to be mastered slowly and painfully. 'In some schools', reported the primary-school-inspector of Uzès in 1897, 'the study of spelling and the recitation of grammar make up almost the totality of the teaching of French.'[25] Little time was left for the self-expression tested by French composition. Again, school-inspectors were ever critical of the rigid and old-fashioned methods employed by many teachers. Above all, teaching was too *livresque*. The book was 'constantly riveted to the hand of the master', who was unable to unshackle himself from the text and launch into personal exposition; pupils were eternally occupied copying passages from the page. Similarly, too much time was devoted to learning by heart and recitation. Children consequently became 'repeat-ing machines, incapable of research on their own' and left school 'their brains full but not formed'.[26] Finally, the

[24] A.D. Gard 1T 814, *instituteur* of Navacelles to primary-school-inspector of Alès, 1 July 1912; school-inspector to Insp. Acad. Nîmes, 3 July 1912.

[25] A.D. Gard 1T 720 *bis*, primary-school-inspector of Uzès to Insp. Acad. Nîmes, 1 June 1897.

[26] A.D. Gard 1T 724, primary-school-inspector of Le Vigan to Insp. Acad. Nîmes, 10 May 1901, 10 May 1902.

certificat d'études concentrated pupils heavily on the subjects required for the examination, largely arithmetic and French, at the expense of history and geography, moral and civic education, and all the sciences, which together accounted for only two marks out of six, and gave rise to a 'veritable epidemic of cramming', a hothouse atmosphere detrimental both to the mind and body of the schoolchild.[27]

Much of what the republican leadership said about education remained at the level of rhetoric, if not of hypocrisy. Like the regimes that had gone before, they were interested in instruction essentially as a means to education. They differed in that their talk involved less direct references to work-discipline, moralization, and cleanliness, and was both more subtle and more sanctimonious. The Opportunists were not, of course, interested in *revanche*, and it was for this defeatism that Radicals and Boulangists berated them. They were interested only in peace, prosperity, and protection. On the other hand, a little sabre-rattling in Tonkin or Tunisia and the liberal distributionof school-prize books on Jean Bart and Joan of Arc, on the war of 1870 and the Dark Continent, could maintain a semblance of national integration, while compulsory gymnastics at least had the benefit of giving youth 'habits of order and discipline'.[28] Again, a civic education that underlined that equality meant above all political equality might conceal deeper social divisions and disguise the fact that power had been concentrated in the hands of a bourgeois oligarchy. Anticlericalism served as a useful smoke-screen, reinforcing class unity by portraying the Catholic Church and a broken-down royalist squirearchy as the real enemies of the Republic, though the limits of anticlericalism became very clear when the social question became more serious in the 1890s. Meanwhile, despite attacks on the obscurantism of the catechism, the republicans were in no way prepared to sacrifice social morality, cemented henceforth not by fear of God but by a sentiment of obligation to the community, the key to which was *devoir*. The emptiness

[27] A.D. Nord 1N 135, Conseil général, report of Insp. Acad. Lille, 1 Aug. 1891, p. 539.
[28] A.N. F17 9162, circular of Jules Ferry to Rectors of Academies, 20 May 1880.

of rhetoric about the 'fusion of classes' was amply demonstrated by the explosion of strikes that greeted the republican coming to power. The miners of Anzin were on strike in 1878, while in the spring of 1880 40,000 textile workers of Lille, Roubaix, and Tourcoing struck in favour of a ten-hour day, dragging out thousands of building and metal-workers in their wake.[29] In the event, republican education concentrated less on success than on hard work and the inculcation of bourgeois values. Books distributed as prizes to the primary schools of Alès in 1879 included such titles as *Conseils aux ouvriers, La Petite Jeanne et le devoir, Maurice ou le travail, Petit Pierre ou le bon cultivateur.*[30] The propagation of *caisses d'épargne scolaires*, or penny banks, based on the example of Britain and Belgium, was an important development at the end of the 1870s, and 577 of them were established in the Nord by 1877. But as far as the Inspector of the Academy of Rennes was concerned, the aim of such banks was not to encourage social promotion but to inculcate the drinking classes with 'habits of order and thrift'.[31]

Faced by the rising threat of socialism in the early 1890s, republican notables developed a coherent social ethic, which, shunning conflict, preached the essential unity of society. This was the ethic of solidarism. Whereas for the Catholics a stable social order was the reflection of the divine order and obedience to earthly magistrates an extension of obedience to God, for the exponents of solidarism the coherence of society was based not on religion but on science. Nascent sociology revealed that men were interdependent by the mechanism of the division of labour, and this social fact reinforced the notion of the subordination of the interests of the individual to those of society.[32] The main function of education became the inculcation of the social instinct and the sentiment of *devoir*. In 1895 the Minister of Public

[29] M. Perrot, *Les Ouvriers en grève, 1871-1890* (Paris-The Hague, 1974), i, 89-91.

[30] A.D. Gard 1T 513, List of prize-day books, 1879.

[31] A.D. I-et-V 1N, Conseil général, report of Insp. Acad. Rennes, 1889, p. 433.

[32] Durkheim's *De la division du travail social* was published in 1893. See M. Ruby, *Le Solidarisme* (Librairie Gedalge, 1971).

Instruction, Poincaré, sent a circular to members of *délégations cantonales, commissions scolaires*, and *caisses des écoles*, asserting that the first phase of the development of primary education, concerned with legislation and administration, was over. The school was 'a sort of national workshop in which the France of tomorrow is being forged', and the second period would see its 'natural radiation over the country by means of a host of voluntary works founded on the initiative of good citizens', school libraries and evening classes, *sociétés du sou* and *patronages laïques*.[33]

The ideology of solidarism manifested itself in the schools after 1895 in three principal areas: mutual insurance, hygiene, and temperance. They were calculated to raise the prosperity, health, and, above all, the morality and dignity of the race. *Mutualités scolaires* were developed on a town or canton level to provide insurance against doctors' bills, funeral expenses, and, it was hoped, to guarantee a capital for retirement. The first rural *mutualité scolaire* was founded in the Gard at Saint-Laurent d'Aigouze, as Protestant as neighbouring Aigues-Mortes was Catholic, by a Protestant *instituteur* of the Cévennes, Émile Lombard, who christened it 'La Ruche'. Set up in March 1896, it was followed by 'l'Enfance prévoyante' at Nîmes. A *mutualité scolaire* was founded at Rennes at the end of 1896, and similar institutions flourished in the textile region of Lille–Roubaix-Tourcoing and in the coal-basin of Douai–Valenciennes after 1897. The Directeur départmental de l'enseignement primaire of the Nord noted that the *mutualités scolaires* were taking over from the *caisse d'épargne scolaires*, which halved in number between 1888 and 1898. But, he reflected, this was no bad thing, since the *mutualités scolaires* 'teach not only thrift, but solidarity. Children learn to give as well as to save, to help themselves and to help others as well.'[34] It was the difference between the bourgeois ethic of thrift and self-help of the 1870s and the liberal collectivism of the 1890s.

Concern for public health also intensified in the last years of the century. The school law of 30 October 1886 provided

[33] Circular of Poincaré, 10 July 1895.
[34] A.D. Nord 1N 143, Conseil général, report of Directeur départemental du Nord, 1899, pp. 254–6.

for medical inspectors, nominated by the prefect and under the supervision of the *conseils départementaux d'hygiène*, to inspect the salubrity of school buildings, both public and private, and the health of the schoolchildren. However, no public funds were allocated for this service, so that while eighty-one doctors were appointed in the Gard for school inspection in 1887, by 1906 only one doctor was still active, at Nîmes, financed by the municipality.[35] Poincaré issued elaborate instructions in 1893 for maintaining strict standards of hygiene in schools, evidence of an official appreciation of the contagion theory of disease, and efforts were made to ensure airy and salubrious school buildings and that children were protected against disease by revaccination at the age of eight.[36] Some *instituteurs*, such as Roux of Saint-André de Valborgne in the Cévennes or Letouzé of Louvigné-de-Bais, in Ille-et-Vilaine, were active propagandists in the cause of revaccination, the latter called upon to vaccinate a hundred children himself in October 1905.[37] Yet there were limitations to what could be achieved. The condition of many schoolhouses and schoolteachers' lodgings was deplorable, not least because many municipalities refused to undertake building projects,[38] while in the Nord as a whole the rapidly growing population caused massive overcrowding, with 120 children crammed into a space meant for fifty, which could be of no benefit to their health.[39] Schools were often little better than death-traps which had to be closed by the prefect at the outbreak of epidemics. The École maternelle of Aigues-Mortes was particularly unfortunate. There, according to the inspectress, 'epidemics rage continuously', with whooping cough in December, January, and February, measles in February, March, April, and May, and bronchitis and marsh fever added for good measure.[40] The nightmare

[35] A.D. Gard 1T 1193, Insp. Acad. Nîmes, to Directeur de l'Enseignement primaire, 27 July 1906.

[36] A.N. F17 11781, *arrêté* of Min. Instr. Pub., 18 Aug. 1893.

[37] G. Dodu, *L'École laïque en Pays Breton*. Lecture to Union républicaine des étudiants rennais, 15 Mar. 1912, p. 19.

[38] A.D. Nord 1N 127, Conseil général, report of prefect of Nord, April 1883.

[39] A.D. Nord 1N 153, Conseil général, report of Directeur départemental, 19 Aug. 1909.

[40] A.D. Gard, Conseil général, report of Inspectrice des écoles maternelles du Gard, 12 July 1911.

of the turn of the century was, nevertheless, tuberculosis. In the Gard, the children of the *écoles maternelles* were said in 1906 to be 'sickly and pale, their limbs slight, their blood poor, their lungs torpid . . . they are ready victims of tuberculosis, if they are not already infected'.[41] In the Nord, the fear was repeatedly expressed that the anaemic girls recruited to the École normale d'institutrices of Douai carried the germs of tuberculosis, and deaths from the disease, either in the *internat*, or in later employment, were frequent.[42]

The obsession with tuberculosis was paralleled by an obsession with alcoholism at the turn of the century. Indeed, the ravages of alcoholism were said, among other things, to undermine the body and expose it to consumption. The fight against alcoholism, involving as it did the wild expenditure of hard-earned wages at the cabaret, was also a campaign for thriftiness and the solidarity of the family. Children themselves must be fortified against this vice, and not out of any abstract moralizing but because, as one *instituteur* in the Nord calculated in 1898, only one in ten of his pupils had never drunk strong liquor. All means of propaganda were brought to bear in the schools. The *instituteurs* of the Nord, meeting in *conférences pédagogiques* in October 1897, were told to use every opportunity to insist on temperance, whether in lessons on morality, or arithmetic, or natural history.[43] Class-rooms were covered with posters, engravings, and maxims, dry reviews such as *L'Alcool* and *L'Étoile bleue* were subscribed to, and branches of the Société contre l'usage des boissons spiritueuses, the members of which would take the pledge, were set up. It would be rash to suggest that Protestantism was a major influence on this puritan moralizing. Nevertheless, such clubs in the Gard were found at Nîmes, Sauve, Saint-André de Valborgne, Aulas, near Le Vigan, Saint-Laurent d'Aigouze, and in the Vaunage at Vergèze and Congénies, all dominated by Protestants. In the Catholic mining commune of La Vernarède, the *instituteur*

[41] A.D. Gard, Conseil général, report of Inspectrice des écoles maternelles du Gard, 31 July 1906.

[42] A.D. Nord 2T 1163[3], *directrice* of École normale d'institutrices to Directeur départemental, 8 June 1902, and 1163[4], 15 May 1906.

[43] A.D. Nord 1N 142, Conseil général, report of Directeur départemental, 1 Aug. 1898, pp. 232–3.

was powerless to fight against endemic drunkenness and the gaming and rowdiness that went with it.[44]

A special role was singled out for women in this enterprise. Not that women had anything to gain from the republicans: republicans, like conservatives, looked to women as a force for continuity and stability in society, and, while countenancing female labour from an early age, preached that the proper place for women was in the home. In 1902 a motion placed before the Conseil général of the Nord by the doctors and industrialists representing the cantons of Valenciennes, Condé, and Saint-Amand demanded that greater time be given to *enseignement ménager* or domestic science in the schools. Girls should be taught to sew, mend and make clothes, to wash and iron linen, to cook, to be 'thrifty and tidy . . . and to realize that well-being and order that makes a house into a home, enticing the husband to remain by the hearth'.[45] The demands of thrift and hygiene would be satisfied and, according to the common assumption of the time, the husband provided with a happy home would not feel the necessity to desert his family for the cabaret. If industrialization had torn the family apart, education, it was felt, could reconstitute it and, with that, promote the solidarity of society. *Enseignement ménager*, reflected the Inspecteur général de l'enseignement technique in 1909, 'by putting the woman in her rightful place, by demonstrating to her her duties as housewife, wife and mother, will contribute to the reign of harmony in the family'. And this education, 'by strengthening the family, will increase the worth of society, which is only the sum total of the energies of families, "true social molecules" '.[46]

A study of the social ideology of the Third Republic must take into account not only the evidence of the class-room but also, since the evils of premature school-leaving had not been exorcised, that of a whole congeries of post-school institutions that made up a veritable Republican Church. Again, the

[44] A.D. Gard 1T 753, *instituteur* of La Vernarède to primary-school-inspector of Alès, 2 Nov. 1904.
[45] A.D. Nord 1N 145, Conseil général, motion of councillors of Valenciennes, N., E., S., Condé and Saint-Amand R.D., 19 Aug. 1901.
[46] A.D. Nord 1N 153, Conseil général, report of Inspecteur général de l'Enseignement technique, 23 July 1909.

revival in this field did not occur until 1895. For the first twenty years of the republican dominance, post-school *œuvres*, the school libraries and evening classes, were continuing under the momentum given to them by Victor Duruy. Even then, the achievement of these engines of war on *mauvaises lectures* and the cabaret was limited. School libraries, intended as stocks of improving reading for rural populations, too often contained old and tatty volumes totally lacking in appeal, while the lots sent out by the Ministry of Public Instruction—on ethics, science, agriculture, and political economy—had a habit of remaining unread on the shelves. Popular taste was for novels, those of Mayne-Reid, Canon Schmidt, Erckmann-Chatrian, and Jules Verne, to be found more plentifully in the *bibliothèques populaires* founded by private societies. Such libraries, which often used *estaminets* as their base and inclined to republican principles, tended to invite the wrath of the Moral Order regime, notably during the hard-fought elections of October 1877, but multiplied rapidly after this, enjoying the patronage of masonic lodges, Protestant pastors, and republican circles. In 1881, a Société Diderot was founded under the presidency of Alfred Giard, Professor of the Science Faculty of Lille, who would be elected as radical-socialist deputy for Valenciennes the following year and would show Zola round the Anzin mines in 1884. Its task was to propagate popular libraries in the countryside and towns around Lille, fighting the *bons livres* of Paul Féval, Henri Conscience, and the Comtesse de Ségur which combined a 'pompous ignorance of the needs of our modern society' with a 'missionary hatred of our republican institutions', by spreading 'a few works conceived in a broader, more scientific and more liberal spirit'.[47]

Evening classes continued to multiply, especially after the end of the Moral Order, and in large centres such as Lille and Rennes popular universities, modelled on the Association philotechnique founded in Paris in 1848, were set up. In these Mechanics' Institutes, academics, *lycée* masters, and members of the liberal professions lectured on mathematics, architecture, natural history, political economy, commercial

[47] B.M. Lille, Fonds Humbert 37, press cutting, July 1881.

law, and elementary medicine. The audience of the Société d'instruction populaire of Rennes was said in 1883 to belong to every class of society and every age-group, including 'young people and workers'.[48] It is nevertheless clear that the level of the courses exceeded the capacity of all but a knowledgeable élite of the working class. Meanwhile, a governmental decree of 22 July 1884 delivered a death-blow to rural evening classes. By requiring that the courses should run for five months in the winter and that every enrolled member should attend at least fifty times before any government funds would be forthcoming, it made the courses unviable. In 1894, the Inspector of the Academy of Rennes reported that 'these courses have all but disappeared in the department of Ille-et-Vilaine'.[49] In the Nord, where municipal funds were greater, there were still 354 courses that year, but in Ille-et-Vilaine there were forty-five and in the Gard only thirty-one.

A decree of 11 January 1895 reopened the flow of state credits for evening classes and within a year they numbered eighty-one in the Gard, 151 in Ille-et-Vilaine, and 757 in the Nord. This was only the beginning of a growth that would continue unabated for the next five or ten years. A survey of 1895 in fact revealed many doubts in the localities about the viability of such courses, and repeatedly the advantages of towns over the countryside were demonstrated.[50] One problem was financial: communes were asked to pay the heating and lighting costs of evening classes, but many rural communes did not have resources available and any attempt to require a subscription from the clientele had the effect of reducing attendance, so engrained had the concept of free education become. Again, in the towns, it was possible to draw on the support not only of *instituteurs* but also of college and *lycée* masters—though tension could and did arise between the two levels of education—academics, magistrates, civil servants, army officers, and doctors. In the

---

[48] A.N. Rennes R93, report of Chatel, Professor of Law Faculty, President of Committee of Société d'instruction populaire to mayor of Rennes, 1883.

[49] A.N. F17 11824, Insp. Acad. to Rector of Rennes, 5 Sept. 1894.

[50] The replies to the inquiry of 1895 are in A.N. F17 11920 for the Nord, 11921 for the Gard, and 11926 for Ille-et-Vilaine.

countryside the *instituteur* was obliged to shoulder the burden of evening classes alone, and that, on top of seven hours'
teaching a day, two hours' preparation, the business of the
mairie and so on, could be overwhelming. Beyond such
problems of organization was the problem of access to a poor
and often semi-nomadic population, absorbed by the neces-
sities of labour. Farm-hands who lived in the 'gilded servi-
tude' of being 'logé, nourri, blanchi' were not released by
their employers to attend evening classes. Attendance was
affected also by the seasonality of work. The cod-fishers of
the Rance estuary did not return from the banks until
November, nor did the *camberlots* of Iwuy come back from
the wheat-plains of Beauce to manufacture chairs and knives
until after the harvest. The masons of Wannehain on the
Belgian border left for Lille every week and did not return
to their families before Sunday. The populations of the urban
villages of French Flanders, who had once been employed as
hand-loom weavers for the great mill-towns, were now
obliged to travel daily to the textile centres, returning late
in the evening. In the mining and metallurgical centres of the
Nord and the Gard, one week's night work alternated with
one week's day work, making regular attendance impossible.
Moreover, released from work and journeying several kilo-
metres home, most of this floating population could think of
nothing other than repairing to the cabaret for the rest of the
evening. The wine-growing proletariat of the Rhône Valley
was no different. The *instituteur* of Roquemaure predicted
that no evening class would succeed there, 'the *roquemauroise*
youth preferring before study, the café, dancing, and other
pleasures'. Finally, there was often a political opposition to
evening classes in clerical or reactionary areas. Conservative
communes could frustrate classes by refusing to vote credits,
while at Saint-Erblon, south of Rennes, the curate told the
children attending catechism in 1896 that 'since the evening
class opened, Saint-Erblon has become quite barbarous. I
am therefore keen to warn you that if any one of you attends
that class, he will be pitilessly expelled from catechism and
will not take his communion.'[51] The geography of evening

[51] A.A. Rennes C5, cited in prefect of Ille-et-Vilaine to archbishop of Rennes,
17 Mar. 1897.

classes in Ille-et-Vilaine in 1895–6 was essentially that of the 'blue' communes.

Where evening classes were satisfactorily launched, there was no guarantee that they would reach the right audience. They served, in the first instance, as adult literacy classes, providing for adolescents an elementary education that had been forgotten since school-leaving at an early age or that had never really been grasped. Yet there was a *bas-fonds* of illiterates who remained outside the pale of literate culture precisely because they were so illiterate. The primary-school-inspector of the Breton *arrondissement* of Montfort noted that most of the recruits of the evening classes either had the *certificat d'études*, or were of a level to obtain it, but 'almost all those whose education is notoriously inadequate do not even think of using them'.[52] At Flines-les-Raches, near Douai, the *instituteur* reported that 'the population is ignorant and cares little about instruction'. The revelation of a conscript's illiteracy by the military commission might prompt him to study, if only because the army would drill him into reading. The army itself was in fact anxious to take advantage of the *instituteurs'* expertise. The *élèves-maîtres* of the École normale of Nîmes were engaged to teach illiterate soldiers in the barracks opposite in 1899, while the general commanding the Tenth Army Corps enlisted *instituteurs* into the garrisons of Ille-et-Vilaine—Rennes, Saint-Malo, Saint-Servan, and Vitré—in 1905.[53] A basic threshold of literacy among the labouring population was required before it would take any interest in evening classes. Moreover, it was the eminently practical function of the classes that was attractive. At the simplest level, it offered an initiation into the mysteries of legal–bureaucratic society: the study of farm leases and rent receipts, the art of writing to request employment, tax-rebates, insurance, or exemption from military service, the way to calculate surface areas and profits. Besides that, classes in towns provided a training in technical drawing, surveying, and accountancy. Lastly, although the Breton

---

[52] A.N. F17 11825, report of primary-school-inspector of Montfort, 16 April 1896.

[53] A.D. Gard 1T 750, director of École normale d'instituteurs of Nîmes to Insp. Acad. Nîmes, 31 Jan. 1899; A.D. I-et-V. 1N, Conseil général, report of Insp. Acad. Rennes, 1905, p. 826.

peasant was said to be indifferent to the evening class because it 'does not train accountants or notaries or even ordinary clerks',[54] in fact, classes at the school of the Halle aux Toiles in Rennes and other large centres did coach for professional exams giving access to the customs, *octroi*, Post Office, the Ponts et Chaussés, the master mariner's ticket, *gendarmerie* and military school of Saint-Maixent, through which NCOs could be promoted to the rank of officer. On the other hand, evening classes were not the only institution that provided such practical facilities. In the Nord, for example, similar courses were offered by *écoles profession-nelles* such as that of Fourmies, the railway workshops of Hellemmes–Lille for their employees, and the *cours acadé-miques* of drawing, architecture, modelling, sculpture, and painting of Douai.[55] Just as school libraries never secured a monopoly of popular reading habits, so evening classes never conquered the field of post-school training.

Many evening classes concerned with promoting basic literacy ended with a reading or informal talk by the *institu-teur*: education followed instruction. With the revival of evening classes after 1895, a concerted attempt was made to develop these talks in their own right, as *conférences popu-laires*. Through these lectures, more moralizing than practical, it is possible to grasp the ideology of the Third Republic as it was transmitted to the peasants, artisans, and *employés* who, together with elements of the middle class in towns, constituted the bulk of the clientele. The creation of national unity around the republican standard was the *arrière-pensée* of the lectures. Much was said about the war of 1870, the brutal behaviour of the Prussians, and the heroic siege of Belfort. The Ancien Régime was compared unfavourably to the post-revolutionary era under such titles as 'Peasant life under the feudal regime' and in histories of the Bastille, while it was the national rather than the divisive aspect of the Revolution that was emphasized: the Tennis Court Oath, the Fête de la Féderation, Valmy, and Hoche. In a period of rapid

[54] A.D. I-et-V. 2T *Cours d'adultes*, mayor of Saint-Ouen-la-Rouerie to prefect of Ille-et-Vilaine, 26 May 1896.

[55] A.N. F17 11920; A.D. Nord 2T 1608, report to general assembly of *professeurs* of Douai, 10 Jan. 1895.

imperial expansion, a taste for exotic places, voyages, and discovery combined with sentiments of jingoism to rally populations around the French Empire: lectures on Tonkin and Tunisia, the Franco-Russian alliance and Turkish atrocities, Madagascar in the light of the expedition of 1895, and China at the time of the Boxer rising attracted enthusiastic audiences, even in the depths of the countryside. And it was the need to create a race fit for imperial rule that gave meaning to lectures on thrift, hygiene, and temperance, sometimes dressed up as vulgarized science about Pasteur and germs. National efficiency was as important a concept in France as in Britain.

Popular lectures required a capital of subject-matter, speakers, and, not least to swell the audience, slide-projectors. Some *instituteurs*, like Théodore Chalmel of Saint-Père-Marc-en-Poulet, by the Rance estuary in Brittany, or Henri Roux of Sauve in the lower Cévennes, were talented and indefatigable lecturers, and local historians in their own right to boot.[56] Other *instituteurs* needed more props. For this reason the Ligue de l'Enseignement founded a Société nationale des conférences populaires in 1890 which prepared printed lectures and sent out speakers from local sections of the Ligue, not only to nearby towns but also into the countryside.[57] On a more reduced level, there sprang up societies such as the Comité des conférences populaires of Douai, a group of municipal councillors, publicists, barristers, and *professeurs* set up in 1896 to make projectors, slides, and lecturers available to *instituteurs* of the town and surrounding region.[58] In larger centres, a more ambitious campaign was launched in 1899 by noted academics and professional men to establish *sociétés pour l'éducation populaire* or *universités populaires*, along the lines of the Association philotechnique. As before, however, the nature of the lectures was too elevated to attract working-class

[56] A.D. I-et-V. 1F 1768, Fonds Chalmel, *dossier d'un instituteur*, 1896-1901; A.D. Gard 1T 751, list of lectures by Henri Roux, 1896-1901.

[57] C. Mora, 'La diffusion de la culture dans la jeunesse des classes populaires depuis un siècle: l'action de la Ligue de l'Enseignement', in *Niveaux de culture et groupes sociaux* (Paris–The Hague, 1967), p. 257.

[58] A.D. Nord 2T 1608, Desmarets, municipal councillor of Douai, to Directeur départemental, 3 Jan. 1897.

audiences. The Université populaire of Alès was attended mainly by *instituteurs*, pupils from the Lycée, teachers from the Collège de Jeunes Filles, and girls and ladies of the bourgeoisie.[59] As the Inspector of the Academy of Nîmes reflected in 1903, 'neither the expositions of the lecturers, nor their talent or good will have lacked, but the people have not come to hear them. The lectures were attended by a group of people who had absolutely no need of them. The working man revealed himself more sceptical and increasingly confined to drinking and the tavern.'[60]

Besides failing to make contact with working people, the *universités populaires* tended, on occasions, to become involved in politics. At Lille, Charles Debierre, Professor of the Medical Faculty and radical-socialist leader, created scandal when he lectured in 1900 on 'The Brain and Thought', denying the immortality of the soul. A local Protestant pastor was quick to denounce such 'materialist and socialist propaganda . . . the revolutionary and masonic character of the enterprise'.[61] Again, when the Université populaire organized lectures in the *arrondissement* of Dunkerque in 1904, it was accused of seeking to stir up a radical-socialist current that would overthrow first the conservative municipality of Dunkerque, and then the conservative deputy.[62] This clearly had nothing to do with the official aim of the *universités populaires*, which was to restore class harmony and social solidarity. According to Gaston Dodu, Inspector of the Academy of Rennes, success would be achieved by 'the lowering of barriers raised up by egoism and that most deplorable of misunderstandings between those who work with their hands and those who work with their minds'; all would now participate in 'the common religion of humanity'.[63] In Ille-et-Vilaine, the main instrument of popular education was the Cercle républicain départemental de l'enseignement laïque, founded in 1903 by various officials

[59] A.D. Gard 1T 751, Poisson, *professeur* of Lycée of Alès to Insp. Acad. Nîmes, 9 July 1899.
[60] A.D. Gard 1T 752, Insp. Acad. Nîmes to primary-school-inspector of Alès, 8 May 1903.   [61] *La Dépêche du Nord*, 21 Dec. 1900.
[62] A.D. Nord 1T 110, sub-prefect of Dunkerque to prefect of Nord, 27 Jan. 1904.
[63] G. Dodu, *Guide pratique de l'éducateur populaire* (Paris, 1901), p. 27.

of the Academy, academics, magistrates, journalists, and employers in order to provide speakers, slides, and circulating libraries for the Breton countryside and towns. Poitrineau, Dodu's predecessor at the Inspection Académique, summed up the task of the Cercle républicain as being 'to draw together social classes by reducing the differences of intellectual culture, and to fortify in men's souls sentiments of solidarity, human fraternity and a deep love of *la patrie*'.[64] It was of the essence of solidarism that it should see inequality of instruction as the most intolerable of all equalities, and imagine that class differences could be bridged by propagating an ideology of unity.

Both school libraries and evening classes were old institutions taken up and developed by the notables of the Third Republic. The third element of the post-school apparatus, the *patronage*, had not previously been exploited by the state; on the other hand, it was an organization that had been pioneered since the Restoration by the Catholic Church.[65] It was nevertheless during the last years of the century that both Catholics and republicans set about multiplying *patronages* to an unprecedented degree. In 1888, it could be reported that 'this word *patronage* intrigues the population of Le Cateau'. The clergy explained that this youth club, run on Sunday evenings by the parish curate and the Frères des Écoles Chrétiennes, bringing together sixty-five former pupils of the Frères with five from the municipal college and twenty from the lay primary schools, was to 'preserve the child from the influences of the street, to keep him within reach'.[66] The concern of the Church to safeguard the morals of adolescents during the dangerous period between First Communion and military service (itself fraught with peril) was traditional. In addition, the *patronage* was a powerful weapon of propaganda, a means of recruitment to Catholic circles of various descriptions in later life, and a means of countering the influence of the 'école sans Dieu'.

---

[64] *Bulletin du Cercle républicain départemental de l'enseignement laïque*, speech of Poitrineau at Banquet of 19 June 1904, p. 20.

[65] See above, pp. 244–9.

[66] A.D. Nord 2T 283/7, principal of College of Le Cateau to Insp. Acad. Douai, 14 Mar. 1888.

The dominance of the Catholic Church in the realm of *patronages* was difficult to dispute. At Houplines, adjoining Armentières, it was said in 1895 that it would be impossible to found an evening class because 'the clergy, all powerful here, has created *patronages* where they monopolize young people of all ages'. At Nîmes, there were three Catholic *patronages* in 1897, one run by the secular clergy, one by the Jesuits, and a third by the Frères des Écoles Chrétiennes, complete with study-group and lectures, sport and gymnastics, and a weekly publication, *Les Tablettes de la Jeunesse Catholique Nîmoise.*[67] In La Grand'Combe and its suburbs, the mining company came to subsidize not only *patronages* such as the Cercle de la Jeunesse Catholique and the girls' Patronage 'Jeanne d'Arc' but also choral societies, gymnastic societies, shooting societies, and the Œuvre des Enfants à la Montagne that annually sent two hundred children of company employees on holidays to the mountains of the Lozère or the Ardèche or to the seaside et La Grau-du-Roi.[68]

It was 1895 before any adequate response was drawn from the republicans. In that year Léon Bourgeois, former Minister of Public Instruction and shortly to form a cabinet himself, as President of the Ligue de l'Enseignement called for the organization throughout France of democratic *patronages scolaires*. As he pointed out, 'the moment has come to organize and encourage throughout France the task of democratic solidarity that has been desired for so long by the friends of the Republic'.[69] The anticlerical nature of the enterprise soon became clear. The first lay *patronage* founded at Lille in 1899 by a group of employees and workers was designed to 'induce the children of municipal schools not to attend the *patronages* of the religious congregations any longer and to remove them from clerical influence'.[70] The Comité Lillois de patronages laïques set up in the same year

---

[67] A.D. Gard 1T 750, M. Fourier to mayor of Nîmes, 3 Nov. 1897; A.F.E.C., NC 716/1, 'Histoire des Écoles des Frères, Maison de la Providence, Nîmes, 1817–1928' (M.S., n.d.).

[68] A.D. Gard 18J 801–805, Compagnie houillère de la Grand'Combe, registres de dépenses de prévoyance patronale, 1910–14.

[69] A.D. Gard 1T 749, circular of Ligue Française de l'Enseignement, 3 Apr. 1895.

[70] A.D. Nord 1T 68/10, report of Commissaire spécial of Lille, 8 April 1902.

was said by the socialist mayor, Gustave Delory, to be composed of devoted republicans,[71] and it was Delory, along with
Édouard Delesalle, *négociant* and editor of the socialist *Réveil
du Nord*, and Charles Debierre, who headed the Société des
Patronages du Nord de la France, founded in 1900.[72]

The lay *patronages*, usually the *association amicale
d'anciens élèves* of municipal schools, provided a framework
for the evening classes, talks, and lectures that had previously
taken place on a more or less regular basis. The leisure activities
at which the Catholics were so proficient provided an additional attraction: games, gymnastics, shooting, the staging of
plays and concerts, excursions and holidays, as resources
permitted. From 1912 the federation of Associations des
anciens élèves of Roubaix organized *colonies de vacances*
under the rubric 'Grand Air pour les Petits' for the grimy
children of the mill-town. At best, the *patronage* included
a *mutualité* that provided for members in need and acted
as 'a small-scale Bourse du Travail' using contacts with
employers in order to find jobs.[73] The ethic of the lay
patronages was summed up by the titles of the associations
set up at Le Vigan and at Sommières in the Gard in 1898:
one was called 'La Solidarité' and the other 'Le Devoir'.
The membership was made up of apprentices, shop assistants,
employés, and accountants: wage-earners and peasants were
underrepresented.[74] The moral teaching of the class-room
was carried over into the *patronages*: mutual aid, hygiene,
and sobriety were still the underlining concepts. Indeed, the
Association coopérative et fraternelle of Aulas, near Le
Vigan, set up in 1896, was both a club for *anciens élèves*
and a temperance society, incorporating lectures, a reading-
room, dominoes and lotto, and the provision of 'healthy
and harmless drinks at reduced prices'.[75] The republican
conception of the role of women in society, cast in the most
conservative of moulds, was underlined in the *associations*

[71] A.D. Nord M 222/791, Delory to prefect of Nord, 28 June 1899.
[72] A.D. Nord 1T 68/10, the Société des Patronages du Nord de la France was
constituted on 23 Feb. 1900.
[73] G. Dodu, *Guide pratique*, p. 35.
[74] A.D. Gard 6M 1188, list of members of 'La Solidarité' of Le Vigan and
'Le Devoir' of Sommières, 1898.
[75] A.D. Gard 1T 750, *instituteur* of Aulas to Insp. Acad. Nîmes, 27 Dec. 1896.

d'anciennes élèves. One of the more successful, that of Saint-Hippolyte-du-Fort in the Gard, meeting twice a week under the *institutrice*, offered improving talks and projection evenings, needlework, singing, and games. But the *institutrice* never ceased to preach that the true place of the woman was in the home as wife and mother, and that her rights could be expressed in three words: 'love, devotion, sacrifice'.[76] The *association amicale* of one of the lay girls' schools of Lille was appropriately named 'Les Futures Ménagères'.[77] And the equivalent of the employment service provided by some boys' *patronages* was the accumulation of the effects of a bottom drawer, to be handed over to the girl on her wedding day. At Roubaix, such a Mutuelle dotale pour les jeunes filles was founded in 1908, while at Valleraugue, the Cévenol mountain community, an Œuvre du trousseau was started in 1912 in order to 'stimulate and maintain love of the home'.[78]

The republicans could never, in the end, catch up with the trail blazed by the Catholic Church. Whatever they did, the Church had done it before, and better. The Directeur départemental of the Nord, reflecting on the success of the fifty-eight Catholic *patronages*, twenty Catholic gymnastic and shooting societies, libraries, and *colonies de vacances* that honeycombed Lille in 1913, noted ruefully that 'our post-school works, more serious, less entertaining and less well supported have a much lower attendance'.[79] It was the difference between the Catholic clergyman and the republican notable: the one adept, boyish, with a twinkle in his eye, the other rigid, self-important, and eminently Victorian.

The Third Republic was the golden age of primary education in France. But though the network of public primary schools was complete, there were many fields in which local

---

[76] A.D. Gard 1T 750, speech of *directrice* of girls' school of Saint-Hippolyte-du-Fort, 13 Dec. 1897.

[77] A.D. Nord M 222/770, the Association amicale, 'Les Futures Ménagères', of Lille, constituted 8 Dec. 1900.

[78] A.D. Gard 1T 706, mayor of Valleraugue to sub-prefect of Le Vigan, 22 Dec. 1912.

[79] A.N. F17 10366, Directeur départemental du Nord to Min. Instr. Pub., 12 Feb. 1913.

private initiative counted for much. This was the case with the campaign to support free schooling with the provision of school equipment for poor pupils, and with the campaign to compensate for early leaving by developing evening classes, popular lectures, and youth clubs. Though these initiatives later came under the auspices of the state, they were essentially the work of the republican petty bourgeoisie and arose out of their world of societies, circles, and lodges, and of the Ligue de l'Enseignement.

It was nevertheless the Churches, both Catholic and Protestant, that had blazed the trail. *Patronages*, to safeguard and moralize adolescents by a combination of learning, devotion, and innocent amusement, were very much an extension of the charitable and social works of the Churches. The Republic imitated the way in which they enveloped and controlled the population in order to form its own Republican Church. For the first time the utilitarian approach to social discipline broke through, preaching education as an all-purpose answer to social problems. The self-help of the 1870s, based on the illusion that poverty could be eliminated by hard work, was replaced by the solidarism of the 1890s, seeking to mend the rifts in industrial society by a pseudo-scientific ethic of social interdependence. But evening classes and popular lectures could be multiplied *ad nauseam*. They were attended by the *classes moyennes* and artisan élite rather than by the labouring masses, and what appealed was not moralization but professional qualifications and sheer entertainment.

The Republican Church was a church of little republican notables, self-righteous and philistine. If the bulk of the population did come within its sphere it was thanks not to the implementation of free compulsory education in 1881–2, but to the economic boom of the Second Empire. And compulsory education coexisted with unreformed labour laws, which did little to prohibit the employment of child workers still necessary for industrialists.

8

The Limits of Ambition

In the last years of the nineteenth century important economic and social change altered on many fronts the pattern of ambition. Agricultural depression and industrial recession between 1873 and 1895 darkened prospects in the productive sector. But the shift from the self-made entrepreneur of the era of cut-throat competition to the large firm of the era of protection, cartels, and empire gave rise to a new middle class, that of the cadres or managers. It was nevertheless at the doors to the liberal professions, the bureaucracy, and the army that pressure built up to the greatest extent, creating once more the problem of an excess of educated men. Further down the social scale, economic difficulty similarly provoked a movement towards occupations that were more secure, if not better paid. White-collar posts were multiplying in commerce, industry, and the railways, as were openings for bank or office clerks, shop assistants, draughtsmen, and accountants. Above all, there was *la fonction publique*, whether in primary-school teaching or in the Post Office, customs, tax administration, or Ponts et Chaussées. Work done on Lille shows that between 1873-5 and 1908-10 the proportion of the working class declined from 47.0 per cent to 37.7 per cent of the population, while amongst the *classes moyennes*, that of artisans and shop-keepers rose from 14.2 per cent to 18.9 per cent, and that of clerks and *petits fonctionnaires* increased from 7.5 per cent to 10.3 per cent. But while the *classes moyennes* expanded numerically from 23.4 per cent to 30.6 per cent of the population, they grew no wealthier. Indeed, their share in the total wealth of the city fell from 9.9 per cent to 7.5 per cent.[1]

The school system was called upon to make some sense of these social pressures. Any notion that the triumph of the republicans would result in the victory of the career open to

[1] F. P. Codaccioni, op. cit., pp. 126, 141, 360, 385.

the talents, the throwing open of secondary education, and an orgy of social mobility was quite misplaced. Entry into the élite remained carefully controlled. Secondary education remained fee-paying while elementary education was made free. The medical profession not only continued to demand Latin and Greek for entry into the faculty but also obtained legislation in 1892 reinforcing its monopoly of medical practice.[2] But while the education of the élite became no more open, primary education had begun its golden age, developing not only at the elementary level but also affording a higher primary education, up to the age of sixteen if need be, that was short, practical, and also free. Seeking access to the right technical school or branch of the administration, the *classes moyennes* who had once surged into the small-town colleges now looked increasingly to the E.P.S. for their needs.

In competition with the E.P.S. on one front, the *lycées* and colleges were faced on another by the challenge of the Catholic school system. Far from being mown out of existence by the lay laws, Catholic education proved hardy, even flourishing. Its most notable achievement was an ability to adapt to the requirements of each class. For the *classes moyennes* it developed a whole range of schools at the intermediary level between elementary and classical education, for the bourgeoisie it provided convent schools and 'aristocratic' colleges. Catholic and royalist nobles may have been excluded from the front rank of political power, but they still set the tone in society, maintained the connections necessary for success in many a career, and made the Catholic colleges they patronized more sought-after by the ambitious middle class than most provincial *lycées*.

At the turn of the century therefore the University was struggling. On the one hand there was the incentive to make its education more compact, practical, and accessible to the *classes moyennes*, for the sake of the survival of the smaller colleges. On the other, there was the need to preserve all the educative pretensions and *snobisme* of the classical course, in order to retain a hold on the bourgeoisie. By 1899 the problem had become so critical that a parliamentary inquiry

[2] J. Léonard, op. cit. pp. 293–8.

was held, the basis both of the education reform of 1902 and of much of the material of this chapter, a discussion of the struggle for pupils waged between different establishments, the imperialism of the schools.

It has been said that special education was a fairly success-ful response to those of the *classes moyennes* whose cultural background and family resources put the long preparation of the classical *baccalauréat* beyond their reach. The Inspector of the Academy of Douai noted in 1874 that such ambitions were anyway misplaced and that 'this wave of pupils should be diverted towards studies that are more humble, more modest, more down to earth, more practical and if you like, more useful, secondary special education'.[3] Special education indeed flourished either in small towns with a prosperous agricultural, industrial, or trading base, such as Condé, Le Quesnoy, Avesnes, and Maubeuge in the Nord, or in towns where competition from a Catholic institution was fierce, as in Flanders at Hazebrouck and Tourcoing, whose Lycée d'enseignement spécial was founded in 1881. Yet special education also had its drawbacks. It was very often a *pis aller*, a course frequented by unlicked primary-school pupils, rejects from classical studies, foreigners, and those short of the time and money necessary for a liberal education. Moreover, it never was a practical, let alone a professional, training and during the 1880s became less rather than more so. The syllabus was lengthened in 1881 to five years, leading to a *baccalauréat ès arts* (but not the *ès lettres*, which mat-tered), and extended again in 1886 to a six-year course, in which the *surménage* of two modern languages, sciences, history, and geography could scarcely be disguised. For the Rector of Douai, this was no disadvantage: special education now provided 'a culture of secondary spirit without the help of Latin or Greek'.[4] But those closer to the turning wheels of the economy insisted on the disadvantage of the reform. The municipal council of Roubaix, which maintained the Catholic school of Notre-Dame des Victoires where special education was strong, noted that the 'average pupil required a practical, applied training', not 'a luxury education designed

---

[3] A.D. Nord 2T 64/11, report of Inspector of Academy, Douai, June 1874.
[4] A.N. F17 6877, report of Rector of Douai, 9 July 1885.

to embellish his mind rather than to arm him against the difficulties of material life'.[5] The principal of the College of Condé was more candid still: graduates from the special course were not finding employment with the industrial concerns of the region, because their education was anything but professional.[6]

As special education increasingly adopted the airs of a new humanism, so the *classes moyennes* turned away in order to find a more practical training that would at least leave them with a job. As in so many instances, it was not public education that was first in the field, but the schools of the Catholic Church, sensitive as ever to the demands of the age. The advantages of the Catholic school system were several. Firstly, it was not tied to the restrictions of the University and could float in the uncertain area between primary and secondary education. At the top level, it could offer the diploma of special education and the certificate of grammar, a passport to the professions of pharmacist and *officier de santé*, both examinations taken at the age of fourteen, and these capped the primary-school system rather than broached the classical culture of the colleges. Secondly, the establishment of free education in all *écoles communales* in 1881 eliminated the benefit of segregation afforded by school fees and played into the hands of the Catholic primary schools. As early as 1879, Jonglez de Ligne, director of the Catholic school committee of Lille, noted 'the repulsion felt by the *petite bourgeoisie* for schools that are entirely free'. This was explained not only by sentiments of vanity: 'it must be recognized that standards of education decline singularly in free schools that are open to the lowest strata of the population'.[7] Twenty years later, when the republican press of the Nord attacked minor functionaries for patronizing the schools of religious congregations, the sub-prefect of Hazebrouck reported that the fees payable in such schools constituted the only available means of selection.[8] Thirdly,

[5] A.D. Nord 2T XI-21, report to municipal council of Roubaix, 16 Jan. 1885.
[6] A.D. Nord 2T 287/2, Insp. Acad. to Rector of Lille, 24 May 1888.
[7] B.M. Lille, Fonds Mahieu, A21, report of Jonglez de Ligne, 29 Dec. 1879.
[8] A.D. Nord 1T 123/10, sub-prefect of Hazebrouck to prefect of Nord, 14 Dec. 1901.

the Catholic *patronat*, promised employees who would know their place in the organization, favoured the development of the Church's intermediary schools.

In the Nord, it was the Marists, promoted by local industrialists, who made the running after 1870. The irascible Georges Colombier founded the Pensionnat Saint-Joseph at Haubourdin, near Lille, in 1871, and was not happy when the spinner, Philibert Vrau, bought out a *maître de pension* in the Lille suburb of Esquermes to establish the Marists there the following year.[9] In the end, however, it was the Pensionnat of Beaucamps, reduced to a purely rural clientele, that suffered. Lille itself acquired the important Pensionnat of Saint-Pierre, founded in 1889, and run by the Frères des Écoles Chrétiennes. It was these Frères who made the most impressive mark in the Gard. The Pensionnat Saint-Louis de Gonzague at Alès, set up in 1880, had among its other enterprises a monopoly of preparation for the local École des maîtres-mineurs. No municipal scheme existed until 1899, when one was adjoined to the *cours complémentaire* of the Quai Neuf, but never did it obtain more than three or four of the twenty places open every year at the mines' school.[10] In the backward reaches of Ille-et-Vilaine it was the Frères de Plöermel, adding *cours supérieurs* and *pensionnats* to their existing schools, who brought intermediary education to Brittany: Redon in 1871, La Guerche in 1875, Fougères in 1877. Redon did not have an *école primaire supérieure* till 1903; Fougères had to wait until 1908.

There was no effective riposte from the state system to this Catholic initiative before about 1880. The *classes moyennes* were not provided for to any great extent in the primary-school sector until the decree of 15 January 1881 on *écoles primaires supérieures*. There now came into being a network of higher primary schools, autonomous in the case of the E.P.S., tacked on to existing primary schools in the case of *cours complémentaires*, that provided a completion of elementary studies including mathematics, book-keeping, technical drawing, practical scientific work, and

[9] A.F.M. BEA 660, Haubourdin, Colombier to superior-general, 4 Jan. 1873.
[10] A.F.E.C. 337/1, 'Historique du Pensionnat Saint-Louis de Gonzague' (M.S., 1925).

manual training in woodwork and metalwork shops. Private initiative was not excluded: the town of Beaucaire benefited from the legacy of a Protestant merchant to set up its E.P.S. in 1882, while the E.P.S. of Fournes, near Lille, had before 1883 been the *pensionnat* Gombert, offering 'a practical, immediately useful education'.[11] But, unlike in 1833, funds were now forthcoming from the authorities and, boarding fees apart, higher primary education was free. Again, levels of training differed from one end of the country to another. The Institut Turgot, founded at Roubaix in 1879 as an *école primaire supérieure et professionnelle* was a model of its kind. On the other hand, the municipal council of Saint-Georges-de-Reintembault, on the Norman border of Ille-et-Vilaine, which had never produced anything but farmers and *instituteurs*, asked in 1897 whether it could be excused from building a workshop and procure a field for agricultural experiments instead.[12] It is nevertheless clear that the higher primary-school system was competing for the same clientele as the colleges, especially those which concentrated on special education. And this time, by contrast with the July Monarchy, it was the colleges that suffered. The principal of the College of Dunkerque complained in 1883 that the *instituteurs* of the Flemish *bourgs*, setting up *pensionnats* and keeping on pupils till the age of fifteen, were starving him of recruits.[13] That of Le Quesnoy accused local *instituteurs* of 'a veritable horse-dealing unworthy of members of a teaching body salaried by the state', of scouring the canton from morning to night, offering education at knock-down prices.[14] In some small towns, colleges whose Latin was not appreciated by the local trading and farming populations and could not compete with colleges in larger centres nearby succumbed altogether. Republican municipalities elected in January 1881 were all too happy to replace them by E.P.S., as happened at Dol, behind Saint-Malo, in 1888 and at Bagnols, near Uzès, in 1892.

[11] A.D. Nord 2T 278/3, Prospectus of *pensionnat* Gombert, 11 Aug. 1883.

[12] A.D. I-et-V, 2T Dossiers communaux, Saint-Georges-de-Reintembault, municipal council, 28 Feb. 1897.

[13] A.D. Nord 2T 277/5, principal of College of Dunkerque to Insp. Acad. Douai, 19 Oct. 1883.

[14] A.D. Nord 2T 278/3, principal of College of Le Quesnoy to Insp. Acad. Douai, 12 Oct. 1884.

The University was not only under attack from developing primary education; it also faced a challenge from the Catholic colleges. In each case, the clientele competed for was different: in the first, the *classes moyennes*; in the second, the bourgeoisie. As far as the education of girls was concerned, the convent schools had a virtual monopoly of the daughters of the bourgeoisie. The elaboration of a state system of secondary education for girls was a bid to break into this market but it could be sure only of a clientele of modest means. The municipality of Lille opened an E.P.S. for girls in October 1870 and the free education it offered proved attractive to the lower middle class. This, however, was precisely what made bourgeois mothers reluctant to subscribe, 'fearing to see their children sitting next to children who do not have the same upbringing and the same language'.[15] In 1877, accordingly, a fee-paying E.P.S., given the more exalted title of the Institut Fénelon, was opened by the municipality in an attempt to seduce them. The battle had only just begun. The *cours secondaires* envisaged by the Camille Sée law, also largely free and little more than lecture courses held in town halls, drew on the girls of the *écoles communales* as much as on those of the lay *pensionnats* and convent schools. Charles Zévort, Directeur de l'Enseignement secondaire, suggested to the Rector of Douai in 1882, that 'far from hindering admissions to the secondary courses, the payment of a fee has the opposite effect of attracting girls of the *classe moyenne* who would be turned away by absolutely free education'.[16] After 1883, colleges for young ladies replaced the *cours* at Lille, Cambrai, Armentières, and Valenciennes in the Nord, charging between 500 and 1,000 francs for boarding fees but often little more than 100 francs a year for day-schooling. Such moves were fiercely denounced by radicals like Émile Moreau, municipal councillor of Roubaix, who insisted that 'advanced education should not be the privilege of the wealthy classes'.[17] There seemed in fact little chance of that, at least as far as the

[15] A.D. Nord 2T 910, Insp. Acad. to Rector of Douai, 28 Feb. 1879.

[16] A.D. Nord 2T 917/3, Charles Zévort to Rector of Douai, 5 Dec. 1882.

[17] A.D. Nord 2T 917/5, Émile Moreau at municipal council of Roubaix, 30 Mar. 1883.

education of girls was concerned. The *Nouvelliste du Nord et du Pas-de-Calais*, admittedly seeking to defend the convent schools, attacked the girls' College of Valenciennes in 1888 not only for breeding freethinking *lycéennes* recognizable by their direct gaze and their slightly masculine bearing but also for 'sacrificing the task of education to the conquest of the sixteen *brevets*, the climbing of an Eiffel tower of parchment', aimed at scholarship girls and *professeurs*' daughters, manufacturing *femmes savantes*, and *déclassées*.[18] In fact, the records for the Nord do little to belie such assertions. Recruitment to girls' colleges in 1889 was more modest than that to boys' colleges in the same town, involving fewer children of professional men or rich industrialists, about the same of *petits patrons* and more of *fonctionnaires*—including secondary-school masters—and *employés*.[19]

The ascendancy of the Catholic colleges over bourgeois families unquestionably increased during the last years of the century. In Lille, for example, the secular clergy organized the foundation of a second college, the Institution Jeanne d'Arc, in 1895, which 'was to share the clerical clientele with the Jesuits of Saint-Joseph'.[20] Both colleges indeed were expanding in the years 1895–1900. Such success was explained in part by political considerations. The Catholic colleges represented an 'eternal France', higher than the discredited Republic with which the *lycées* were identified. On the eve of the Dreyfus Affair, indeed, the *lycées* could not count on the support of the state's own employees: senior bureaucrats, magistrates, and above all, army officers. The military sent its sons to the Catholic colleges, which in turn prepared them for Saint-Cyr: it was here that the alliance of 'le sabre et le goupillon' was forged. In addition, the presence of aristocratic elements in the Catholic colleges exercised a powerful influence on the aspiring middle classes. The Catholic *patronat* could not be relied upon to favour the Lycée of Tourcoing rather than the Institution libre du Sacré-

---

[18] *Nouvelliste du Nord et du Pas-de-Calais*, 23 Mar. 1888. Cutting in A.D. Nord 2T 939.

[19] A.D. Nord 2T 285/2. The categories are fixed by the survey. The sample includes 362 girls, 1,019 boys.

[20] A.D. Nord 2T XI-20, report of Insp. Acad. Lille, 28 April 1897.

Fig. 9 Intake of Municipal Colleges in the Nord, 1889

Profession of Father	Cambrai		Douai		Valenciennes	
	boys	girls	boys	girls	boys	girls
substantial rentiers	5.5	5.9	4.6	1.7	3.1	3.0
small rentiers	4.5	2.5	5.4	6.8	7.5	—
liberal professions	6.9	3.8	11.9	6.8	7.5	1.5
public service	15.6	30.5	19.5	30.5	15.8	38.8
rich industrialists and merchants	17.0	6.8	8.6	3.4	4.4	6.0
small employers	26.6	31.4	29.7	27.1	34.4	29.8
employés	9.3	10.2	10.5	18.6	20.6	17.9
substantial peasants	6.6	5.9	3.0	3.4	3.9	3.0
small peasants	3.1	2.1	5.9	—	—	—
wage-earners	4.8	0.8	0.8	1.7	2.8	—

Cœur.[21] The Association des anciens élèves of the Lycée of Lille noted in 1899 that 'it is *bien porté*, it is *de bon ton*, in certain circles that were once loyal to the Lycée, to entrust the Jesuits or other religious congregations with their children's upbringing'. There they would not only receive 'a more finished education, but they would create more aristocratic relationships', with a view to social contacts, marriage, and business connections.[22]

A study of enrolment registers of the Lycée and Catholic colleges of Rennes for the period 1875–90 permits some concrete assessment of the situation.[23] The Institution Saint-Vincent stands as a typical 'aristocratic' Catholic college, drawing almost exclusively on the Breton landowning class, army officers, magistrates, *hauts fonctionnaires*, lawyers, and bankers and virtually excluding the *classes moyennes*. Comparison with the Institution Saint-Martin, run by Eudist priests, nevertheless reveals once again that there was no single type of Catholic college. Saint-Martin remained faithful

[21] A.N. F17 13940, minutes of masters' meeting, Lycée of Tourcoing, 23 Jan. 1899.
[22] A.D. F17 13940, *Association des Anciens Élèves du Collège et du Lycée de Lille* to *proviseur* of Lycée, 17 April 1899.
[23] A.D. I-et-V 10T for Lycée of Rennes. The enrolment registers of the Institution Saint-Martin are in the school archives. Those of Saint-Vincent are more disappointing and the statistics are gleaned from the bursary accounts. The samples are respectively 969, 557, 151.

Fig. 10. Recruitment to Schools of Rennes, 1875–1890

| | Saint-Vincent | Lycée of Rennes | | Saint-Martin | Cours Complémentaire 1894–8 |
		classical	special		
landowners, rentiers	21.7	19.9	12.8	9.5	1.8
liberal professions	33.0	13.5	3.7	20.8	0.9
management	7.8	9.7	9.0	3.8	10.9
public service	30.4	22.6	4.9	5.1	—
industrialists, entrepreneurs	0.9	4.1	7.4	3.4	4.5
merchants, bankers	4.4	15.0	17.7	15.1	2.7
shopkeepers	0.9	2.5	10.3	11.5	10.0
artisans	—	2.9	9.1	11.3	12.7
employés, petits fonctionnaires	—	7.6	16.9	7.4	21.8
peasants	—	0.4	3.3	10.4	1.8
wage-earners	—	1.4	3.7	1.8	30.0
servants	—	0.4	1.2	—	2.7

to its *petit séminaire* origins, offering reduced rates below its nominal boarding fee of 475 francs to needy families and recruiting heavily to the priesthood. The *classes moyennes*, taken here as tradesmen, artisans, *employés*, and *petits fonctionnaires*, provided 30 per cent of its intake, while another 10 per cent were drawn from the solid peasant families of the basin of Rennes. At the Lycée, the classical side included fewer of the élite and more industrialists, merchants, and *classes moyennes* than Saint-Vincent, while the special side had proportionately less of the liberal professions and peasantry, more employers and petty bourgeois than Saint-Martin. Yet it would not be true to say that the *classes moyennes* had surged into the Lycée since the Second Empire. Comparison with the figures for 1863–70[24] indicates that the proportion of *classes moyennes* on the classical side had increased from 12.4 per cent merely to 13.0 per cent, while that on the special course had risen from 34.5 per cent to 36.3 per cent. If republican France stood for the heyday of the *nouvelles couches sociales*, for 'individuals in a process of rising' from one class to another by dint of instruction and hard work, it is not evident from these figures. By contrast, the *classes moyennes* together with the working

[24] See above, p. 201.

class undoubtedly favoured the *cours complémentaire* (Rennes did not have an E.P.S. till 1945), as the sample for 1894–8 demonstrates.[25] It is possible that the *classes moyennes* sensed the futility of a long secondary education, given 'the overcrowding of the liberal professions'. The entrenchment of the bourgeois dynasties in the liberal professions cannot be doubted. Though the calculation must remain impressionistic, given the number of cases studied, career patterns established by tracing individuals from the enrolment registers of the secondary schools of Rennes to the records of the respective *Associations des anciens élèves* suggest that 45 per cent of sons of *avocats* and notaries in turn became lawyers while 33 per cent of sons of doctors, pharmacists, vets, and *officiers de santé* entered the medical profession.[26]

Confronted by challenges both from the E.P.S. and from Catholic colleges, the University could choose between alternatives. Either it could seek to become more attractive to the *classes moyennes* by offering a more practical alternative to the classics, or it could appeal to the bourgeoisie by upgrading special education and eliminating any associations it had with manual labour. The reform proposed by the Minister of Public Instruction, Léon Bourgeois, in 1891, opted for the latter course, Special education was replaced by modern education, claiming to offer the same culture as Latin and Greek but founded instead on modern languages and science.

Modern education nevertheless fell between two stools. On the one hand, it could not offer the élitist, humanistic training claimed for the classics. Where weight was attached to the duration of study, its course lasted for six years instead of the seven required for the classical *baccalauréat*. Academic administrators, university professors, and headmasters questioned in the parliamentary inquiry of 1899 doubted whether modern education had the same 'vertu éducatrice' as classics or imparted the same 'haute culture intellectuelle'. It was a 'disastrous intrusion of dangerous

[25] A.N. F17 11683, the samples is of 110 cases.
[26] R. N. Gildea, 'Education in Nineteenth-Century Brittany. A Study of Ille-et-Vilaine, 1800–1914' (Oxford D.Phil. thesis, 1977), p. 247.

overwork' which sought to substitute an encyclopaedic cramming of botany, zoology, cosmography, and hygiene for the slow osmosis of genuine culture.[27] A future *universitaire* who came up through modern education at the Lycée of Rennes after working, like his father, in the shoe factories of Fougères, later confessed that his lack of classical culture weighed heavily against him. He was a 'strange *khâgneux* who knew neither Latin nor Greek'. 'I suffered', he wrote later, 'to feel myself so stupid and so clumsy in the midst of those whose tempered and agile minds served them so well in all our exercises . . . I had studied a few textbooks, swotted, but the interconnection, the essence, the sense of things eluded me completely.'[28] It was such weaknesses that closed the Faculties and hence the liberal professions to holders of a mere modern *baccalauréat*. Modern studies gave access to *grandes écoles* such as Polytechnique, Saint-Cyr, and Centrale, but the Faculties refused to open their doors, arguing that the classics were necessary to understand Roman law or the Greek roots of medical terms, that classical education was 'the mark of a certain culture', and that the classical *baccalauréat* served as a barrier against the ambitions of those who sought to gain entry into professions that were already oversubscribed. The Law Faculties of Lille and Rennes were adamant; the Arts Faculties scarcely less conservative. The only exceptions were progressive voices in the Medical Faculties who argued that modern languages to keep abreast of scientific movements were more valuable than Greek grammar.[29] Figures for the Nord in 1891–2 show that liberal professions and *hauts fonctionnaires* were heavily represented on the classical side of *lycées* and colleges, while the sons of industrialists and merchants, shopkeepers, artisans, and farmers gravitated to modern education.[30]

---

[27] Open letter of Abbé Léon, superior of Saint-Sauveur, Redon, to Alexandre Ribot, published in *Le Redonnais*, 4 Feb. 1899. Cutting in A.N. F17 13945.

[28] Jean Guéhenno, *Changer la vie: mon enfance et ma jeunesse* (Paris, 1961), pp. 195, 202. (A *khâgneux* was a pupil of the *lycée* class preparing for entry to the École Normale Supérieure.)

[29] A.N. F17 13945, submission of Joubin, professor of Science Faculty, Rennes, Jan. 1899.

[30] A.D. Nord 290/6. The sample is of 342 cases.

Fig. 11. Classical and Modern Education at *Lycées* and Colleges of
Nord, 1891–1892

	Classical	Modern
landowners, rentiers	12.3	12.3
liberal professions	13.4	2.3
public servants	14.0	6.4
industrialists, merchants	23.4	30.4
shopkeepers	7.6	16.4
employés, petits *fonctionnaires*	18.7	15.2
peasants	8.8	13.5
artisans, workers	1.8	2.9
others	—	0.6

On the other hand, modern education was clearly inap-
propriate for the mass of the *classes moyennes* who intended
to enter technical schools like the Arts et Métiers, return to
agriculture, trade, or industry, or find a niche in the lower
reaches of the administration. It was far too long, abstract,
and—in a period of economic depression—expensive to
command favour among the *petite bourgeoisie*. What they
required was an education that was short, practical, cheap,
and, above all, left them with employment. The suffering
of the colleges in this respect was exposed in the parliamentary inquiry of 1899. In industrial areas, there was a feeling
that special education should never have been tampered
with. The Bureau d'administration of Tourcoing recom-
mended a more practical approach: applied mathematics,
language teaching geared to the spoken word, the return of
accountancy and draughtsmanship.[31] The *proviseur* of the
Lycée of Alès reported that those who completed the
modern course became useless *déclassés* and those who never
intended to stay for six years found the *cours complémentaire* of Saint-Louis de Gonzague more to their taste.[32]
A variety of remedies was tried. In 1892, the college of
Maubeuge established a preparatory course for the Arts et
Métiers alongside its modern classes, designed also to appeal
to those seeking to enter the Post Office and Ponts et

[31] A.N. F17 13940, Bureau d'administration of Lycée of Tourcoing, 11 Mar.
1899.
[32] A.N. F17 13941, report of *proviseur* of Lycée of Alès, 14 Jan. 1899.

Chaussées, including bookkeeping, technical drawing, and manual work. So successful was this enterprise that the Ministry of Public Instruction set about selling such professional courses to other colleges, and they caught on at Condé, Le Cateau, Avesnes, Armentières, and even Catholic institutions like Notre-Dame des Dunes at Dunkerque.[33] The principal of Le Cateau noted in 1893 that now 'instruction for the modern *baccalauréat* will be the luxury, the exception, the boast; it is the *brevets* and the little professional examinations that will attract a clientele'.[34]

As the battle for the same clientele between primary and secondary education grew fiercer, so each side tried to eliminate the rivalry of the other. One proposal in the secondary schools was that the E.P.S. should be fused with modern education. The E.P.S. teachers vigorously attacked such a plan, arguing that modern education was general and theoretical, while higher primary education was practical and professional. An E.P.S. teacher of Landrecies in the Nord asserted that the E.P.S. 'is not the college degenerated, but the school perfected',[35] while the director of the E.P.S. of Dol claimed that the E.P.S. 'springs from the primary-school sector as the mushroom springs from the earth'.[36] Many University administrators were in favour of harmonizing syllabuses to make possible a transition from the E.P.S. to the higher classes of modern courses, but at the Institut Turgot of Roubaix it was argued that 'the day this establishment becomes the entrance-hall of the *lycée*, it will lose three-quarters of its pupils'.[37] On the other hand, the demand from many E.P.S. that the primary classes of *lycées* and colleges be suppressed and that all pupils, including those destined for secondary school, pass through the *communale* provoked the riposte that middle-class families would simply send their children to the private schools of religious congregations in order to avoid 'mauvais contacts'. As a last

---

[33] Circular of superior of Notre-Dame des Dunes, July 1898, cited by R. Flahault, *Notes et documents*, pp. 230–1.
[34] A.D. Nord 2T 294/2, report of Insp. Acad. Lille, 20 June 1893.
[35] A.N. F17 13940, *instituteur* of E.P.S. of Landrecies, 23 Jan. 1899.
[36] A.N. F17 13945, director of E.P.S. of Dol, 22 Jan. 1899.
[37] A.N. F17 13940, director of Institut Turgot, Roubaix, 25 Jan. 1899.

resort, the schoolmasters of the college of Dunkerque suggested that higher primary education be abolished, but this view was countered by the teachers of the E.P.S. of Valenciennes, who asserted that vegetating colleges should be suppressed.[38]

The saving grace of the colleges was of course that they were colleges, not schools. As the principal of the college of Saint-Servan, on the Breton coast, said, families 'come to us to receive a *secondary* education, not for a primary education dressed up and disguised. They find the former, what shall I say? more aristocratic.' His schoolmasters were not quite so optimistic, pointing out that 'the overcrowding of the liberal professions, the campaign waged against Latin, Greek and the "vides et pâles bacheliers", together with the development of professional schools, have shrouded secondary education with a sort of disfavour that has made many young people scorn it.'[39] The advantage of higher primary education, as its protagonists never ceased to claim, was that it trained men of action who might pull the country out of economic recession, rather than parasites, *déclassés*, and 'bacheliers sans emploi'. The case that they made out, however, was never wholly convincing and as a network of properly professional schools developed,[40] so the E.P.S. attracted the same criticisms that their supporters had levelled against modern education. The main problem was a combination of the economic situation, which did not encourage those with an education to try to make a living in farming or business, and the nature of higher primary schooling, which was general rather than a specific apprenticeship training. Graduates from the E.P.S. were all too readily attracted by white-collar posts, teaching, and *la fonction publique*. *Cours complémentaires* and E.P.S. for girls were no exception to this tendency. The girls' E.P.S. of Nîmes, for example, founded in 1886, initially did little but turn out *institutrices* who could not be employed in view

---

[38] A.N. F17 13940, masters of College of Dunkerque, 23 Jan. 1899; teachers of E.P.S. of Valenciennes, 25 Jan. 1899.

[39] A.N. F17 13945, principal of College of Saint-Servan, 10 Feb. 1899, and submission of *fonctionnaires* of the College, n.d.

[40] See below, Chap. 10.

of the domination of female education by the religious congregations. An 'industrial course' was added in 1892 to train seamstresses and *blanchisseuses*, but it was never very popular.[41] In 1899 the Chamber of Commerce of Nîmes was announced that the E.P.S. was no different from modern education, except in one respect: that it was free and therefore available not only to the *petite bourgeoisie*, but also to the working classes.[42]

Inspectors of Académies and, in the Nord, the Directeur départemental, energetically resisted such assertions. They pointed out that higher primary education inculcated manual dexterity and an aptitude for economic life but was not intended to teach the skills required for particular trades. The Inspector of the Academy of Rennes provided figures to prove that only a tenth of school-leavers from E.P.S. entered any administration. 'It cannot therefore be said', he concluded, 'that the school produces *fonctionnaires*, neither can it be said that it produces *déclassés*.'[43] Much evidence of this kind was produced, not all of it supporting such verdicts. Information on graduates of E.P.S. in the Nord between 1900 and 1905, to take one survey, indicates that while 9.7 per cent of school-leavers entered technical schools and 30.7 per cent were apprenticed to trades or returned to their families to take up industrial or commercial occupations, 46.0 per cent became *instituteurs* or *employés* in banks, private concerns, railway companies, or administrative services from the Post Office and tax offices to the customs and Ponts et Chaussées.[44] The only significant figures, however, must be those gathered from the inquiry which established, for every pupil of the *cours complémentaires* between 1894 and 1898, the occupation of their parents and the activity embraced after leaving the course.[45] For each of five social categories, the proportion which succeeded to the parental occupation, that which undertook

[41] A.D. Gard 1T 782, *directrice* of girls' E.P.S. of Nîmes to Insp. Acad., April 1897.

[42] A.D. Gard 1T 781, Chamber of Commerce of Nîmes, 8 Mar. 1899.

[43] A.D. I-et-V 1N. Conseil général, report of Insp. Acad. Rennes, 1901, p. 560.

[44] A.D; Nord 1N 150, Conseil général, report of Insp. Acad. Lille, 4 Aug. 1906. The sample is of 3,099 cases.

[45] A.N. F17 11683 for Gard and Ille-et-Vilaine, F17 11685 for Nord. The number of cases is 386 for Ille-et-Vilaine, 408 for Gard, and 963 for Nord.

Fig. 12 Career Patterns of Leavers from *Cours Complémentaires* 1894–1898

		Reproduction	Manual Occupations	E.P.S. professional schools	Écoles normales White-collar	Secondary schools
Shopkeepers	I-et-V	12.8	42.5	6.4	29.8	8.5
	N	33.6	18.2	9.1	34.5	4.5
	G	33.3	20.0	4.4	35.6	6.7
Artisans	I-et-V	53.0	21.2	6.1	13.6	6.1
	N	40.2	16.3	6.5	33.7	3.3
	G	40.3	12.5	6.9	38.9	1.4
Employés *Petits* *fonctionnaires*	I-et-V	14.7	18.9	11.6	43.2	11.6
	N	17.9	15.4	14.5	42.7	9.4
	G	12.8	13.7	13.7	54.9	4.9
Peasants	I-et-V	74.2	11.2	2.2	11.2	1.1
	N	60.8	8.2	2.3	22.2	6.4
	G	67.4	9.6	3.0	19.3	0.7
Workers	I-et-V	6.7	41.6	7.9	39.3	4.5
	N	31.8	28.0	6.1	33.3	0.8
	G	16.7	38.9	5.5	37.0	1.9

other manual occupations, that which graduated to E.P.S. or technical schools, that which entered *écoles normales* or secured white-collar posts (including all forms of *employé* and *petit fonctionnaire*), and that which went on to secondary school has been calculated.

The table may be analysed in three ways: by social occupation, by department, and, adducing evidence (albeit scantier) for girls' *cours complémentaires*, by sex. The most mobile category were clearly the *employés* and *petits fonctionnaires* including teachers, *gendarmes*, and office-clerks, followed by shopkeepers and other small tradesmen. Already separated by at least one generation from the farm or workshop, these groups tended to suffer the same contradiction between bourgeois aspirations and proletarian incomes and to exhibit the same combination of anxiety and family solidarity, which had the effect of sharpening ambition.[46] The least mobile groups were the peasantry, two-thirds of whom returned to the land, and the *artisanat*. By contrast, those of the working class who continued their studies in the *cours complémentaire* tended to desert their origins. Here, of course, a distinction must be drawn between large towns like Rennes and closed, working-class communities.[47] At Aniche, near Douai, in the coal-basin of northern France, 61 per cent of the sons of miners and glass-workers who attended the *cours complémentaire* followed their fathers into the pits or furnaces.

The geographical distinction is of the first importance. There was clearly an enormous difference between backward, socially stratified areas like Brittany and industrial, complex societies like the Nord. Nevertheless, social mobility was greatest in the Gard, whether because of economic crisis in the wine-growing and silk industry at the end of the century or because of cultural factors such as high literacy rates, Protestantism, or radicalism. Taking all our social categories, the proportion of those who became *instituteurs, employés*, or *fonctionnaires* was 37.1 per cent in the Gard, 33.3 per

[46] On this, see C. Lévy-Leboyer, *L'Ambition professionnelle et la mobilité sociale* (Paris, 1971), pp. 159–80.

[47] M. Gillet, *Les Charbonnages du Nord de la France au XIXe siècle*, pp. 325–6, confirms this view.

cent in the Nord, and 27.4 per cent in Ille-et-Vilaine. But if the proportion of those who reverted to manual occupations is compared with the proportion that progressed to secondary school, an interesting paradox emerges: Ille-et-Vilaine is at the head of both lists. One possible explanation is that in the west, where the *couches moyennes* formed a relatively narrow stratum between the élite and the masses, the alternatives between *immobilisme* and escape by way of extended education were very few.

A final comparison between boys' and girls' *cours complementaires* indicates that the possibilities of social mobility for girls were far behind those that obtained for boys.[48] In the first place, 26.1 per cent were recorded as returning to their families to help in the home. Then, the fraction that moved on to teaching or white-collar occupations was 22.2 per cent in the case of girls against 32.6 per cent for boys. The proportions which went on to secondary school were 3.5 per cent for girls against 4.8 per cent for boys, while those who turned to manual labour represented 28.7 per cent in the first case, 21.1 per cent in the second. Economically, as well as ideologically, women still stood for conservatism in society.

There was in fact a broadening of prospects for girls in higher primary education after 1900. One reason was political: the offensive against the female congregations whose *pensionnats* had dominated the education of middle- and lower-middle-class girls for so long. The legislation of 1901 and 1904 created a void that had to be filled by lay schools. Progress was made in the traditionally Catholic west, where Fougères acquired a girls' E.P.S. before that for boys was set up, while the radical mayor of Rennes, Jean Janvier, obtained acceptance for a girls' E.P.S. in 1911, though it did not open until 1917.[49] Moreover, it was in 1906 that secondary education for girls was put on a secure base in Ille-et-Vilaine with the opening of the girls' Lycée at Rennes. This new enthusiasm for the education of girls reflected a multiplication of career opportunities. Until 1900, there were

---

[48] Figures for girls' *cours complémentaires* and E.P.S. in the Gard have been completed with information in A.D. Gard 1T 787.

[49] A.M. Rennes IDI/76, municipal council of Rennes, 15 Dec. 1911.

few alternatives to teaching, assistance in shops, and manual work as seamstresses or launderers. Subsequently, openings appeared for girls in the Post Office, nursing, and banks. At the girls' E.P.S. of Nîmes, a course in typing and shorthand was opened in 1907, while the inauguration of an *école d'infirmières* at a local hospital, pioneered by English nurses who replaced the religious congregations, attracted graduates from the E.P.S. after 1908. Comparison of the careers of leavers from the girls' E.P.S. of Nîmes in 1900–10,[50] as against 1894–8, bears witness to a new social mobility. The proportion of those who went into white-collar occupations such as teaching, nursing, clerkships, or secretarial work rose in the case of daughters of shopkeepers from 26.3 per cent to 32.1 per cent, in that of daughters of artisans from 20.0 per cent to 39.0 per cent, and in that of daughters of working men from 14.3 per cent to 43.8 per cent.

How far this flurry of activity in the primary-school sector in the *Belle Époque* was carried over into secondary education must be gauged from a study of the reform of 1902, towards which the parliamentary inquiry of 1899 had led. While many members of the University, *conseils généraux*, and even the Chamber of Commerce of Ille-et-Vilaine had favoured retaining the classical *baccalauréat* as the sole route into the élite, others had pointed out that this only resulted in an overcrowding of the liberal professions and the abandonment of the productive sectors of the economy by those required to take them in hand. On the Conseil général of the Nord, Henri Sculfort, machine-tool manufacturer of Maubeuge, argued that France's poor performance in the struggle for world markets could be explained largely by the stultifying effect of the classics: 'we have neither the urge to action, nor the gift of enterprise', he reflected. From a rather different angle, the socialist councillor Henri Ghesquière pointed out that 'the dead languages, the whole Greco-Latin system of education have made an entirely reactionary ruling class of our French bourgeoisie'.[51] The education reform of 1902 was some attempt to take account of such criticisms, to reply to the

---

[50] A.D. Gard, 1T 793.
[51] A.D. Nord 1N 143, Conseil général, 29 Aug., pp. 472–81.

upsurge of democracy and the demands of the modern age. In the first place, the privileges attributed to the classical *baccalauréats* were removed. Instead, four courses were now available to the *lycéen* or college boy entering the second form (age 15), all with equal value: Latin–Greek (known as option A), Latin–Modern Languages (B), Latin–Sciences (C), and Sciences–Modern Languages (D). Secondly, secondary and primary education were for the first time co-ordinated to make easier the transition between the two. The primary-school pupil could enter secondary school either after the *cours moyen*, starting in the sixth form (age 11) or after the *cours supérieur*, starting in the second form.[52]

Jean Janvier, Radical mayor of Rennes, himself a product of the primary-school system, could tell *lycéens* and their parents in 1909 that 'because of the role of intelligence, the *lycée* is the highest bond between the wealthy classes and others in our modern society'.[53] Yet it must be doubted whether any notable 'democratization' of the secondary-school system had taken place. The co-ordination of primary and secondary schooling remained of little importance while the first was free and the second fee-paying, with no great extension of the number of colleges. Considering secondary education in 1907 one *instituteur* of Nîmes observed that 'it would perhaps be wise to warn parents that the studies will be long, the costs high and the results uncertain'.[54] Subsequent research has demonstrated that the intake of *classes moyennes* and popular elements into the *lycées* remained substantially unchanged until after the Second World War.[55] In addition, the 'equality of sanctions' remained formal, for the distinction between classical and non-classical education continued to be made by families. It has been said that 'unfortunately, there exists no statistics by section of the

[52] In general, see J.-B. Piobetta, *Le Baccalauréat*, pp. 241–4, C. Falcucci, *L'Humanisme dans l'enseignement secondaire* (Toulouse, 1939), pp. 505–8; F. Vial, *Trois Siècles d'histoire de l'enseignement secondaire* (Paris, 1936), pp. 245–56.

[53] *Avenir hebdomadaire*, 1 Aug. 1909.

[54] A.D. Gard, Périodiques 115, *Bulletin d l'Amicale des Institutrices et Instituteurs du Gard*, report of Raous, *instituteur* of Nîmes, 1907, first trim., 11–14.

[55] A. Girard, 'L'Origine social des élèves des classes de 6e', *Population*, 17, Jan–Mar. 1962, 10–23.

the socio-professional origin of the pupils'.[56] This may be true as far as official surveys are concerned, but enrolment registers can provide some guidance, even if none survive for the period in any of our three departments. However, the registers of the Lycée of Rouen, extant for the years 1905–6 and 1908–17, make it possible to draw several conclusions.[57] Firstly, pupils were not distributed equally between the four sections, 59 per cent of the sample opting for courses A–C while 41 per cent opted for D. Next, the clearest social differences are evident in the intake of the Latin-based courses on the one hand, and the non-Latin course on the other. The proportion of sons of landowners, professional men, and civil servants ranged from 58.6 per cent to 68.8 per cent in the first, against 40.3 per cent in the second, while the proportion of shopkeepers, artisans, and *employés* was between 8.8 per cent and 13.8 per cent in the Latin courses but 29.3 per cent in course D. In this connection, it is interesting to note that Françoise Mayeur dates the *embourgeoisement* of women's colleges and *lycées* to the period after 1900.[58]

The cumbersome nature of the University was still evident at the turn of the century. It seemed as ever incapable of adapting its courses to the specific needs of the market. The *classes moyennes* had at last found a 'commercial' course in the colleges and *lycées* in the 1860s in the shape of special education, but the University insisted on upgrading this to a 'modern humanism'. The middling sorts were driven into private *pensionnats* or E.P.S. which now rivalled the University with increasing success. On the other hand, modern education was not the classical education that gave access to the liberal professions, and even when the distinction between the two courses was broken down officially in 1902, it remained in terms of class snobbery. Furthermore, the bourgeoisie in search of aristocratic connections did not patronize the municipal colleges or *lycées* at all, but sent

[56] V. Isambert-Jamatie, 'Une réforme des lycées et collèges. Essai d'analyse sociologique de la réforme de 1902', *L'Année sociologique*, 20, 1969, p. 50, note.

[57] A.D. Seine-Maritime, 10T 84, 85, 86. Lycée de Rouen, registres des élèves, 1905–6, 1908–9, 1910–17. The sample is of 440 cases.

[58] F. Mayeur, *L'Enseignement secondaire des jeunes filles*, pp. 187–91.

Fig. 13. Intake of Lycée of Rouen, 1905–1917

Profession of Father	A: Latin–Greek	B: Latin–Languages	C: Latin–Sciences	D: Languages–Sciences
landowners, rentiers	6.9	7.3	5.2	11.0
liberal professions	17.2	31.3	17.8	14.9
management	10.3	6.2	13.3	13.3
public service	34.5	30.2	43.0	14.4
industrialists, entrepreneurs	13.8	6.2	3.7	8.8
bankers, merchants	3.5	5.2	6.7	5.5
shopkeepers	3.5	6.2	2.9	13.8
artisans	—	—	—	3.9
employés, petits fonctionnaires	10.3	6.2	5.9	11.6
peasants	—	1.0	1.5	1.6
wage-earners	—	—	—	0.6
servants	—	—	—	0.6

its children to the great Catholic colleges and convent schools. Though the forces of conservatism had been defeated politically, they still set the tone socially.

Catholic education was as usual far more adaptable than the University. It offered a distinct institution for each social category. For the élite, the colleges. For those seeking a cheap, quick Latin education the *petits séminaires* or (since they were becoming strictly training academies for priests) minor colleges that had a great deal of the seminary about them. And for the *classes moyennes* who would return to farming, trade, or industries, the *pensionnats* of the religious congregations provided a sophisticated education and guarantees for a conservative *patronat*, a decade before the republican government replied with remodelled E.P.S.

One of the functions of the E.P.S. was of course to attract those of the *classes moyennes* whose ambitions would be directed towards the desperately overcrowded professions if they attended the colleges but who might become the shock troops of economic recovery if they remained within the primary-school sector. They were a response to the propertied classes' fear of *déclassement*. By the turn of the century even the E.P.S. were accused of turning out useless *fonctionnaires*, but the notables need not have worried. In the last resort social mobility was determined not by institutions of education but by the social and economic structure of the country. The overcrowding of the professions and the multiplication of opportunities at the managerial and employee levels of business regulated the flow of labour towards those sectors. And since the occupational structure varied from one part of the country to the next, it was clear that backward regions like the west would combine the social *immobilisme* of the lower classes with a passion for secondary education in order to move on, while the economic development of regions like the north would encourage school-leavers to remain in the productive sector. But more than any other region it was the Midi that claimed pride of place as a manufacturer of *déclassés*.

PART III
THE ECONOMY AND EDUCATION

9

The Pioneering Decades

Over the course of the nineteenth century, European economies faced the challenge both of a rising population and of international competition for markets. Frequently it was the more backward countries that placed a heavy emphasis on professional education, in order to make up artificially the ground that had been lost naturally. One of these countries was France. The extent to which various forms of technical education succeeded in France may conveniently be studied at the provincial level. Indeed, more than any other study of educational institutions, those of technical education require a local or regional base. In the first place, it must be ascertained whether professional education was more important in areas that were agriculturally and industrially advanced, such as the Nord, or whether it was of greater value in poorer, less developed regions like Brittany. Again, it should be discovered whether it was better adapted to large-scale industry, such as textiles or coal-mining or to small-scale, artisanal crafts. It is necessary to examine the level at which it was pitched, whether it was the landowners and entrepreneurs who profited most from technical education, or the mass of the labourers, or the intermediary groups, the farmers, managers, foremen, and technicians. The regional analysis permits an analysis of whether specific professional schools were effectively geared to the economic needs of the locality, or whether they functioned with very little sense of what was required by employers and remained marginal to economic developments. An extension of this point is the degree to which these schools were part of a national blueprint drawn up by the central government or, on the other hand, were devised in the locality either privately, by landowners and industrialists, or publicly, by municipalities and departments. Lastly, since the imparting of technical expertise was combined with more general education, it must be discovered how far professional formation did train vanguard fighters for the economy and how far it tended only to stimulate ambition and to increase the very *déclassement* that it was supposed to prevent.

As far as industrial education was concerned, there were two main strands of continuity with the Ancien Régime: military apprenticeship and the *écoles académiques*. The *écoles académiques* were schools set up in the larger towns after 1750 to compensate for the inadequacy of the training provided by the *corps de métier* and maintain the artistic qualities of the French artisan élite. The *échevins* of Douai noted in 1777 that in the guilds 'the years of apprenticeship have degenerated into years of idleness', while the masters, 'using the pretext of the ignorance of the young apprentices, employ them only to run errands'.[1] Architecture courses were added to the drawing-schools in towns such as Douai and Lille, while public lectures in mathematics and botany funded by the municipalities transferred some of the ideas of the Enlightenment to a wider audience. When the *écoles centrales* were founded, these public courses tended to be absorbed into the drawing, mathematics, and natural-history departments of those schools; whether they would be lost or salvaged after the abolition of the *écoles centrales* in 1802 depended very much on the municipalities and prefects.[2] At Lille, for example, the drawing-school had become attached to the *école secondaire communale* after 1802 but was re-founded separately in 1806 on the initiative of the prefect, General Pommereul, who noted that not only painters and sculptors but also joiners and cabinet-makers, jewellers and goldsmiths, stone-cutters and turners required expertise in drawing for the perfection of their skills.[3]

The second strand, which impinges indirectly rather than directly on our departments, illustrates the way in which technical education was bound up with military and strategic concerns. In 1788, the Duc de La Rochefoucald-Liancourt founded a school on his estate near Paris for the children of soldiers of his dragoon regiment, who would be taught

---

[1] B.M. Lille, Fonds Lemaire, II, 73, ordinance of *échevins* of Douai, 16 Jan. 1777.　　　[2] See above, pp. 50.

[3] B.M. Lille, Fonds Lemaire, II, 57, prize-day speech of prefect of Nord, Pommereul, 25 Aug. 1808.

by NCOs to read, write, count, and to exercise the trades of tailor and cobbler, so useful to the army. La Rochefoucald was forced to emigrate in 1791 and the school moved to Compiègne, but he returned in 1799 to save it and in 1803 Bonaparte as First Consul reorganized it. Classical studies would be confined to the *lycées*; the College of Compiègne would educate 'orphans of the Patrie' in its workshops to become smiths, founders, turners and fitters, carpenters, joiners, and wheelwrights, training them also in French, drawing, arithmetic, geometry, and mechanics.[4] Moved again in 1806 to Châlons-sur-Marne, this became the first École d'Arts et Métiers; the second, founded at Beaupréau (Maine-et-Loire) to 'colonize' the *chouan* west, was transferred to Angers during the Hundred Days. At the Restoration, under the ordinance of 26 February 1817, the military regime was phased out, and the government agreed to finance 500 scholarships at the Arts et Métiers. The practical bent of the institution was strongly underlined. As colleges and seminaries, as well as mutual schools and those of the Frères des Écoles Chrétiennes, began to expand under the pressure of popular ambition, it was stated that the Arts et Métiers would train 'foremen and workers exercised in the practice of industrial arts'.

The defeat of Napoleon, the dismantling of the blockades and the Continental System, the exposure of France to the rigours of competitive trading, were all powerful forces behind the development of professional education at the Restoration. Since all nations, reflected La Rochefoucauld, were seeking to import as few industrial goods as possible and to export the maximum, 'we have to manufacture more cheaply and better. The intelligence of the work-force and techniques of production must be improved.'[5] Baron Charles Dupin (1784-1873), a *polytechnicien* who studied under Monge and was fascinated by the success of Britain, visiting it repeatedly and writing three volumes on its *Force militaire, Force navale*, and *Force commerciale* (1820-4), urged that in the face of such competition every effort must be made to increase 'the enlightened and thinking portion

---

[4] A.N. F17 14317, memorandum of La Rochefoucauld to Min. Int., 3 June 1814.
[5] A.N. F17 14317, memorandum of La Rochefoucauld to Min. Int., 1815-16.

of the labouring classes' replacing brawn by brain, time by economy, routine by science.[6]

The impact of renewed competition was evident in both of the traditions of professional education under review. A fillip was given to drawing-schools intended for the *artisanat*, with foundations at Rennes and Saint-Malo in 1817,[7] and at Nîmes in 1820. There the moving spirit was another *polytechnicien* and *cadastre* official, Simon Durant, and the cause was taken up by the Académie royale de Nîmes and the prefect in order to improve the quality of silk products to rival those of Lyon. The École gratuite de dessin was aimed at 'the children of merchants or industrialists who intend to undertake the manufacture of silk garments, and the children of master-masons and other workers distinguished by their skill in the mechanical arts'.[8] Records show that in 1821 29 per cent of entrants were sons of textile merchants or manufacturers, and another 28 per cent represented an artisan élite of jewellers, china-makers, and printers; there were very few silk-weavers or building workers.[9] The course was in fact highly artistic, run by the painter, Jean Vignaud (1775–1826), who had been drawing-master at the École Centrale at Nîmes and then gone to Paris to work at the studio of David. The ritual of drawing heads, 'académies' (studies from the nude), and bosses was orientated exclusively in the interest of the luxury crafts and silk industry. This also benefited from a course in applied chemistry, to train dyers, in 1820.

Parallel to such developments was a campaign organized by Charles Dupin to vulgarize a knowledge of geometry and applied mechanics among the working classes. He had lectured on this subject from 1824 at the Paris Conservatoire royal des Arts et Métiers, but in November 1825 he managed to press the Minister of the Interior into sending out a

---

[6] Charles Dupin, *Effets de l'enseignement populaire de la lecture, de l'écriture, de l'arithmétique, de la géometrie et de la mécanique appliquées aux arts sur les prospérités de la France* (Paris, 1826).

[7] A.N. F17 10213, Comte d'Allonville, prefect of Ille-et-Vilaine, to Min. Int. 31 May 1817.

[8] A.D. Gard 4T 44, Regulation of Cavalier, mayor of Nîmes, 26 Sept. 1820.

[9] A.D. Gard 4T 45, List drawn up by Commission directrice of École gratuite de dessin, Nîmes, 23 Jan. 1821. The sample is of 103.

circular to prefects, urging them to set up similar courses in their departments, to be funded by the municipalities. The mayor or Rennes, Louis de Lorgeril, responded at once, and a course was opened under Pierre Legrand, *professeur* at the Collège royal. In the Nord, Douai and Dunkerque went into action long before Lille, which did not found a course until 1828. Even then, the mayor, the Comte de Muyssart, complained that it was far too theoretical to be of any use to workers and artisans of limited educational achievements.[10]

That the working class might be infected by theory was indeed a recurrent fear among the supporters of the existing order. Theoretical education excited pride and ambition, and weakened resistance to revolutionary ideas. When a wave of unrest affected the Écoles d'Arts et Métiers at Châlons and Angers in 1826, the director of Angers was told by the Minister of the Interior that too much theory and not enough hard grind in the workshops, together with a tendency of accepting boys as old as sixteen and seventeen, not a few of them of prosperous families but rejected by colleges for stupidity or misbehaviour, had combined to provoke the disturbances.[11] There was clearly a misunderstanding as to the purpose of the school. The director replied that 'almost all the youths who populate our schools have been born to a situation that places them above the class of ordinary workmen'; that without the Arts et Métiers three-quarters of them would be learning Greek at dusty municipal colleges with greater dangers of 'vanity'; and that there was nothing impractical about training surveyors, draughtsmen for architects, and *conducteurs des Ponts et Chaussées*, as well as masons and carpenters.[12] Despite his protests, an official commission reported that numbers at the schools should be limited, the age of entrance reduced to between thirteen and fifteen, that manual work be increased to fill two-thirds of the timetable, and that 'an insurmountable barrier' be erected

---

[10] A.D. Nord 1T 158/1, Baron Dupin to prefect of Nord, 30 Dec. 1825; 1T 56, Comte de Muyssart to prefect of Nord, 28 Feb. 1829.

[11] A.N. F17 14318, Min. Int. to director of École d'Arts et Métiers, Angers, 10 May 1826.

[12] A.N. F17 14318, Billet, director of École d'Arts et Métiers. Angers, to Min. Int. 27 June 1826.

against useless subjects such as figure-drawing, literature, and history.[13] After the Revolution of 1830, moreover, the course was reduced from four years to three, and candidates were required to have spent one year apprenticed to a trade before entering the school.[14]

A more real problem in technical education was betrayed by the term *arts industriels* or *arts mécaniques*. With the exception of the Arts et Métiers which had some concept of engineering, professional education concentrated on perfecting the skills of the artisan, and that at the artistic, luxury end of the market. A new awareness of this shortcoming made itself felt after 1830. At the drawing-school of Nîmes the emphasis was on ornamental and figure-drawing, not line-drawing. Yet there was a demand for competent draughtsmen, even in the silk industry, which was having to summon them from Lyon, while nothing was meeting the needs of stone-masons and architects, joiners and cabinet-makers, locksmiths and mechanics. As a result of a report by a municipal commission in 1834, line-drawing was introduced into the drawing-school of Nîmes,[15] and industrial drawing formed an important part of the course at the weaving-school set up in 1836.

At the other end of France, where the growth of a large-scale textile industry was in full spate, mill-owners were realizing the difficulties posed by a shortage of skilled foremen and managers, those men said by a director of one of the Écoles d'Arts et Métiers to be 'situated between the engineer who invents and the practician who executes, capable of understanding the one and supervising the other'.[16] In 1837, the municipal council of Lille, dominated by the textile *patronat*, began its campaign for the establishment of a third École d'Arts et Métiers there.[17] Told that there

---

[13] A.N. F17 14318, report of commission on reform of Écoles d'Arts et Métiers to Min. Int. 22 Sept. 1826. The royal ordinance embodying their proposals is of 31 Dec. 1826.

[14] A.N. 14317, royal ordinance of 23 Sept. 1832.

[15] A.M. Nîmes R271, report of commission of municipal council of Nîmes, 30 Jan. 1834.

[16] A.M. Angers 26R2, Dauban, director of École d'Arts et Métiers, Angers, to mayor of Bordeaux, 1 Dec. 1836.

[17] A.D. Nord, 1T 156/1, municipal council of Lille, 13 Dec. 1837.

would be more room at Châlons when a third school was opened in the Midi, they had the chagrin of witnessing the foundation of that school in a third non-industrial area, at Aix-en-Provence, in 1843. Indeed, 1843 was a good year for such large-scale industry as existed in the Midi: the mayor of Alès, the *ingénieurs* of the Corps des Mines, and the directors of the mining companies of La Grand'Combe, Rochebelle, and Bessèges in the Gard obtained the foundation of a school to train *maîtres-ouvriers-mineurs* from candidates who had spent at least a year in the pits, in one of the wings of the municipal college of Alès.[18]

At the beginning of the railway age, the government took an interest in those spheres of professional education that were linked with engineering and mining industries, but abandoned responsibility for perfecting the *artisanat* to the towns and manifested no concern about the state of agriculture. The movement to eliminate agricultural backwardness and increase food production, though a precondition of an expanding industrial economy, was essentially local and private. Agricultural improvement was confronted by the structural problem of a gulf between landlords who were interested only in rents or, at best, had the amateur, bookish interest in agronomy of the agricultural societies, and peasants hamstrung by lack of capital and short leases.[19] The *comices agricoles*, associations of like-minded, improving farmers, offered some possibility of bridging the gap, and in Ille-et-Vilaine Louis de Lorgeril founded one in the canton of Tinténiac as early as 1817.[20] Landowners of the Gard, on the other hand, could only reflect on the want of understanding between non-resident proprietors and the poverty and insecurity of tenure of their *métayers*.[21]

Improvement in the early nineteenth century was initiated

[18] A. Mauban, *L'École des mines d'Alès et son histoire* (Alès, 1950).

[19] The intellectual, paternalistic 'despotisme éclairée de l'agriculture' is described by A.-J. Bourde, *Agronomie et agronomes en France au XVIIIe siècle* (Paris, 1967), pp. 987-96.

[20] M. Goyet, 'Les Comices agricoles et les sociétés d'agriculture en Ille-et-Vilaine, 1831-1905' (*mémoire de maîtrise*, Rennes, 1968), pp. 8-12.

[21] A.D. Gard 13M1, mayor of Aramon to prefect of Gard, 7 June 1822; de Lacour-Lagardiolle, landowner and mayor of Conqueyrac to prefect of Gard, 8 Jan. 1822.

by an élite of landowners who undertook agriculture not as
rentiers but as practical farmers, developing model estates
to which they annexed institutions of education and workshops for the production of agricultural implements and from
which they propagated literature to vulgarize new scientific
methods. Their doyen was Alexandre Mathieu de Dombasle
(1777–1843), grandson of an ennobled *officier* of the duchy
of Lorraine, who in 1822 leased the estate of Roville near
Nancy, developing it as a model farm with the aid of subscriptions from local landowners gathered into an association
by the prefect of Meurthe, Comte Alban de Villeneuve-
Bargemon.[22] The Institut de Roville opened with forty-five
pupils, a workshop was founded in 1823 to build machines
like the light, manœuvrable Dombasle plough, while treatises
like the *Calendrier du Bon Cultivateur* broadcast the latest
methods. The movement fanned out across the country. One
of Dombasle's pupils, Colonel Auguste Bella, rented an estate
from Charles X at Grignon, near Versailles, in 1827, with the
help of a similar company, adding workshop and school,
while another pupil, Jules Rieffel, was put in charge of a vast
tract of Breton moorland at Grandjouan (Loire-Inférieure)
by a *négociant* of Nantes, raised capital by means of a company formed in 1833, and set up a farm-school in 1833.

The penetration of these new practices through to the mass
of the peasantry seemed an insuperable problem. Yet in Ille-
et-Vilaine landowners were concerned lest the combined
effects of routine in farming, recruitment under the Gouvion
Saint-Cyr law of 1818, and education which would habituate
the solid peasants' son to 'laziness and disgust with the
labours of the fields' might not empty the countryside
to the benefit of the growing towns.[23] It was as a result
of the initiative of Pierre Legrand, promoted Rector of the
Academy of Rennes after the 1830 Revolution, that Ille-
et-Vilaine moved into a pioneering position. Legrand not
only founded the École normale d'instituteurs;[24] he also

[22] Édouard Bécus, *Mathieu de Dombasle, sa vie et ses œuvres* (Paris, 1874),
passim.

[23] A.D. I-et-V 26M1, Jullien du Plessis, landowner of Argentré, to prefect of
Ille-et-Vilaine, 21 Dec. 1829. See also letter of Bonapartist colonel Félix Guim-
berteau de Lamalotière to prefect, 29 Jan. 1830.

[24] See above, p. 95.

put forward a plan to annex to it an agricultural school. The young teachers, who should for the most part be sons of 'prosperous farmers', would learn enough about modern agriculture not to make 'consummate agriculturists' of their pupils, but to spread through the Breton countryside 'firstly, ideas of improvement and subsequently, the improvements themselves'. In addition, in a period when the authorities were seeking to destroy the myth that teaching was a means of *déclassement* and, having provided for the towns, to develop rural schools, agricultural training would serve to reconcile the *instituteurs* to their rural mission.[25]

The threefold task of the agricultural school, launched early in 1833, was therefore to impart some agricultural knowledge to pupil-masters, to train a certain number of 'young peasants', and to provide a terrain for the experiments of the departmental Agricultural Society, refounded after a sixty-year interval in 1831.[26] The government provided an initial grant of 3,000 francs and in 1835 founded six scholarships at 300 francs each, though these were later financed by the department. The Conseil général made available 1,500 francs to rent a farm, but this became inadequate after the agricultural school moved to the larger estate of Trois-Croix in 1837, the difference having to come from the farm profits.[27] The school was thus in an ambiguous position, as an educational establishment partly financed out of the public purse, and as a farm (and later workshop for agricultural machinery) run at the 'risques et périls' of the director.

In Jean-Jules Bodin, Legrand found a director of the first order. A Sarthois, the son of an *officier de santé* who had gone to Bella's school at Grignon after failing his medical studies at Tours, Bodin was a convert to the cause of agricultural progress through education. In some ways his approach was intellectual. Like Olivier de Serres, he was steeped in Virgil's *Eclogues* and the classical rustic tradition while, like the Physiocrats, he

[25] A.D. I-et-V 2T, École normale de Rennes, Legrand to Conseil d'administration of École normale, Rennes, 30 April 1832.

[26] A.D. I-et-V 26M1, Prospectus of École d'agronomie, Rennes, 26 Jan. 1833.

[27] See H. Le Gall and A. Bessec, 'L'École des Trois-Croix: un établissement au service du progrès agricole, 1835-1889' (*mémoire de maîtrise*, Rennes, 1972).

believed that land was the only source of public wealth; commerce and industry only made it 'change place'. But he had a deep sense that progress would only be ensured once the gap between the discoveries of science and the everyday practice of farming had been bridged by 'a reasoned practice'. Though only ten or fifteen youths entered the agricultural school every year, Bodin's policy of selection was intractable. The candidates should be 'sons of farmers, intending to become farmers themselves'.[28] Unlike himself, they should not be drop-outs from colleges or *pensions*, bringing with them the evil influence of the city and trying agriculture as a *pis aller*, but have a rural background. They should be neither landowners who did not intend to farm themselves nor agricultural labourers, farm-hands, or waggoners who had not the property or capital to put their training to use. He wanted solid *laboureurs*, 'the active and influential class',[29] an élite of enlightened peasants that would carry the gospel of agricultural progress back to the villages and propagate it there.

Despite the low standards of Bodin's recruits—their primary education was 'very neglected and often non-existent'[30]—he did manage to draw on the sons of peasant-farmers of the basin of Rennes and ensure that after the two years' training they returned to act as a leaven on the community. Bodin's influence on the École normale was probably less, inculcating 'a taste for the countryside' into youths who wanted only to leave it behind, and withdrawing his services in 1840 after the authorities reduced his salary. On the other hand, a model farm, lying off the main Rennes to Saint-Malo road, Trois-Croix, became 'the rendezvous of good farmers',[31] who studied his war on the 'infernal circle' of cereals and fallow by integrating artificial meadows and new root-crops, purchased guano fertilizer without the risk of fraud, and paid a small fee for the 'sallies' of his prize Durham and Ayrshire bulls. The workshop was turning out steam-powered threshing-machines by 1846, and in 1840

---

[28] A.D. I-et-V 25M1, Bodin to prefect of Ille-et-Vilaine, 1 Feb. 1838.
[29] J.-J. Bodin, *École d'Agriculture de Rennes, rapport au Conseil général,* 1849, p. 3.
[30] Ibid., 1841, p. 5.
[31] Ibid., 1844, p. 4.

Bodin's *Éléments d'agriculture* won the prize offered by the Société royale et centrale d'agriculture for the best elementary treatise, and was widely distributed with the aid of a state subsidy.

The success of Bodin at Rennes may best be gauged by examining the record of the Nord and the Gard. Here, there were no local initiatives, and the pointlessness of government initiatives operating in a vacuum was amply demonstrated. In 1838 the Minister of Public Instruction, Salvandy, required *écoles normales* to run courses on the grafting and pruning of trees, as a gesture towards agriculture education. Douai and Nîmes both responded dutifully, but at Nîmes the director reported that drought had struck: 'the lessons were thus confined to theory, and do not seem to have produced much effect'.[32] It was not until the Second Republic that the government planned a nation-wide structure of Farm Schools in every department, then in every *arrondissement*, regional agricultural schools, and an Institut national agronomique to serve as the keystone of the edifice at Versailles. The Institut closed in 1852, Grignon and Grandjouan were assimilated as regional schools, and most of the Farm Schools, despite the state's payment of staff, boarding fees, and an end-of-course premium of seventy-five francs, came rapidly to grief. In Ille-et-Vilaine, Trois-Croix was actually threatened by the bids of other landowners, notably in the heathlands of the west of the department, to have Farm Schools annexed to their properties for the purposes of *défrichement*.[33] 1849 saw the foundation of Farm Schools in French Flanders at Templeuve and in the Gard at Mas-le-Conte, between Nîmes and Alès. The first, however, closed in 1851, the second in 1853. The reasons are not difficult to discover. Whereas Bodin was a dedicated agronomist and improver in the tradition of Dombasle, the directors at Templeuve and Mas-le-Conte were merely the owners of the properties and that of Templeuve, at least, was accused of

---

[32] A.D. Gard 1T45, director of École normale, Nîmes, to *conseil d'administration*, 11 Sept. 1839.

[33] A.N. F10 2589, Eric du Béru was supported by legitimist deputy Audren de Kerdrel, de Rochefort by the sub-prefect of Redon (letters to Min. Agric., 1 April 1849, 8 Oct. 1849).

using his charges as a source of unremunerated labour.[34] In addition, the prosperous farmers of Flanders, practising some of the most advanced agriculture in the country, saw no need for training. One proprietor who bid for the Farm School and then withdrew thought 'this institution more desirable in those departments where agriculture is more backward and labour at less of a premium than our own'.[35] His last words give the clue to a final problem, that of the dependence of peasant families on the free labour of their children, who, at the entrance age of sixteen, were capable of a man's work, either on their own properties, or hired out elsewhere. This dependence would be all the greater in the *garrigue* scrubland around Mas-le-Conte, the director of which reflected that even if the training were free, poor peasant families could not afford the *trousseau* so that the few inmates were going around in rage.[36]

If the inauguration of the Second Empire saw agricultural education in the doldrums, it witnessed an important attempt to launch a new form of industrial education. The moment was propitious. There was a sentiment that the revolutionary leaders of 1848–52 had been young men rendered ambitious by over-education in Greek and Latin, disappointed to find the liberal professions overcrowded and their path to public office blocked, who vented their rage by attempting to destroy the social order.[37] But this same ambition had carried middle-class talent away from the productive sectors of the economy. If this could be remedied, it was said, then revolutionary discontent would be replaced by national prosperity. The most significant initiative came from a private individual, César Fichet, director of a private industrial school at Menars (Loir-et-Cher), who in November 1852

[34] A.D. Nord 1T 175/3, reports of JPs of La Bassée and Saint-Amand, 18 and 22 Aug. 1851.

[35] A.D. Nord 1T 175/3, Darche, proprietor of Gognies-Chausée to president of Société d'Agriculture of Avesnes, 13 Nov. 1848.

[36] A.D. Gard 6M 91, Baron-Vigne, director of Mas-le-Conte, to prefect of Gard, 24 Aug. 1849.

[37] See, for example, A. Audiganne, 'Du Mouvement intellectuel parmi les populations ouvrières', *Revue des deux mondes*, 10, 1851, 862. On the background, L. O'Boyle, art. cit., and G. D. Sussman, 'The glut of doctors in mid-nineteenth-century France', *Comparative Studies in Society and History* vol. 9, 1977, 281–304.

submitted a *Mémoire sur l'enseignement professionnel* to the Emperor. Deeply influenced by Saint-Simonian ideas, he attacked French education as having 'more taste for Romantic literature and politics than for practical skills', and argued that a training that united theory and practice would promote harmony between the productive classes and impart 'a movement ever onwards and upwards' to commerce and industry. Bifurcation was an important reform, but education in the University was by definition impractical. The E.P.S. had produced only 'mediocre copyists of works of art or quasi-*littérateurs* . . . bound to bring trouble into the workshops'. He proposed twenty new Écoles d'Arts et Métiers for boys of between twelve and fifteen, who would become 'théoriciens-practiciens' trained in drawing, mathematics, and the workshop, ready to become foremen and managers, engineers and architects, accountants and surveyors.[38] Fichet's disciple, Victor Denniée, a civil engineer who ran a similar industrial school in the rue Beaumarchais, Paris, put forward a similar case in his pamphlet, *De l'enseignement professionnel*, published in 1852. He quoted a recent statement of J.-B. Dumas that 'instead of these unemployed *bacheliers* embittered by their impotence, born petitioners of every public office, disturbing the state by their pretensions, we will see emerge from our *lycées* generations vigorously trained for the combats of production', and publicized a professional education that would train 'according to the degree of their intelligence and skill, competent workers, foremen, teachers, managers of factories, etc.'[39]

Fichet was successful in convincing Hippolyte Fortoul, Minister of Public Instruction and author of bifurcation, of the need for his *écoles professionnelles*, and Lille along with Mulhouse and Toulon were selected as sites.[40] At Lille, where Victor Denniée was appointed director on the advice of Fichet, two difficulties rapidly became apparent. Firstly, it was unclear how far the school was private, and how far it

[38] A.N. F17 11708, César Fichet, *Mémoire sur l'enseignement professionnel,* Nov. 1852.

[39] V. Denniée, *De l'enseignement professionnel, 1762–1852* (Paris, 1852), pp. 8, 16. See also R. Anderson, *Education in France, 1848–1870,* pp. 90-4.

[40] On the École professionnelle of Mulhouse, see R. Oberlé, *L'Enseignement à Mulhouse de 1798 à 1870* (Paris, 1961), pp. 173-93.

would be financed out of public funds. Secondly, it was never certain at what stratum of society the school was aimed, and whether it was designed to train skilled workers, foremen, or managers. As far as the first issue was concerned, Fortoul instructed the Rector of Douai in August 1853 to find premises where a professional school could be set up. The Emperor, he pointed out, was anxious that the lacuna between primary education, training 'the greatest number' for apprenticeship, and the *lycées*, training the bourgeoisie for the liberal professions, should be filled by schools preparing 'the intermediary class' for commerce and industry.[41] That November, César Fichet appeared before a commission of the municipal council of Lille to explain his plans,[42] and in January 1854 the mayor of Lille, Auguste Richebé, managed to convince a reluctant council to associate itself with the government's plans, financing the original establishment of the school in the old Mont-de-Piété and maintaining the building, so long as the state paid the salaries of the staff.[43] The confusion was in fact complete. As the École professionnelle du Nord revealed itself a fiasco, the municipality of Lille refused to have anything more to do with it, arguing that, apart from an initial injection of public money, it was a private school, that the town had acceded to the government's project only under pressure, and that the needs of the *classes moyennes* were amply satisfied by the *écoles académiques*, the E.P.S., the commercial classes of the Lycée, and the course in applied science which opened at the Faculty in 1854.[44] On the other hand, Denniée, who told Fichet '. . . really, I am starting to regret having got mixed up in this mess', argued that he had only accepted the post because he believed he had been appointed 'official of the University', and was calling the school the École impériale professionnelle du Nord.[45] As the creditors closed in, the

[41] A.D. Nord 2T 311/1, Fortoul to Rector of Douai, 29 Aug. 1853.

[42] A.N. F17 11708, commission of municipal council of Lille, 25 Nov. 1853.

[43] A.D. Nord 2T 311/1, municipal council of Lille, 20 Jan. 1854; Richebé to Rector of Douai, 23 Jan. 1854.

[44] A.D. Nord 1T 139/1, Richebé to Min. Instr. Pub. Rouland, 18 Oct. 1856.

[45] A.D. Nord 1T 139/1, Denniée to Fichet, 5 Sept. 1855; Denniée to Min. Instr. Pub., 12 June 1856, A.N. F17 11708, Inspecteur-général Magin to Min. Instr. Pub., 13 Dec. 1856.

government was indeed obliged to take over the school, appointing a *commission de surveillance* and making the primary-school-inspector Bernot the director of what was now the École des arts industriels of Lille.[46]

The second problem concerned the level of training offered by the École professionelle. Fichet told the municipal council of Lille in 1853 that it would turn out 'young industrial chiefs'. Richebé told the same body that it would be 'a nursery of trained foremen'. The prospectus published by a deflated Denniée in 1856 tended to the more modest view: the school would train 'foremen', 'qualified workers', and 'clerks' for the mines, railways, and Ponts et Chaussées.[47] The only point at which the industrialists of the Nord saw a use for school was in 1860, when the Free Trade treaty with England swept away the protectionist wall behind which the textile *patronat* had been sheltering. A petition signed by two hundred of them, headed by the brothers Scrive, Wallaert, and Bernard, demanded that the sons of their workers and foremen be adequately trained as foremen and production managers in an improved professional school at Lille.[48] Unfortunately, Bernot had set his sights on raising the status of the school, attracting 'the children of the industrial bourgeoisie who can pay the cost of their education'.[49] The prospectus that he published about 1864[50] announced that the school was training engineers and managers for the textile, machine-building, mining, chemical, sugar-refining, distilling and brewing industries, preferred candidates with a *baccalauréat* or diploma of special education, and required a boarding fee of 1,000 francs. Although the school population rose to ninety in 1867, the École des arts industriels was not a success, as Bernot admitted just before the fall of the Empire. The services of the school were not appreciated by the industrialists. In the Nord, it was possible to get rich

---

[46] A.D. Nord 1T 139/1, Min. Instr. Pub. to prefect of Nord, 2 Feb. 1857.

[47] A.D. Nord 1T 139/1, Prospectus of École impériale professionnelle du Nord, 30 Aug. 1856.

[48] A.D. Nord 1T 159/1, Petition of industrialists to prefect of Nord, 15 June 1860.

[49] A.D. Nord 1T 159/1, Bernot to prefect of Nord, 17 July 1860.

[50] A.D. Nord 1T 159/1, Prospectus of École spéciale des arts industriels et des mines, of Lille, c.1864.

without an education: 'there are still mill-owners who cannot read' and an engineer could be sacrificed for skilful salesmen and 'a foreman hard enough to extract the maximum amount of labour from the worker'. Bernot sought the patronage of the 'aristocracy of fortune', but the sons of manufacturers who had completed their studies at Marcq or some other college then went off to perfect their knowledge in some model factory or abroad, but not at Lille. Those of the *classes moyennes* who did want a technical education were put off by the inadequate workshops and poor teaching. The Commission de surveillance of the École des arts indus-triels, which included reputed employers such as Édouard Scrive, Auguste Wallaert, and Kuhlmann, president of the Chamber of Commerce, noted that one-third of the candi-dates rejected nationally from Châlons or the École Centrale at Paris came from the industrial north but that they were not interested in Lille as an alternative. Bernot knew that they received only 'college pupils in whom a dislike of study seemed to have revealed an aptitude to wield a hammer'.[51]

One shortcoming of professional education that emerges clearly is that it tended to function at the margins of the economic world, blind to real needs. If the professional aspect was wanting, then the educational function could go awry, provoking *déclassement* instead of countering it. This problem was already evident in the École des maître-mineurs of Alès. In the first place, the recruitment of the school was not limited to miners. Indeed, between 1851 and 1864 only 29 per cent were miners, and although 42 per cent in the widest sense came from the mining sector, 26 per cent were the sons of peasant-farmers, 11 per cent sons of artisans, and another 29 per cent were sons of managers, entrepreneurs, *employés*, and tradesmen.[52] On top of that, one of the directors of the mines of La Grand'Combe complained in 1859–60 that the training given by the school was too theoretical, so that what was being turned out was a half-breed of supervisors, somewhere between master-miners and

---

[51] A.D. Nord 1T 160/2, Bernot to prefect of Nord, 10 July 1870; report of commission de surveillance of École des arts industriels to prefect, 18 July 1867.

[52] École nationale des mines d'Alès, Cahiers des notes particulières des élèves 1851–64, 114 cases.

engineers.[53] A commission was set up by the school's Conseil d'administration with the result that in 1861 the course was simplified. Lastly, the École des maîtres-mineurs at Alès suffered from the problem of isolation from the mainstream of the economy. Though miners came to it from the coalfields of Carmaux, Décazeville, Rive-de-Gier, and Commentry, it remained a school for the centre and south of France, not for the north, where a third of France's coal was produced. In the Nord, opinion grew increasingly impatient with this state of affairs. In 1864 the Conseil général refused to finance candidates to go to Alès and sent them instead to the École des arts industriels.[54] On the coalfield, the Conseil d'arrondissement of Valenciennes campaigned for its own mining school, although after February 1870 it seemed that Douai would win the appeal, since it had the support of the Comité des houillères of the Nord and Pas-de-Calais, on the border between which it lay.[55]

Such agricultural education as existed under the Second Empire was a response, like industrial education, to the problem of the *déclassement* and *déracinement* of the rural population that was making a shortage of labour, a *manque de bras*, sharply felt in the countryside. In this sense, professional education, though ostensibly on the side of progress, had a conservative function, the shoring-up of the existing social order, and this function would become more pronounced as the century advanced.

To explain this depopulation of the countryside, various scapegoats were brought forward. The Free Trade treaty of 1860, that 'abolitionist' measure which removed France's Corn Laws, could be singled out by those who regretted the opening-up of the national market to cheap foodstuffs. Military service was another target. In 1866, Ange de Léon, a former royalist mayor of Rennes, stated that 'the rural population, being the healthiest and the least well off, provides the greatest number of them for the call-up and buys itself out most rarely'.[56] Recent research has indeed shown

[53] École nationale des mines d'Alès, Procès-verbaux des séances du conseil d'administration, 31 Oct. 1859, 19 July 1860.

[54] A.D. Nord 1N 106, Conseil général, 24 Aug. 1864.

[55] A.D. Nord 1T 164/1, mayor of Douai, to Min. Trav. Pub., 21 Feb. 1870.

[56] *Enquête agricole, 1866* (Paris, 1868), ii, 3.

that poor regions like Brittany tended to make a living by the 'substitution trade', while rich ones like Flanders managed to buy out their sons.[57] It was generally supposed by agrarians that youths corrupted by an urban, barrack life would not return to the villages. Lastly, the development of elementary education was itself seen to be a motor of *déclassement*. Peasants over-educated in schools were then sent on to college, cherished ambitions to enter the liberal professions, and, suffering 'bitter disappointments', became 'people who were a danger to society, or at the very least peasants of extreme indolence'.[58] The *instituteur*, that prime example of the rural *déclassé*, had a marked inclination towards republican politics; every effort should be made to 'preserve simple and modest tastes among the *instituteurs*, and to attach them by positive interest to the soil of the communes'.[59]

The answer to these dangers was agricultural education. Materially, by building up the prosperity of the countryside, and morally, by inculcating rustic and Christian values into the populations of the villages, the drift towards perdition in the towns could be checked. As Eugène Bodin, who was to succeed his father as director of Trois-Croix, put it, 'our goal is essentially moral'.[60] Trois-Croix reached a peak of prosperity around 1860, but by the end of the decade Eugène Bodin was noting, 'the shortage and expense of labour in the countryside makes being deprived of their children more and more difficult for peasants. So for a long time, we have seen applications for bursaries become rare.'[61] When after the death of Jean-Jules Bodin Trois-Croix was brought under the regime of the Farm Schools (1869) a premium of 300 francs was promised to those who completed the two-year course in an attempt to attract recruits. Even so, its record until then had been surprisingly

---

[57] B. Schnapper, *Le Remplacement militaire en France. Quelques aspects politiques, économiques et sociaux du recrutement au XIXe siècle* (Paris, 1968), pp. 79, 82.

[58] A.N. F17 9325, report of primary-school-inspector of Le Vigan, April 1856.

[59] A.D. Gard 1T 929, Rector of Montpellier to Insp. Acad. Nîmes, 23 April, 1865.

[60] A.D. I-et-V, 26M2, report of E. Bodin, 15 April 1873.

[61] E. Bodin, *École d'agriculture de Rennes, rapports au Conseil général*, 1869.

good. Of 343 scholars who passed through Trois-Croix between 1835 and 1869, 76 per cent returned to agricultural pursuits rather than moved on to become tax employees, railway clerks, tradesmen, soldiers, or *instituteurs*.[62]

Whatever the value of agricultural schools like Trois-Croix, a wider audience would be reached only by promoting agricultural education in the *écoles normales*. The conservative syllabus of 31 July 1851 which came near to turning *écoles normales* into seminaries[63] included this feature, but was not until 1855 that the combined forces of the Minister of Public Instruction, the Rector of Rennes, and the prefect of Ille-et-Vilaine, together with a grant from the Conseil général, managed to persuade Jean-Jules Bodin to resume his lecture course at the École normale of Rennes. By 1862 Bodin was announcing that 'the *instituteurs* on one side, the trainee farmers on the other, will destroy routine and propagate that solid knowledge without which the countryside will become deserted'.[64] The focus of attention, however, was not so much arable and animal husbandry as market gardening. The railways had created openings for market gardening, horticultural societies were springing up and denouncing the parlous state of the garden plots that peasants maintained alongside their main exploitation, and it was argued that lessons in the cultivation of fruit-trees and vegetable patches were especially appropriate for *écoles normales*.[65]

Rather than the government and *conseils généraux*, it was the agricultural and horticultural societies and the *comices agricoles* that provided the patronage for agricultural education in the schools during the Second Empire. The Horticultural Society of the Gard was responsible for founding a course at the École normale of Nîmes in 1864; the lack of such a society in the Nord delayed the appearance of any effective agricultural education there until 1867.[66]

---

[62] H. Le Gall and A. Bessec, op. cit., p. 153.   [63] See above, pp. 47, 102.

[64] J.-J. Bodin, *École d'agriculture de Rennes, rapports au Conseil général*, 1862, p. 2.

[65] A.N. F17 11713, President of Société Centrale d'Horticulture d'Ille-et-Vilaine to Min. Agric. Comm., 23 April 1856.

[66] A.D. Gard, 1T 929, director of École normale to Insp. Acad. Nîmes, 8 Feb. 1865; A.D. Nord 1T 183/1, Girardin, Dean of Faculty of Sciences of Lille, to prefect of Nord, 22 June 1865.

The horticultural societies awarded prizes and medals to the masters and pupils of the *écoles normales.* At Nîmes, the local society organized a course for *instituteurs* already teaching in the villages, along the lines of the *conférences pédagogiques.*[67] In the Nord, the agricultural society of Douai was giving lectures to *instituteurs* in 1858, while the Comice agricole of Cambrai distributed elementary treatises on agriculture and seed for the kitchen gardens attached to schools.[68]

The possession of a kitchen garden or experimental field was essential if agricultural education in he villages was to be in any way successful. A circular of Victor Duruy in 1867 urged communes to provide plots of at least ten *ares* for *instituteurs,* but few did. Apart from the cost involved it was often felt that cultivation would remove the teacher from his essential work in the class-room. The usefulness of a garden was not only pedagogic: the lower incomes of rural *instituteurs* and the high price of foodstuffs could also be guarded against by growing vegetables and keeping a cow.[69] Yet without a plot of land, agricultural education would remain fatally theoretical. The primary-school-inspector of Cambrai reported in 1867 that twenty *instituteurs* in his *arrondissement* had claimed to have given practical instruction in agriculture but he added, 'one must be careful, for that means that the pupils have read in class a book called *A Course in Practical Agriculture'.*[70] Even theoretical lessons could be rendered pointless by the lack of a good textbook. In Ille-et-Vilaine the agricultural society commissioned Bodin to write a little treatise for schools, and his *Lectures et promenades agricoles* appeared in 1857. Too often, however, the texts were heavy going and did not apply to local agricultural conditions. What text, inquired the primary-

[67] A.D. Gard 1T 929, primary-school-inspector of Nîmes to Insp. Acad., Feb. 1867.

[68] A.D. Nord 1T 67/14, reports of primary-school-inspectors of Douai and Cambrai, 30 June 1868.

[69] A.N. F17 10792, statements of Vallée, *adjoint* at municipal school, Rennes, 2 Feb. 1861, and F17 10776, Lavoisier, *instituteur* of Radinghem, Nord, 31 Jan. 1861.

[70] A.D. Nord 1T 67/30, primary-school-inspector of Cambrai to Insp. Acad. Douai, 9 Feb. 1867.

school-inspector of Uzès, ever dealt with the cultivation of the mulberry tree, so important for the silk industry of the Cévennes?[71] School libraries were often replete with agricultural treatises, their distribution subsidized by the government, but it was reported from Douai that 'no one reads them. If, by some extraordinary whim, some peasants do read, it is rare that they choose works on agriculture.'[72] In Brittany the agronomists tried to leap over this obstacle by writing in the popular almanacs which the *colporteurs* carried in their bundles to hamlets and farms. Jules Rieffel published advice under the pseudonym, *Maître Jacques*; Eugène Bodin followed his example in almanacs published in Dinan.[73]

The propagation of agricultural education could easily be defended for a backward region like Brittany, but in northeast France no one could doubt that an agricultural revolution had taken place. The Inspector of the Academy of Douai saw it as superfluous 'in a region as fertile and well cultivated as this where the best agricultural methods were in use before the development of primary education'.[74] In the *arrondissement* of Cambrai, where 'many farmers are *too* capable', *instituteurs* who were 'as foreign to the cultivation of wheat as they were to that of oranges and olives' excited only ridicule by lecturing on agriculture.[75] Since knowledge of agriculture was required only for a *brevet complet*, a qualification that few teachers had, even their theoretical grounding was severely limited. In such circumstances, the true purpose of agricultural education became clear. It was not so much to improve farming as to check the tide of rural depopulation, and that not by science but by the perpetuation of a mystique of rural life. An *instituteur* of the outskirts of Nîmes demonstrated that he had understood this mission when he suggested that informal talks and readings on

---

[71] A.D. Gard 1T 929, primary-school-inspector of Uzès to Insp. Acad. Nîmes, 27 Sept. 1867.

[72] A.D. Nord 1T 67/30, primary-school-inspector of Douai to Insp. Acad. Douai, 8 Feb. 1867.

[73] F. Foliard, 'Les Almanachs populaires bretons au XIXe siècle' (*mémoire de maîtrise*, Rennes, 1970), pp. 67–71; writings of Bodin in *Le Bon Almanach Chantant de la Bretagne*, 1879, 1880.

[74] A.D. Nord 1T 67/30, Insp. Acad. to Rector of Douai, 19 Feb. 1867.

[75] A.D. Nord 2T 299/14, report of *instituteur* of Maretz, 26 July, 1868; 1T67/30, primary-school-inspector of Douai to Insp. Acad. Douai, 9 Feb. 1867.

agriculture 'portraying rustic life and its pleasures would be most appropriate to inspire children with a taste for the countryside and to make them sense the happiness of the simple and hard-working life of the peasant'.[76] Unfortunately, this Rousseauistic view of the countryside was entirely bourgeois and in no way corresponded to the attitude of the peasant. For the peasant, work in the fields was arduous, nature was capricious, landlords were brutal. Towns offered one temptation which outweighed all others: higher wages. The Inspector of the Academy of Douai was obliged to admit in 1867 that 'primary schools . . . have not . . . contributed to spread and strengthen a love for work in the fields, and have not prevented emigration towards the centres of industry'.[77] Once again, professional education was functioning on the margins of economic reality and was failing to prevent the *déclassement* that was its object.

It is clear then that professional education was organized entirely on a local basis before 1880. The initiative was taken by municipalities, mining companies, or, in the case of agricultural education, by a small group of dedicated physiocrats. Neither departments nor the state gave much assistance; when they did, it was as often as not ill-directed. The Arts et Métiers were set up in all the wrong places, the Farm School project of 1848 all but wrecked flourishing local schemes, and half-hearted support of the *écoles professionnelles* after 1852 condemned them to impotence unless they could be revived by local authorities.

The relationship between technical education and the locality was extremely delicate. It was of great significance for backward regions like the west, less important for the prosperous agricultural–industrial economy of the north of France. Directed towards the artisan élite or skilled workers it might be of great value, but there was no point during the early phase of the industrial revolution, when the self-taught entrepreneur and family firm were so important, in attempting to train technocrats, nor was it possible to reach the mass of workers and peasants. At the end of the day, the

---

[76] A.N. F17 10781, statement of Tarron, *instituteur* of Milhaud, Gard, 12 Dec. 1860.

[77] A.D. Nord 1T 67/30, Insp. Acad. to Rector of Douai, 19 Feb. 1867.

raison dêtre of professional education too often seemed not
to stimulate growth areas but to prevent *déracinement.*
A peasant mentality was inculcated into trainee *instituteurs,*
as in the Swiss seminaries, in order to blunt their ambition,
so that they in turn could prevent their pupils from drifting
to the towns in search of work. The discontented Latinists
whose generation were responsible for the Revolution of
1848 were guided into technical colleges by Saint-Simonian
eccentrics. But the danger lay in dispensing any form of
education; the tendency of students was to make of it what
they would.

10

The Agonies of Modernization

In its strategy of transmitting scientific developments to economic practice, professional education gave the impression of being on the side of progress. In fact, its basic function was to counter the stresses and strains to which France was subjected as a result of rapid economic change, to defend the existing social order against disintegration. Never was this more true than at the end of the nineteenth century when the pressure of international competition hounded France into the depression of 1873–95 and exposed three major social problems: rural depopulation, the crisis of apprenticeship—age-old but now particularly acute—, and the mushrooming of the service sector.

The rural economy faced intense competition on two counts: agricultural and industrial. The opening of the American wheat plains and the development of steamships flooded European markets with cheap cereals, so that between 1873 and 1894 it has been calculated that agricultural prices fell by 37 per cent, production by 11 per cent and incomes by 10 per cent.[1] Land values declined, farmers were unable to pay rents and taxes and fell into debt. At the same time, the destruction of rural industry, in competition with that of urban centres which drew on world markets for raw materials and used new technological methods, was completed. Ille-et-Vilaine lost its forest ironworks of Pléchâtel, Martigné-Ferchaud, and Paimpont during the 1860s. Cottage hemp- and linen-weaving disappeared: Amanlis, once known as the 'city of weavers', had fifty-three weavers in 1872 but only one in 1896.[2] On the Channel coast, the two-masted schooners which sailed to the cod-banks of Newfoundland and Iceland had largely been ousted by the steam-trawlers of Boulogne by the First World War. In the Nord,

[1] F. Caron, *Economic History*, p. 130.
[2] A.D. I-et-V 23 MC 177, Amanlis, census, 1872; 23 MC 234, Amanlis, census, 1896.

the cottage weavers of the Cambrésis and Lys valley disappeared as the textile conglomeration of Lille–Roubaix–Tourcoing expanded, sucking in cotton from America, linen from Russia, and wool from Argentina and Australia. In the Midi the silk industry of the Cévennes suffered in competition with Lyon, now drawing on the cheaper and better eggs and cocoons of the Far East. In some instances the rural economy, weaning itself away from cereals and adopting industrial techniques, demonstrated a capability of resisting such challenges. Wine-growing in the south of France, struck down by phylloxera in the 1870s, recovered with the planting of American vines until the following crisis, that of over-production after 1903. In the Nord endless fields of beet were put to use in the sugar-refining, brewing, and distilling industries. There, and in Brittany also, arable was turned over to pastureland and dairy herds kept to supply urban markets with milk, butter, and cheese. Even so, rural depopulation took its toll. Within departments, populations shifted to the expanding single-industry towns: the woollen town of Fourmies and the steel centres of Denain in the Nord, the shoe-manufacturing phenomenon of Fougères in Ille-et-Vilaine, and in the Gard, Alès, 'which looked like a mushroom town of the New World, a sort of Chicago growing out of a traditional landscape'.[3] Frequently, where there was insufficient industrial capacity to take up the surplus population, as in Brittany, the movement of populations was towards Paris, albeit to the occupations of the bottom of the social scale: unskilled labour on the railways, domestic service, and the sweat-shops of the clothing trade.[4] Clergy and notables such as Viscount Charles de Lorgeril, secretary-general of the Association bretonne, might protest,[5] but the population of Ille-et-Vilaine declined absolutely after 1891 and in the five-year period 1896–1901 alone it lost twenty-one of every thousand of its inhabitants through emigration.

[3] André Chamson, *Le Chiffre de nos jours*, p. 18.

[4] J. Choleau, *Questions bretonnes des temps présents* (Vitré, 1942), i, 124. He prints an occupational analysis of 10,600 Breton emigrants in the department of the Seine in 1911.

[5] C. de Lorgeril, *La Propriété foncière en face de l'abandon de la culture par les populations rurales* (Saint-Brieuc, 1886).

The urban predicament was no less severe. Industrialists met competition by increasing productivity, but this entailed greater mechanization and a more intense division of labour. Not only was the provision of apprenticeship training un-economic but the destruction of artisanal skills also made it unnecessary. Industries, particularly heavy industries such as textiles, mining, metallurgy, and chemicals, were concentrating in larger units, each employing a mass of wage-workers. In Lille, the proportion of textile establishments employing over a hundred workers increased from 44 per cent in 1891 to 66 per cent in 1911.[6] At the same time, however, the proletariat was declining in size relative to the rest of the population.[7] As small entrepreneurs gave way to large firms, so between the owners and the manual workers new strata of employees multiplied, from clerks, accountants, and foremen to production managers and engineers. It was this sector, the cadres, that formed such an important element in the late nineteenth-century economy.

The heavy industry of the Nord and, to a lesser extent, the Gard, was one area in which the demand for cadres was evident. The exodus of young men from the north of France to the industrial schools of Belgium brought home to the Conseil général of the Nord the need to reorganize the École des arts industriels et des mines of Lille. A commission of industrialists including a chief engineer of the Ponts et Chaussées, Masquelez, was appointed in 1872 to visit industrial schools at Amiens, Mulhouse, Liège, Ghent, and Antwerp. Bernot, threatened with redundancy, now claimed that manufacturers were looking not for a 'learned initiation' but a practical training that could be acquired only in the workshop. The École des arts industriels should therefore be stripped down to a one-year technical course, 'a methodical transition between the college and the factory'.[8] The conclusions of Masquelez, on behalf of the commission, were quite different, proposing a two-year course for foremen crowned by a three-year course serving as 'a sort of regional École Centrale' funded partly by the municipality of Lille, partly

[6] L. Trénard, ed., *Histoire des Pays-Bas français*, pp. 442-3.
[7] See above, p. 282.
[8] A.D. Nord 1T 159/1, report of Bernot, 13 May 1872.

by a *société industrielle* set up by the Chamber of Commerce, modelled on that set up by Dollfus at Mulhouse.[9] The new school, the Institut industriel, agronomique et commercial which opened under the directorship of Masquelez in the autumn of 1872, duly presented this two-tier combination. It was partly for the sons of landowners, industrialists, bankers, and *négociants* intending to continue the family business, partly for 'a few chosen individuals fitted to become the precious auxiliaries of the heads of large establishments'. The courses in agriculture, commerce, and technology (mechanics, textiles, and chemistry) were for two years, while the engineering division, subdivided into machine-building, mining, and spinning and weaving, was for three.[10]

The Institut industriel was not exempt from the three difficulties faced by professional schools: marginality, *déclassement*, and conflict with the central government. It rapidly became clear that the *négociants* of the Nord saw no advantage in having graduates trained in the commercial section, and this was closed in 1881.[11] The agricultural section, which seemed to serve no purpose, was split between the technology and engineering sections in 1883, and phased out altogether in 1889. Within the industrial sections, the mining option was empty in 1884 and suppressed, while the new director noted that 'the spinning and weaving section includes only a few pupils who, when they leave, will have the greatest difficulty in finding jobs'.[12] It is true that the depression of the 1880s was a difficult time for finding any employment, but it was also a question of responding to new areas of industrial development. Those who graduated with engineering qualifications were immediately taken on in construction workshops, by the railway companies or the navy; and the school miscalculated badly by failing to turn out chemists at a time when the chemical industry was expanding.

[9] A.D. Nord 1T 139/1, report of Masquelez to mayor of Lille, 6 Aug. 1872.

[10] A.D. Nord 1T 159/1, Prospectus of Institut industriel, agronomique et commercial de Lille, 1872.

[11] A.D. Nord IT 159/3, reports of Masquelez, 22 June 1878, 18 June 1879.

[12] A.D. Nord 1T 138/1, Director of Institut industriel, Olry, to prefect, n.d. [1884 or 1885].

The second problem related to the fact that, among the cadres, there were too many officers and not enough NCOs. The École Centrale in Paris was supposed to turn out engineers and the Arts et Métiers foremen, but just as in 1826 Baron Dupin had prophesied that the mechanical arts would 'cease to belong to the purely mechanical class and rise to the level of liberal arts or fine arts',[13] so the graduates of the Arts et Métiers, if they started out as skilled workers, foremen, and draughtsmen, rose in the course of their career to the higher echelons of industry.[14] In 1879 a manufacturer of Lille complained that 'instead of remaining foremen, most of these young men rapidly become overseers, managers and employers',[15] and the same accusation was levelled against the Institut industriel. Any student worth his salt hurried to leave the technology division, which was weaker than the Arts et Métiers, to enrol in the engineering division, acquiring after three years 'a scientific or technical knowledge that was similar to that of graduates of the École Centrale'.[16] When a campaign was launched in Lille to annex an École d'Arts et Métiers—the country's fourth—to the Institut industriel, allowing some movement between them, Masquelez had to defend himself against criticisms that working-class boys would rush to become engineers by insisting that 'we can limit *déclassement* as much as we like, by reducing the number of scholarships'.[17]

The question of France's fourth École d'Arts et Métiers laid bare the third problem, that of tensions between the locality and central government. Characteristically, it was the Catholic *patronat* of Lille that took the first initiative, setting up a limited-liability company in 1877 to found a 'Catholic Châlons', entrusted to the Frères des Écoles Chrétiennes.[18]

---

[13] C. Dupin, *Effets de l'enseignement populaire*, p. 16.

[14] C. R. Day, 'The Third Republic and the development of intermediate education in France 1870-1914', *Proceedings of the Fourth Annual Meeting of the Western Society for French History*, 1976 (Santa Barbara, 1977), pp. 348-9.

[15] A.N. F17 14337, Carlos Delattre, president of the Commission des écoles académiques et des cours techniques of Roubaix, 31 May 1879.

[16] A.D. Nord 1T 159/7, Olry, director of Institut industriel, 1885.

[17] A.D. Nord 1T 156/1, memorandum of Masquelez for municipal council of Lille and Conseil général of Nord, 7 May 1879.

[18] A.F.E.C. 654/6, notarial act, 19 Oct. 1877; L. Baunard, *Les Deux Frères*, pp. 78-80.

On the lay side, municipalities and associations of industrialists throughout the north of France, orchestrated by Pierre Legrand, *avocat* and deputy of Lille, mobilized to obtain the fourth school at Lille, and Lille and the department of the Nord offered over a million francs in land and subsidies towards the foundation. But though a law of 10 March 1881 conceded that the fourth school would be at Lille, it took Pierre Legrand two periods as Minister of Commerce in 1882-3 and 1885 to get the National Assembly to vote its share of the budget.[19] Further problems delayed the building of the school and Lille's Arts et Métiers was not opened until 1900, two years after the archbishop had blessed that of the Catholics, run in the event by the Jesuits.

In another branch of heavy industry, that of coal-mining, the positive contribution of the state was even less. Douai acquired its own École des maître-mineurs in 1878, but it was financed by the town, the departments of the Nord and Pas-de-Calais, and interested mining companies. Even there, the original function of the school seemed to be perverted. Designed to train miners in their mid-twenties with several years' experience in the pits, the school had only nineteen students in 1881. Most miners of that age, it was said, were married with families, and could not sacrifice the wages required to maintain them.[20] In addition, there seemed little point in returning to the school-bench for two years, since it was the policy in many mines to recruit master-miners from the better workers available. Lastly, at a time of economic depression, intensified in the mining industry by the rising importation of British coal, there could be no guarantee that even a qualified graduate of Douai or Alès would not have to work as a common miner. In the period 1873-9, only 2.7 per cent of graduates from Alès returned to work as miners; in the period 1880-90, the proportion rose to 13.7 per cent, and by way of compensation there was a considerable emigration of qualified miners to the colonies.[21]

---

[19] A.D. Nord 1NI 23, Conseil général, 28 Aug. 1879, pp. 203-18; P. Arnous, *Pierre Legrand, un parlementaire français de 1876 à 1895* (Paris, 1807), pp. 332-8.

[20] A.D. Nord 1N 128, report of director of École des maître-mineurs, Douai, 30 July 1881.

[21] A.D. Gard, Conseil général, report of director of Écoles des maîtres-mineurs,

If the older miners felt little inclination to attend the schools of Alès or Douai, this was not the case among younger men of eighteen or twenty who had some expertise in drawing and geometry and had worked in the offices of the mining companies rather than underground. The tendency to recruit younger candidates was reinforced by the universal military-service law of 27 July 1872 which created an incentive to complete training before service fell due at the age of twenty to avoid the interruption of studies.[22] These younger men were destined not to become master-miners but mine-surveyors, and as the proportion of graduates of Alès who became master-miners fell between 1873–9 and 1880–90 from 29.2 per cent to 14.3 per cent, so that of those who became surveyors rose from 23.0 per cent to 25.2 per cent. The tendency to produce higher cadres was reinforced by a decree of 2 January 1883 which opened the post of *garde-mines* (called *contrôleur des mines* after 1894) to the top three graduates of Douai and Alès in any year. By the end of the decade, a controversy flared up on this subject between the Minister of Public Works on the one hand and the *conseils d'administration* of the mining schools, dominated by directors of mining companies, on the other. Whereas the mining companies demanded 'a general improvement of scholarship' in order to supply more and better cadres, the government feared that the mining schools were being shifted away from their original function of training master-miners, and clipped the syllabus into shape under a decree of 18 July 1890. The Conseil d'administration of Alès protested in 1894, pointing out that most master-miners were trained on the job, while France, unlike Germany and Austria, had no special *Markscheiderschulen* for the training of surveyors.[23] It was simple for the government to blame the Écoles des maîtres-mineurs for producing *déclassés*. In fact, the developing technology of extractive industries

Alès, 20 June 1885. The figures are collated from the annual reports of the director.

[22] École nationale des mines d'Alès, Procès-verbaux des séances du Conseil d'administration, 25 Nov. 1872, 25 Nov. 1873.

[23] École nationale des mines d'Alès, Procès des séances du Conseil d'administration, 19 Dec. 1894.

made it inevitable that the trained minority would rise to higher positions. Of 508 graduates of Douai in the period 1880-1908, 25.7 per cent became surveyors and chief surveyors, 35.9 per cent overseers, *porions*, and *chefs-porions*, while 5.9 per cent became *controleurs*, and 14.4 per cent engineers, managers, and directors.[24] The Écoles des maîtres-mineurs had gone the same way as the Écoles d'Arts et Métiers.

In those branches of industry where artisanal methods predominated, the problems were somewhat different. Mechanization and the division of labour required to compete in the open market were undermining the skills of the artisan and eliminating apprenticeship in the workshop. The *chambres syndicales* of metal-workers and of joiners and cabinet-makers told the mayor of Nîmes in 1883 that 'apprenticeship has been profoundly changed by the ever increasing use of improved machinery, so that the time during which the apprentice receives training from the employer has been considerably reduced' and higher wages could not compensate for the suffering of professional education.[25] These developments were not the only tragic cost of industrialization. The proportion of luxury goods among the manufactured goods exported from France, especially in textiles, was very high, so that the disappearance of fine craftsmanship, the unique blend of art and trade, the *arts industriels*, would have serious consequences for the economy.

Alongside the tradition of the Arts et Métiers there was also, of course, that of the municipal drawing-schools, the *écoles académiques*, and these continued to develop with the times. In the Nord, the Écoles académiques of Douai, where 'nothing was done but art for art's sake', joined forces early in 1870 with the E.P.S., where the pupils did not have 'the taste and feeling for the beautiful', in order that the teaching should have 'a double tendency, industrial and artistic'.[26] At Roubaix, the Écoles académiques acquired a painting and drawing-master who would train 'not artists,

[24] M. Mettrier, 'L'École des maîtres-mineurs de Douai', from *Lille et la Région du Nord* (Lille, 1909), pp. 6-9.

[25] A.M. Nîmes R 271, petition of July 1883.

[26] A.D. Nord 1T 140, Écoles académiques of Douai, prize-giving of 16 Jan. 1870.

but artisans with taste' in 1863,[27] while in 1878 they were completed by a weaving and dyeing school, along the lines of those at Amiens and Mulhouse, debated by the municipal council since 1870.[28]

The completion of apprenticeship in the evenings, outside the workshop, was thus a well-established practice before the Third Republic. Nevertheless, it was the Opportunist Republic which organized a basic change towards training in the school. A speech of Jules Ferry on 23 April 1881 called for a revival of artistic education among the people, not to encourage 'the production of *déclassés*' or to aggravate 'the unfortunate scourge of artistic mendicity', but to achieve 'the magnificent union of craft and art'. 'We want you to bring art into trades, we want you to awaken the artist in every artisan'.[29] Accordingly he proposed a system of artistic education on three levels, the municipal drawing-schools, *écoles régionales des Beaux-Arts*, and *écoles d'art décoratif*, such as those to be set up at Limoges, specializing in ceramics, and at Roubaix, reviving Flemish tapestries.

At Nîmes and at Douai, there were no formal changes: the *écoles académiques* continued to render service. Indeed, Douai amply demonstrated the success of an evening- and Sunday-class system for the working populations. In the first place, for those apprenticed to the trades of mechanic, turner, fitter, joiner, and carpenter, there were courses in technical drawing and draughtsmanship, together with woodwork and metalwork shops for practical application. Secondly, for the *ouvriers d'art*, painters and decorators, sculptors and cabinet-makers, there were courses in decorative composition. Ornamental drawing for future architects and calligraphy and accounting for the city's clerks were also offered. Thirdly, there was training for candidates for the Beaux-Arts or (after 1894) the École de Sèvres, together with *cours normaux* including tuition in anatomical drawing and the history of art for would-be drawing-masters. Rennes, under

---

[27] A.-J. Lestienne, *L'Instruction publique depuis la Révolution à Roubaix: les écoles académiques* (Roubaix, 1933), p. 17.

[28] A.D. Nord CC 144, report of commission of municipal council of Roubaix, 8 April 1870.

[29] A.M. Rennes R54, copy of speech of Jules Ferry at Sorbonne, 23 April 1881.

its energetic Radical mayor, Edgar Le Bastard, did manage to obtain an upgrading of its *écoles académiques* to the status of an École régionale des Beaux-Arts in 1881, but its structure was no different. Its apprenticeship courses catered for workers from the Arsenal, the artillery foundry, and the railway workshops, among others. Its *ouvriers artistes* included, as well as painters, sculptors, and cabinet-makers, jewellers and lithographers who might find employment with the local printing firm of Oberthür. And its teacher-training course added public lectures for *instituteurs* and *institutrices*.

The promotion of the Écoles académiques of Roubaix to become, under the law of 5 August 1881, the École nationale des arts industriels of Roubaix, was a further notch up the scale. In the whole enterprise, there seemed to be shades of France's mercantilist past. Just as the Administration des Beaux-Arts, attached to the Ministry of Public Instruction, presided over the Sèvres factory and, by extension, endowed Limoges with an École nationale for ceramics, by the same token, its direction of the Manufactures royales of Gobelins extended to the endowment of Roubaix with an École nationale for Flemish tapestries.[30] Again, there was the sentiment that the rich and colourful woollen cloths, crossed with silk, that had been the glory of Flanders and displaced in the early nineteenth century by the cotton fever that transformed Roubaix into a mushrooming mill-town, must once again see the light. Finally, there was the competition with Britain, and the knowledge that in Yorkshire woollen towns like Bradford technical education was forging far ahead.[31] Émile Moreau remarked at the municipal council of Roubaix that the École nationale resembled the Arts et Métiers in its workshops and drawing courses, chemistry and mechanics, while its tapestry-making based on models from the local museum annexed to weaving and dyeing represented 'the teaching of fine arts applied to industry'.[32]

Ferry's network of artistic schools clearly faced difficulties. Any tampering with the *écoles académiques* signified the

---

[30] A.D. Nord CC 180, Carlos Delattre to mayor of Roubaix, 15 April 1881.
[31] A.N. F17 14359, Dutert, Inspecteur-général to Min. Instr. Pub., 7 June 1881.
[32] A.N. F17 14359, municipal council of Roubaix, 28 May 1881.

encroachment of the central government on municipal institutions that were in some cases over a century old. But
whereas the École régionale at Rennes remained very much
under the control of the mayor, the École nationale at
Roubaix was to a much greater extent at the mercy of
Paris. And just as the Catholic *patronat* and clergy expressed their particularism over the Arts et Métiers, so
they answered the École nationale at Roubaix with an
Institut technique, founded by secular priests, and the
École Saint-Luc at Lille, run by a Frère des Écoles Chrétiennes
who had studied the artistic traditions of the Low Countries
at Ghent.[33]

A further problem, all too common to professional
schools, was that local industrialists found the apprenticeship offered in the class-room inadequate for what was
required in the workshops. The Chamber of Commerce of
Douai reported in 1884 that the *écoles académiques* offered
'a significant contribution to apprenticeship' but in no
sense turned out skilled workers, let alone foremen.[34]
A member of the municipal council of Rennes complained in
1891 that housepainters emerging from the École des Beaux
Arts had to undergo a new apprenticeship on site, only to be
told by the director of the school, 'today, apprenticeship
is no longer provided by employers . . . the training of
apprentices is the responsibility of schools'.[35]

Lastly, the artistic schools could fall into the trap of
being art schools divorced from practical considerations, producing only *déclassés*. The socialist municipality of Gustave
Delory at Lille reorganized the Écoles académiques as an
École régionale des Beaux Arts in 1897, but in 1900 municipal criticism that the students needed to do more than 'know
how to splash paint on to canvas' provoked the director to
break away and found a separate studio, until the municipality

---

[33] J. Balmont, *L'École nationale des Arts industriels de Roubaix* (Lille, 1909),
p. 13. G. Rigault, *Histoire générale de l'Institut des Frères des Écoles Chrétiennes*,
vii (Paris, 1949), pp. 389–90.

[34] A.D. Nord 1T 135/3, Chamber of Commerce of Douai to prefect of Nord,
22 Feb. 1884.

[35] A.M. Rennes R54, municipal council of Rennes, 16 Sept. 1891; director of École régionale des Beaux-Arts, Rennes, to mayor of Rennes, 17 Oct.
1891.

apologized five years later.[36] Nearby, the École nationale des arts industriels at Roubaix did not really prosper until 1902, when it was reorganized by a director who insisted that 'Art must have above all a utilitarian function, training for gainful trades.'[37] Artistic design for textiles was enhanced, but the engineering side of the industry was developed to include the training of mechanical engineers and electricians and the dyeing and printing aspects refined to train for the chemical industry in general. On the other hand, the study of architecture, for so long—at least outside the École des Beaux Arts of Paris—connected with the *arts industriels,* finally broke away as a liberal profession with foundation of *écoles régionales d'architecture* in the provinces. Both Lille and Rennes acquired them in 1905, although the syllabus was so elevated that at the latter scarcely any candidates were eligible to follow the course as full-blown students for several years.[38]

It was clear that if the republican notables were to combat the decline of apprenticeship effectively, they would have to cast their net wider than the old *écoles académiques.* The inauguration of compulsory, free education in the years 1881-2 .made the need for a broader-based professional education all the more pressing because universal education in generalities only served to increase the risk of popular *déclassement.* Besides, as ever, the Catholic clergy and *patronat* were blazing the trail. In 1880 they founded a network of technical schools at the higher primary level, such as the École de La Salle at Lyon[39] and the École Catholique de commerce et d'industrie at Lille. The school at Lille, an extension of the *classe d'honneur* of the rue des Urbanistes, was conceived by a group of employers under Philibert Vrau, elaborated with the help of Bernot who had turned to the Catholics after his dismissal from the École des arts industriels in 1872, and installed in *La Monnaie* under the Frères des Écoles Chrétiennes.[40]

---

[36] *Progrès du Nord,* 31 Dec. 1900; Z. de Winter, *Pharaon de Winter, sa vie, son enseignement, son œuvre 1849-1924* (Lille, 1926), pp. 54-9.

[37] J. Balmont, op. cit., p. 14.

[38] *Le Bâtiment,* 25 Sept. 1904; A.M. Rennes R63, director of École régional d'architecture to mayor of Rennes, 12 Mar. 1906.

[39] A. Prévot, *L'Enseignement technique chez les Frères des Écoles Chrétiennes au XVIIIe et au XIXe siècles* (Paris, 1964), pp. 170-90.   [40] Rigault, vii, 388.

The response of the government was a law of 11 December 1880 which was only permissive, followed up by a decree of 30 July 1881, which envisaged a twin response: *écoles manuelles d'apprentissage* and *cours complémentaires* with workshops adjoined. In fact, the whole burden of the enterprise fell on those Radical mayors who were vigorous enough to take an initiative. The director of the *cours complémentaire* in the rue Jean Reboul at Nîmes obtained money from the Protestant mayor, Ali Margarot, in 1882 to convert three class-rooms into workshops, arguing that the pupil trained to use a file and a plane would develop a liking for manual labour and would think of becoming a good worker rather than a clerk in some branch of commerce or administration,[41] for white-collar ambitions in the Midi were well known. At Rennes, the main concern of Edgar Le Bastard was slightly different. Fearing that apprenticeship was no longer given in the workshop, that the division of labour reduced the worker to a 'fragmented being', and that the use of machine tools turned the artisan into a labourer, he argued in favour of an *école manuelle d'apprentissage* which would train 'real workers, with a high level of technical knowledge, both theoretical and practical'.[42] The 'school' set up in 1883 was essentially three workshops in the rue d'Échange, for joinery, fitting, and metalwork, run by an ex-foreman and two *chefs d'atelier*. The pupils, 'sons of small shopkeepers, clerks and above all workers' looking for a trade and employment in the Arsenal or railway workshops, worked in the shops from 6 a.m. to 5 p.m., and then did an hour's general education in the municipal school, followed by two or three hours in the evening, drawing at the Beaux-Arts. The workshops were also open to pupils of the municipal schools and to apprentices seeking to perfect their training.[43] The École manuelle was 'purely municipal' and did not receive any support from the state. When a subsidy was required in 1887, the municipality only came up against the tangled

---

[41] A.M. Nîmes R265, Vallat, director of *cours complémentaire* to mayor of Nîmes, 6 Dec. 1882.

[42] A.M. Rennes R33, report of commission of municipal council of Rennes, Oct. 1884.

[43] A.D. I-et-V 1N, Conseil général, report of Martel, Inspecteur-général de l'enseignement professionnel et technique, 1890, pp. 423–8.

condominium of the Ministries of Public Instruction and Commerce, each suspicious of the other and neither prepared to be of real assistance.

The government did plan a series of *écoles nationales professionnelles* to come under its auspices at Vierzon, Voiron, and the mill-town of Armentières. But the dangers of official colonization were made clear over the school at ·Armentières, set up under a decree of 10 March 1882. Ten years later, the huge palace of a building had only 180 pupils because, as Félix Pécaut observed, 'the Catholic or rather clerical employers put the lay school at whatever level strictly on the Index'.[44] The schools founded by local industrialists for their own needs were far more successful. The Chamber of Commerce of Tourcoing, for instance, anxious to train an élite of the working class in the finer skills of spinning and weaving with a view to becoming foremen, set up an *école industrielle* in 1889 which was essentially an evening class for workers of between seventeen and twenty-five, equipped with the latest textile machinery imported from Bolton, Rochdale, and Oldham.[45]

While local industrialists were making their own arrangements, an involved debate was raging in Paris between two schools of thought as to the proper function of professional education. For those at the Ministry of Public Instruction professional training must be part of a general education and confined to imparting a certain manual dexterity and a 'taste for work'. For their opponents at the Ministry of Commerce, professional education should go much further, training an 'élite of workers' who had a scientific knowledge, both theoretical and practical, enabling them to exercise a variety of skills, and, seizing on new techniques, to dominate the machine rather than be enslaved to it.[46] The law of 26 January 1892 decided matters in favour of the Ministry of

[44] A.N. F17 14352, Félix Pécaut, Inspecteur-général to Min. Instr. Pub., 30 June 1893.

[45] A.D. Nord 1T 135/3, president of Chamber of Commerce of Tourcoing to prefect of Nord, 23 Feb. 1884; 1T 14 6/4, report on École industrielle of Tourcoing, 1895.

[46] J.-P. Guinot, *Formation professionelle et travailleurs qualifiés depuis 1789* (Paris, 1948), pp. 154-5; C. R. Day, 'The Third Republic and the development of intermediate education', p. 346.

Commerce and a network of *écoles pratiques de commerce
et d'industrie* were organized under its jurisdiction. The *écoles
pratiques* were not in fact new foundations but relabelled
cours complémentaires, E.P.S., and *écoles manuelles
d'apprentissage*. Some E.P.S. applied for status as *écoles
pratiques* largely because the government undertook to pay
their teaching staff (though not their foremen). Dol and
Bagnols were among these, but as former municipal colleges
which turned out little other than clerks, teachers, and
farmers they saw their applications refused.[47] The École
manuelle d'apprentissage of Rennes, which saw its task as
being to 'train apprentices with a certain theoretical grounding', was included as an *école pratique d'industrie*,[48] as was
the Cours complémentaire at Nîmes. Insisting that he wanted
to offer only 'studies leading to an occupation or to knowledge of a trade', the director of the Cours complémentaire
had in 1889 abolished the agricultural section and the general
section, which trained only *fonctionnaires* for the Post Office
or, 'with its *demi-monsieur* and *demi-savant* character',
candidates for the École normale.[49]

Despite this shift towards a more technical education, the
écoles pratiques did remain somewhat marginal, not really
answering the needs of employers. One aspect of the problem
was political. Alès had no *école pratique*, indeed no E.P.S.
before 1902, largely because the Catholic *patronat* of the
coal and iron companies preferred to draw its recruits from
the Pensionnat of Saint-Louis de Gonzague. In addition,
large companies were running training courses to meet their
specific requirements, and these were often closed to others
than the sons of their own employees. At Rennes, Edgar Le
Bastard entered into a heated argument with the director
of the Compagnie des chemins de fer de l'Ouest in 1888–9
in order to try to obtain guarantees that pupils from the
École pratique would find some openings in the station

---

[47] A.D. I-et-V, 2T34 Dol, primary-school-inspector of Saint-Malo to Insp.
Acad. Rennes, 13 July 1892; A.D. Gard 13M 998, Min. Instr. Pub. to prefect of
Gard, 18 Dec. 1893.
[48] A.M. Rennes R30, Prospectus of École pratique d'industrie, Rennes, 1893.
[49] A.D. Gard 1T 782, Vallat, director of Cours complémentaire, Nîmes to
Insp. Acad. Nîmes, 29 Oct. 1891.

workshops.[50] At worst, it was complained that the *ouvrier achevé* or *ouvrier complet* was of no use in the workshop: the young school-leavers were jacks of all trades and masters of none, they were not accustomed to the rhythm of the production line, and they were pretentious enough to imagine that they would be promoted foremen ahead of the old hands. It was on grounds such as these that the abolition of the École pratique was demanded in the municipal council of Rennes in 1897. As one councillor said, 'we all know that when they leave school, the pupils are incapable of making workers . . . I know a good number of them who are on the street'.[51]

Attacked as training centres of the unemployable, the *écoles pratiques* were not spared the criticism that, in their own way, they produced *déclassés*. Within the working class there was an acknowledged hierarchy of trades. Pupils avoided the joinery shop and even metalwork in order to train as fitters, for that was the training that would lead on to the railways, the job of ship's mechanic, and the Arts et Métiers. At the École nationale professionnelle of Armentières, in the heart of the textile country, there were only six pupils out of 204 training in the spinning and weaving shops in 1894, so sought-after was fitting.[52] At Nîmes, there was no interest in metalwork even though the development of light engineering and the bicycle trade meant that there was a 'shortage of good metal-workers in the various workshops of the region'.[53] The director of the École pratique of Rennes gave away the secret in 1903 when he announced that 'intelligent and hardworking pupils find at school the means to raise themselves, with one leap, above the situation of their parents, by studying energetically for the exams to the Arts et Métiers and the fleet'.[54] Moreover, as commercial

[50] A.M. Rennes R30, Le Bastard to director of Compagnie des chemins de fer de l'Ouest, 13 Dec. 1888; reply, 8 Jan. 1889.

[51] A.M. Rennes R30, municipal council of Rennes, 2 July 1897.

[52] A.N. F17 14352, director of École nationale professionnelle of Armentières to Buisson, Directeur de l'Enseignement primaire, 25 Jan. 1894.

[53] A.M. Nîmes, Conseil de perfectionnement de l'école pratique de commerce et d'industrie, Nîmes (1897–1918), report of director Vallet, 26 July 1899.

[54] A.M. Rennes R30, report of Martin, director of École pratique d'industrie, Rennes, 1903.

houses, banks, and insurance companies developed, a demand was set up for those with qualifications in accountancy, typing, and shorthand. Rennes, which was only an École pratique d'industrie, was showing interest in a commercial section before 1910.[55] At Nîmes, which was a full École pratique de commerce et d'industrie, the director was quick to state that 'our young men are not looking for little places in some administration', but did not consider the movement into other white-collar employment as a form of *déclassement*.[56]

Professional education was only ever of interest to the intermediate levels of the economy. The mass of the working population on the one hand, the capitalist dynasties on the other, saw no purpose in it. How far this was true of the employer class was made clear at the turn of the century by attempts to offer a scientific training for the *patronat* at Lille. Again, it was the Catholic industrialists who took the initiative, noting that 'governmental careers are closed, or at least it is impossible to enter them without bowing one's head', and hoping to do for the captains of industry what the Catholic Faculties were doing for the liberal professions. The École des hautes études industrielles opened in 1885,[57] and it was not until 1892 that a lay response came from the Chamber of Commerce of Lille, which set up a rival École supérieure de commerce. This was designed to provide a training in accountancy, commercial law, commercial geography, political economy, and foreign languages for 'our future *négociants*, directors of companies and higher employees'.[58] Recruitment was feeble, though it is significant that a large proportion of students came from Catholic colleges: 36 per cent in the period 1900-2, as against 22 per cent from the *lycées* and colleges of the north of France. A fierce debate on the reasons for this stagnation took place under the auspices of Charles Petit-Dutaillis, professor at

[55] A.M. Rennes R30, report of director Martin, n.d. but before 1910.

[56] A.M. Nîmes, Conseil de perfectionnement (1897-1918), report of director Clavel, 17 Dec. 1902.

[57] A.D. Nord CC149, Notice on the École des hautes études industrielles et commerciales of Lille, 1907.

[58] A.D. Nord CC178, circular of Trannin, director of École supérieure de commerce, Lille, 1 May 1893.

the University of Lille and director of the École supérieure de commerce after 1899. But the warnings of Petit-Dutaillis that 'the everyday practice of business is an inadequate training for the merchant and the manufacturer', and that they required a 'high economic culture', were unlikely to be appreciated by entrepreneurs who had become rich by means only of their shrewd business sense.[59] Neither was the advice that 'the time of empiricism and of hazardous enterprises is passed' to be heeded as the French economy pulled out of the depression about 1895 and gathered momentum with the new century.[60] About a quarter of the students were scholarship boys from modest rather than bourgeois families, but the training that they were offered was not the technical expertise that they might gain at the Institut industriel and it was not encouraging to be told in 1904, 'as you have seen not all the *anciens élèves* yet have a situation'.[61] There was a reform in 1903 that sought to make the school more industrial, adding sections on textiles and sugar-beet-based industries to the commerce and banking section in the second year, but in 1905 there were only twenty-three applicants for sixty available places and in 1906 the Catholic École des hautes études industrielles dealt a death-blow by opening its own commercial course.

Among the mass of the working population the most striking feature of professional education was its absence. This was the 'crisis of apprenticeship'. Indeed Roubaix, under its mayor Eugène Motte, was the focus in 1911 not only of an international exhibition but also, running concurrently with it, of a Congrès national de l'apprentissage. Many reasons were advanced for the disappearance of apprenticeship from the workshop. The division of labour and 'machinisme' were all old war-horses and clearly applied to large-scale industry. In textiles, the child leaving primary school was taken on as a *rattacheur* in spinning or a winder in

---

[59] A.D. Nord CC177, Petit-Dutaillis to senators and deputies of National Assembly, 31 Oct. 1904, in *Association amicale des anciens élèves de l'École supérieure de commerce de Lille*, Dec. 1904.

[60] A.D. Nord CC188, Petit-Dutaillis to president of Chamber of Commerce, Lille, 31 Dec. 1902.

[61] A.D. Nord CC177, president of Association des anciens élèves to banquet, 22 Oct. 1904.

weaving, and professional training was irrelevant except in the case of draughtsmen, mechanics, and textile designers.[62] A master at the École pratique of Roubaix suggested that children of thirteen or fourteen should go straight into the mills, and that any later training could be done on the basis of the old *école de midi*.[63] The Congress was told that in the mining industry youths started at the age of thirteen, screening for flints, before going underground to work the ventilation doors or load coal until they went up to the coal-face as an *aide-mineur* at the age of eighteen. No professional training was required, except perhaps for master-miners, overseers, and *porions*. In metallurgical works, 'all workers are unskilled labourers' and acquired expertise only 'by practice and a greed for earnings'. In short, technical training was of use only in *métiers bourgeois* for artisans and skilled workers.[64]

But even in small-scale industry apprenticeship was disappearing. One reason was economic, that the small employers, harassed by the competition of larger firms, could not afford to devote time or money to training apprentices. In a metalwork shop, a youth would be taken on as a *gamin*, then a *demi-ouvrier*; working-class parents were happy that he should be earning at once, and the illusion of some sort of apprenticeship was preserved.[65] Just as significantly, the labour laws designed to limit the working day had negative repercussions on apprenticeship. The labour law of 2 November 1892, preserving as it did four categories of workers by age and sex, was felt to be inadequate, and two politicians of the Nord, Maxime Lecomte in the Senate and Gustave Dron, deputy of Tourcoing, campaigned for its revision. The resulting law of 31 March 1900 was a veritable 'Ten Hour Act', reducing the working day of all those under eighteen to $10\frac{1}{2}$ hours in 1902 and to 10 hours in 1904.

---

[62] A.N. F12 7621, report of Cromback, president of Syndicat des filateurs de laine of Fourmies, 16 Oct. 1901.

[63] A.D. Nord CC179, Lecointre, master of École pratique of Roubaix to Congrès national de l'apprentissage, Oct. 1911.

[64] A.D. Nord CC179, Wauthy, metal-founder, Inspecteur départemental de l'enseignement technique to Congress of Roubaix, Oct. 1911.

[65] A.D. Nord 1N 152, Conseil général, report of Labbé, Inspecteur général de l'enseignement technique, 10 Aug. 1908.

Unfortunately, many employers began to consider the work of youths under eighteen uneconomic and lay-offs intensified after 1902. The building trade claimed in 1911 that only 3 per cent of its work-force were apprentices, whereas without the labour legislation the figure would have been 15 or 20 per cent.[66]

The solution offered to the crisis of apprenticeship by technical schools could only be partial. Schools like the Arts et Métiers, the building industry asserted, dealt with the élite, not with 'the majority of children of the people . . . that Pleiad of youths destined to form the battalions of the great army of Labour'.[67] The *écoles pratiques* had a wider appeal and were continually developing. In 1906 *écoles pratiques* were annexed to the E.P.S. of Roubaix and Tourcoing, renamed respectively the Institut Turgot and the Institut Colbert. Three years later, the Institut Turgot added to its commercial section a school to train commercial travellers and to its industrial section a weaving-school, though this had the disadvantage of taking place in the daytime. Meanwhile, 1909 saw the opening at Tourcoing of an École pratique commerciale, industrielle et ménagère pour les jeunes filles. Admittedly, the industrial course concentrated on the dressmaking, embroidery, and lace-making arts required for the *couturière* and the *enseignement ménager* retained the creation of a happy home as a guarantee against alcoholism as an *arrière-pensée*, but the commercial section, with its accounting, commercial law, English, typing, and shorthand reflected the demand for female clerks, shop-assistants, and office-workers.

Even so, the Inspecteur-général de l'enseignement technique observed in 1912, technical education benefited only 'a minority of privileged' and this was to be the case for a long time to come.[68] Two alternative courses of action were available. The first was financial encouragement to professional organizations to offer training schemes of their own. The municipality of Rennes, for example, made a grant of a

---

[66] A.D. Nord CC179, report of building industry to Congress of Roubaix, Oct. 1911.

[67] Ibid.

[68] A.D. Nord 1N 157, Conseil général, report of Labbé, 15 July 1912.

thousand francs in 1907 to the corporation of hairdressers to start their own 'academy' and two grants of five hundred francs in 1912 to the *chambres syndicales* of tailors and printers.[69] The nationalized Chemin de fer de l'Ouest ran an apprenticeship course in the railway workshops, paying its apprentices thirty centimes an hour, the cause of another dispute with the École pratique, whose director asked whether the state confessed to 'lack of confidence in the value of its technical education'.[70] The second alternative was the provision of *cours de perfectionnement* for apprentices, that would take place in the evenings and on Sundays. These were little different from the courses organized for young workers by the *écoles académiques* and *écoles des Beaux-Arts*, except that they were attached to *écoles pratiques* and had the same industrial and commercial sections. After 1908 they were becoming very common. At the Institut Colbert, Tourcoing, the Syndicat des entrepreneurs organized a 'school of building' in 1910, open between 5.30 p.m. and 7.30 p.m., and including training in joinery and carpentry, stone-cutting and masonry, plumbing, plastering, and glazing.[71] The centre-piece of the Institut Turgot of Roubaix, meanwhile, was a course in industrial electricity, dedicated to the training of a new member of the skilled working class in 1911.[72] Whereas the technical schools in the Nord saw 1,903 pupils in 1912-13, the *cours de perfectionnement* dealt with 6,844. These courses were very much in the interest of private employers. They were geared to specific trades, not to the cultivation of the mythical *ouvrier complet*. They did not in general interfere with the working day. And while, in the Nord, a quarter of the bill was paid by industrialists, merchants, trade unions, and chambers of commerce, half was paid by the municipalities and the last quarter by the

[69] A.M. Rennes R94, letters to mayor of Rennes to *coiffeurs*, 16 Sept. 1907, *maîtres tailleurs*, July 1912, *travailleurs du livre*, 15 June 1912.

[70] A.M. Rennes R30, Martin, director of École pratique d'industrie, Rennes, to prefect of Ille-et-Vilaine, 28 July 1909.

[71] A.D. Nord CC170, director of Institut Colbert to Congress of Roubaix, Oct. 1911.

[72] A.D. Nord CC170, director of Institut Turgot to Congress of Roubaix, Oct. 1911.

state.[73] The capitalists had managed to shirk their responsibility, and one more brick was knocked into the edifice of
the *État-Patron*.

If one of the functions of professional education was to
cushion the effects of rapid industrialization, another was to
stem the flow of rural depopulation. As such, it was an
alternative to the protection for which so many agrarian
interests were campaigning in the 1880s. Among our three
departments, the main interest continued to be centred on
Ille-et-Vilaine, where a dairy school and an *école nationale
d'agriculture* were added to the Farm School of Trois-Croix.

It might be true to say that the heyday of Trois-Croix was
over. It came into conflict with the government in 1875,
when the Moral Order regime proposed a network of *écoles
pratiques d'agriculture*. These were catalogued as intended
for peasant-proprietors, while Farm Schools were for 'agricultural labourers' and *écoles nationales* like Grandjouan and
Grignon were for the sons of large landowners. A certain
level of instruction would have to be ensured at the *écoles
pratiques* and the course would last for three years. Eugène
Bodin was incensed by the high-handed behaviour of the
government. Trois-Croix, he pointed out, had always appealed
to the 'sons of small peasant-proprietors or of prosperous
farmers', never to agricultural labourers. Their primary
education might be rudimentary, but the essential thing
was that they should have 'agricultural antecedents' and not
represent 'small commerce, the dry fruit of the colleges';
'with those, one would undoubtedly make *déclassés*, aimless
people'. Finally, three years was far too long a course; even
if a family had a scholarship, it could not afford to sacrifice
the son's labour for that time.[74]

Matters were not improved by the bad harvests at the end
of the 1870s and falling agricultural prices. Not only did the
finances of the farm suffer, but the workshop could not sell
its agricultural machinery. In 1882, on the death of Eugène
Bodin, the enterprise was taken over by a Société agricole et
industrielle des Trois-Croix, and in 1888 the school was
bailed out by the department, which became proprietor of

[73] M. Labbé, *Rapport sur l'enseignement professionnel* (Lille, 1913).
[74] A.D. I-et-V. 26M2, Eugène Bodin to Min. Agric., 17 Aug. 1875.

the farmland, and was transformed at last into an *école pratique*. While the conservative landowners formed themselves into the Société agricole et horticole centrale of Ille-et-Vilaine presided over by Charles de Lorgeril to campaign for a protective tariff, the prefect argued that scientific agricultural education, making French farming more efficient and competitive, was the principal way to fight the depression. In this context Trois-Croix, which was 'scarcely only an apprenticeship school', must become 'a nursery of young agriculturalists, intelligent, hard-working, educated', who would 'assimilate all the sciences that concern agriculture'.[75] By 1894, the Nord had also founded a *école pratique d'agriculture* at Wagnonville, near Douai, in spite of the massed opposition of the Flemish protectionists led by Jean Plichon in the Conseil général.[76]

The contradictions of fighting agricultural depression with science rapidly became apparent. Recruitment in the countryside was difficult, and made no easier by the lack of scholarships, the suppression of the leaving premium, the failure of the school to obtain any exemption under the military-service law of 1889, and the more difficult entrance examination, which favoured the urban candidates. Yet urban candidates were unlikely to move on to the land, while many peasant youths were seeking education as a means to escape from it. Between 1871 and 1904, 37 per cent of graduates from Trois-Croix did not go into farming, becoming artisans, shopkeepers, notaries' clerks, pharmacists, *gendarmes*, schoolteachers, soldiers, railway employees, insurance agents, and accountants instead.[77] Such obvious *déclassement* came under attack in the local press in 1908, when an article in the *Nouvelles Rennaises* demanded why taxpayers should pay 28,000 francs a year for a 'conjuring school' which took in farmers' sons and turned out shop assistants and dish-washers.[78] In his defence, the director

[75] A.D. I-et-V. 1N Conseil général, report of prefect of Ille-et-Vilaine, 1888, p. 505. See also P. David, 'Les origines du syndicalisme agricole en Ille-et-Vilaine, 1885-1918' (*mémoire de maîtrise*, Rennes, 1972), pp. 36-9.

[76] A.D. Nord 1N 137, Conseil général, 11-13 April 1893.

[77] A.D. I-et-V, 1N, Conseil général, report of Hérissant, director of Trois-Croix, 1905.

[78] *Les Nouvelles Rennaises*, 2 Sept. 1908.

argued that he could do very little if sons of *employés* were among the recruits to the school, but though he insisted that agricultural improvement was the only means of fighting the depopulation of Brittany he dutifully resigned his post.[79]

A more successful bid to counter the depression by agricultural science was made at the farm of Coëtlogon, adjoining that of Trois-Croix on the outskirts of Rennes, and acquired by Jean-Jules Bodin in 1848. The challenge of markets for dairy produce in England and the Paris region, articulated by rapid rail transport, was the incentive behind the foundation there in 1888 of an *école de laiterie*. The scheme was proposed as early as 1883 by the Chamber of Commerce of Rennes, was forcefully supported by Edgar Le Bastard, and, despite the opposition of the protectionists led in the Conseil général by de Sallier-Dupin, councillor of the legitimist canton of Argentré, was received warmly by the Ministry of Agriculture. The Bodin dynasty reigned still, for Eugène's widow was appointed to direct the school.

The sphere of influence of Coëtlogon was clearly greater than that of Trois-Croix. Whereas Trois-Croix drew principally on Ille-et-Vilaine and the Côtes-du-Nord, pupils came to Coëtlogon from dairy-farming areas in Brittany, Normandy, and northern France, the Alps and Jura, Massif Central and Pyrenees, from Belgium and Norway. Madame Bodin claimed that most of her recruits came from 'large peasant families';[80] in fact, applications for scholarships from Ille-et-Vilaine between 1896 and 1911 showed that 58 per cent of scholars were daughters of farming families, while 17 per cent were working class, 8 per cent daughters of tradesmen, and 6 per cent of minor professions such as pharmacist and veterinary surgeon.[81] The course included instruction in the manufacture of milk, butter, and cheese, enough veterinary science to be able to raise a dairy herd, pigs, and poultry, and enough horticulture to be able to maintain an orchard and kitchen garden. The hours were long and the rates of production high. The *Nouvelles Rennaises* attacked the school as being a

[79] A.D. I-et-V 1N, Conseil général, report of director Hérissant, 1910, pp. 775–81.

[80] A.D. I-et-V 1N, Conseil général, 26 Aug. 1884.

[81] A.N. F10 2633, scholarship to Coëtlogon, 1896–1911.

penitentiary or a forced-labour camp: 'at the end of one month, the young boarders generally fall sick'.[82] Local industrialists complained that the department was subsidizing a factory that was undercutting the market.[83] But these were signs of the success of the enterprise.

Dairy farming and its industrial subsidiaries were expanding, and there was little *déclassement* from the school. About 52 per cent of the girls returned to family farms, while 44 per cent went into the dairy industry.[84] Madame Bodin had inherited her pedagogy from her father-in-law. The aim of the school, she told a conference in 1900, was to 'fight routine', but the girls would learn 'without aiming to become bluestockings and sedulously avoiding pedantry'. They were taught to love their condition and not to 'dream about the conquest of some higher position'. The prosperity and integrity of the French countryside must be maintained by methods that were both material and moral. Alongside technical education went 'domestic education' and a glorification of 'rustic milieux' where ideas were simpler, families larger, lives longer. Madame Bodin's 'model housewife, hardworking and capable' was Jean-Jules's 'good angel of the hearth'.[85]

The problem of reaching the mass of working population existed for Coëtlogon as for any other professional school. In her report of 1900 Madame Bodin suggested a *cours temporaire d'industrie laitière* to take place in the heart of the countryside during the slack season. It was in fact one of her Belgian pupils who carried the first *école volante* (or, more modestly, *ambulante*) *de laiterie* into practice in her native country, and the idea was taken up in the Nord, financed by the Conseil général.[86] A first *école ambulante* was set up in 1905, a second in 1908, so that one concentrated on the plains of Flanders, the other on the poor clay soils

---

[82] *Les Nouvelles Rennaises*, 4 Nov. 1908.

[83] A.D. I-et-V 1N, Conseil général, 19 Aug. 1909.

[84] A.D. I-et-V 1N, Conseil général, report of Madame Bodin, 1905, pp. 766-7.

[85] A.D. I-et-V 1N, Conseil général, report of Madame Bodin to VIe Congrès international d'agriculture, 1900, reprinted pp. 433-42, J.-J. Bodin, *Conseils aux jeunes filles qui doivent devenir fermières* (1864, new ed. 1880), p. 16.

[86] A.D. Nord 1N 149, report of Ducloux, Professeur départemental d'agriculture, 9 Feb. 1905.

and scrubby forest of Hainaut and Thiérache, where cattle-raising was the main agricultural interest. The development of such schools did pose the first threat of *déclassement* to Coëtlogon: in 1908, Madame Bodin requested funds to set up a *cours supérieur* to train teachers for them, and Ardouin-Dumazet saw a prosperous future for Coëtlogon as an agricultural *école normale*.[87]

The links between agriculture and industry demonstrated by Coëtlogon indicated that the training of 'cadres' would be as necessary as 'cultivators'. In the Nord, beet farming flourished alongside dairy farming as cereal prices declined, used not only for fodder but also for its industrial content, sugar, and alcohol. But sugar-refining, brewing, and distilling required teams of foremen, managers, and chemists, and the notables of Douai exerted pressure on the government to set up an *école nationale des industries agricoles*, a sort of agricultural Arts et Métiers.[88] The school opened in 1893, providing a full training for those destined for the agricultural industries of the north and courses for graduates of the *écoles nationales d'agriculture*.[89] The *écoles nationales* were also becoming more scientific. The pioneering days, when the task of Grandjouan was to tame a vast Breton heathland and bring it under cultivation, were over. Now agricultural education had to be linked to the universities for the sake of research and experiment.[90] The École nationale of Grandjouan was therefore transferred to Rennes in 1893, despite the protests of the deputies and senators of Loire-Inférieure, and the government agreed to set aside 15,000 francs a year towards the repayment of a loan of 650,000 francs from the Crédit Foncier.

The increasing complexity of patterns of employment in the agricultural world is evident from a study of the

---

[87] A.N. F10 2632, Madame Bodin to Min. Agric., April 1908; Ardouin-Dumazet, *Voyages en France, Haute Bretagne intérieure*, 4e ed. (Paris, 1914-17, p. 28. This was written in 1909.

[88] A.D. Nord CC 143, Trannin, vice-president of Chamber of Commerce of Douai to Min. Agric., 10 June 1891.

[89] A.D. Nord 1T, 176/1, Prospectus of École nationale des industries agricoles of Douai, 1893.

[90] A.N. F10 2483, report of Tisserand, director at Min. Agric., 4 Aug. 1892. Tisserand was the driving force behind agricultural education at the Ministry in this period.

careers of the graduates of the École nationale. From the enrolment registers, though they rarely give the profession of the father, and from the *Annuaire des anciens élèves*, it is possible to see that there were two 'streams' of pupils.[91] The first, with secondary-school backgrounds, had a greater tendency to return to the land, especially if they had the classical *baccalauréat*: these were the 'modernizing land-lords'. Those with 'special' or 'modern' qualifications, or those who had come up through the E.P.S., Farm Schools, or *écoles pratiques*, rarely went back to the land but took advantage of the wide range of possibilities now open to them.

Fig. 14. Careers of Graduates of the École nationale d'Agriculture, Granjouan-Rennes

	Grandjouan		Rennes
	1871–84	1885–94	1895–1913
Landowners, managers	46.2	33.3	29.1
Agricultural teaching	24.0	31.2	25.3
Auxiliary to agriculture	11.5	25.0	28.5
(veterinary surgeons)	(0.5)	(11.8)	(12.7)
Outside agriculture	10.6	7.0	11.4
Colonies	7.7	3.5	5.7

It is clear that the proportion of graduates that went into farming was falling, as a result of the agricultural depression, even before the school moved from Grandjouan to Rennes. Agricultural teaching was a first alternative, the proportion reaching nearly a third during the last decade at Grandjouan. But other possibilities linked to agriculture—veterinary science, research, the agricultural industry, agricultural credit, insurance, and journalism—increased after the move to Rennes, as did the 'wastage' into commerce, the professions, or the administration. *Déclassement* was in the nature of the development from a simple stratified society to a complex modern one in which the proportion of intermediary professions was that much greater.

[91] École nationale supérieure agronomique de Rennes, enrolment registers 1871–1913. A.D. I-et-V 2 Per 135, *Annuaire de l'Association amicale des anciens élèves de Grandjouan-Rennes.*

One of the teaching careers open to graduates of the agricultural colleges, intended to bring the farming population into contact with the advances of agricultural science, was that of *professeur départemental d'agriculture*. The post was provided for under a law of 16 June 1879, with the twin responsibilities of teaching in the *école normale d'instituteurs* and 'going to the people' as a nomadic lecturer, but some departments did not appoint such a professor for a decade or more. Of our three departments, the Gard was the first in the field. From 1882, the professor was touring about thirty communes a year, lecturing on those subjects of most concern to peasants: the phylloxera epidemic and the planting of American vines, silk cultivation, and the need to use fertilizers. The Nord held aloof, its prefect considering that 'in the Nord, agriculture has been improved to such an extent that the need for a departmental professor is felt here certainly less than anywhere else'.[92] However, it would not let itself fall behind neighbouring departments, and appropriated the Pas-de-Calais' professor in 1889. Ille-et-Vilaine's professor did not start touring until 1891, but attracted audiences with his lectures on butter-making and the manufacture of cider.

Touring a whole department was obviously beyond the capabilities· of one man, and progressively the *professeur départemental* received support through the appointment of *professeurs spéciaux*, attached to a college or E.P.S. and lecturing within the confines of a single *arrondissement*. In Ille-et-Vilaine, Dol and Redon had *professeurs spéciaux* by 1895, Vitré by 1901, Fougères by 1908. In the Nord, it was the southern *arrondissements* that were provided for first— Avesnes, Cambrai, and Valenciennes in 1894—but Douai had to wait till 1902 and Flanders was ignored until a chair was created in 1908 at Cassel. In the Gard, a *professeur spécial* was appointed to the E.P.S. at Bagnols in 1892, the nearest thing the department had to an *école pratique* (for which young Gardois had to go to Avignon), and to Beaucaire in 1902. But the most interesting development was the foundation in 1897 of a chair of silk cultivation at Alès, with responsibility for the silk-cultivating departments of the

[92] A.D. Nord 1N 125, Conseil général, report of prefect of Nord, 1 Sept. 1881, p. 315.

Gard, Ardèche, Drôme, and Vaucluse, even though, in the face of the opposition of reactionaries on the Conseil général like de Ramel, the burden of maintenance fell almost exclusively on the republican municipality and the state. The task of the new professor, Mozziconacci, as he envisaged it, was to fight the shrinking production of silk, falling despite the premiums offered under a law of 1892, and to carry the war against archaic and detrimental practices of silkworm-raising into 'the most inaccessible parts of the countryside'.[93]

Despite the novelty of the enterprise, this agricultural crusade faced many difficulties. In the first place, the E.P.S. and colleges to which the special chairs were attached provided only limited and grudging audiences. Pupils at E.P.S. in the Gard put a position in the Post Office, tax office, *école normale*, or Arts et Métiers first, and would not consider the agricultural section of the school as other than a *pis aller*.[94] Similarly, youths went to colleges in the Nord 'hoping to win academic qualifications' so that, the departmental professor advised, it was only worth training them in agriculture after they had realized the fruitlessness of their ambitions and returned to the countryside.[95] It was with this in view that *écoles ambulantes agricoles*, on the model of the itinerant dairy schools, were launched for the benefit of young farmers as farmers' sons in the Nord from the winter of 1908-9. By 1911, the course involved instruction about agricultural machinery, although there was little point in sending the mission to Flanders. A course at Merville in 1911-12 was emptied when the Catholic College of Hazebrouck replied with a rival school. In the Gard, the *professeur départemental* thought that fifteen lessons and five practicals at each stop was inadequate and—remembering that no *école pratique* had yet been organized—arranged something more permanent: an *école d'hiver* that was established outside Nîmes in 1913.[96]

---

[93] A.D. Gard 13M 999, report of Mozziconacci to prefect of Gard, 18 Sept. 1897.
[94] A.D. Gard, Conseil général, report of Convergne, Professeur départmental d'agriculture, 25 July 1912.
[95] A.N. F10 2646, Ducloux, Professeur départemental d'agriculture, Nord, to Min. Agric., 3 Feb. 1912.
[96] A.D. Gard, Conseil général, report of Convergne, 31 July 1913.

A second problem concerned leaving an impression in a commune that did not vanish with the departure of the professor. Many peasant communities were suspicious of his coming anyway, taking him in the area around Vitré to be a fertilizer salesman or an insurance agent.[97] In the Nord, the Professeur départemental had decided by 1895 to substitute for his whistle-stop tours longer stays of five or six days, during which he could visit farms, talk with farmers, and organize a *champ d'expérience* for testing the cultivation of oats, potatoes, tobacco, or hops. Increasingly, professors held regular 'surgeries' on market-days and left behind printed material or pamphlets written by themselves. Raoul Blanchard noted, perhaps somewhat surprisingly, that 'the Flemish farmer reads newspapers and practical reviews, and attends the lectures organized by the state professor'.[98] It was not rare, however, especially in Brittany, for a large proportion of the audience to be made up of *instituteurs*. They had often organized the publicity for the meeting; they would be left to supervise the *champ d'expérience*, and in the Gard Mozziconacci looked on them to set up model silkworm-raising units, selecting the eggs, incubating them correctly, arranging a large, well-ventilated cocoonery, and keeping the litter of fern and heather constantly renewed as the worms moulted.[99]

The crucial role of the *instituteurs* was not overlooked by the *professeurs départementaux*. That of Ille-et-Vilaine called them in 1908 'the best placed . . . to help us transmit fundamental notions of agricultural science to the mass of peasants'.[100] Agricultural societies and syndicates and *comices agricoles* awarded prizes to *instituteurs* for talented agricultural teaching. Yet they required *champs de démonstration* and in the Gard in 1902 only a quarter of boys' schools were equipped with them. The primary-school-inspector of Alès reflected that some of these gardens were of

[97] A.D. I-et-V, 1N, Conseil général, report of *professeur spécial*, Vitré, 1901, p. 713.

[98] R. Blanchard, *Le Flandre*, pp. 344-5.

[99] A.D. Gard, Conseil général, report of Mozziconacci, 5 July 1899; reports of 6 July 1913 and 1914.

[100] A.D. I-et-V 1N, Conseil général, report of Pic, Professeur départemental d'agriculture, 1908, p. 717.

a 'derisory size of only a few square metres' and that the
instituteurs were too busy doing the secretarial work of
the *mairie* or giving evening classes to look after them.[101]
Theoretical lessons were usually unsatisfactory, not least
because 'notions of agriculture' were not required for the
brevet supérieur until 1897. An excellent alternative was the
promenade agricole, visiting farms and collecting objects for
leçons de choses. But even in the Gard, where these practices
were current, they were 'too rare'.[102] Finally, it seemed
that the role of the *institutrice* had been universally ignored.
Spending time in urban *pensionnats* to prepare for the *école
normale*, seeing the teaching career as a means of *embour-
goisement*, anxious to marry in the town and not be relegated
to an obscure rural post, most of them had little taste for
agricultural matters. 'To bother about what goes on in the
garden, in the farmyard, with the cows and hens', mocked
the new departmental professor of the Gard in 1911, 'for
shame! Mesdemoiselles, that's fit only for peasants.' And
yet, as he pointed out, the farmer's wife was an indispensable
part of the rural economy and *écoles normales* should be
training 'institutrices of the countryside'.[103]

Thirdly, the activities of the agricultural professors could
only be those of the catalyst: in time, the agricultural populations would have to organize their own affairs. That is
why the ideas of *mutualité* were so often on the professors'
lips. They made enormous strides, in harness with agricultural syndicates like the Syndicat des agriculteurs d'Ille-
et-Vilaine, set up in 1887 and largely run by agricultural
masters, in the organization of co-operative institutions.
There were ço-operatives for buying seed, fertilizer, and
machinery and for selling agricultural produce, mutual-
insurance societies to guard against fire, hail, and the death
of livestock, and there were agricultural credit organizations.
It was, of course, an uphill struggle. The Professeur départemental of the Gard pointed out that 'the particularist spirit

---

[101] A.D. Gard 1T 929, primary-school-inspector of Alès to Insp. Acad. Nîmes,
15 Mar. 1904.
[102] A.D. Gard 1T 929, primary-school-inspector of Nîmes to Insp. Acad.
Nîmes, 18 Aug. 1896.
[103] A.D. Gard 1T 707, report of Convergne to prefect of Gard, 25 Mar. 1911.

is difficult to overcome'.[104] The peasants of Ille-et-Vilaine, it was said, 'are distrustful of each other and it is very difficult to persuade them of the usefulness of agricultural associations'.[105] In addition, the whole question of the organization of the peasantry had political overtones. There was no possibility of the agricultural professors forming mutual-aid societies among the Flemish peasantry because *caisses rurales* were being set up by Christian democratic clergy orchestrated by the abbé Lemire,[106] while an agricultural syndicate set up at Templeuve by its *curé* and the Comte d'Hespel was said to be 'a veritable royalist clerical association, founded under the auspices of the Comité de la Jeunesse Catholique of Lille'.[107] In Ille-et-Vilaine, the conservative, protectionist Syndicat agricole et horticole central was fighting for the solidarity of rural society against urban socialism and favoured mixed syndicates, while the republican Syndicat des agriculteurs had as the editor of its bulletin a master of Trois-Croix, Rey de Boissieu, who argued that ' "à bas les exploiteurs" was a precondition of "vivent les exploitants" '.[108] Education remained inseparable from politics.

It was not until 1880 that the French government was able to develop any national system of technical education. Even then, the networks were usually put together from schools that existed already: *écoles académiques*, municipal schools of apprenticeship, or *cours complémentaires*. Gaps persisted in critical areas: Lille, for example, founded its own Institut industriel in 1872 but was without an École des Arts et Métiers until 1900, by which time the Catholics had already opened a school to rival it. Indeed, in the field of technical education, as in every other field, Catholic schools, the graduates of which were especially favoured by Catholic

[104] A.D. Gard, Conseil général, report of Chauzit, Professeur départemental, 1903.
[105] A.D. I-et-V, 1N, Conseil général, report of Pic, 1901, p. 660.
[106] A.D. Nord 6V 44, Commissaire spécial of Lille to prefect of Nord, 19 Oct. 1899.
[107] A.D. Nord 2V 59, Commissaire spécial of Armentières to prefect of Nord, 26 Oct. 1901.
[108] A.D. I-et-V. 3Per 510, *Bulletin mensuel du Syndicat des agriculteurs d'Ille-et-Vilaine*, May 1898.

employers, were established either ahead of state initiatives or very rapidly in response to them, replying to what was seen as state imperialism threatening to breach the integrity of the local economic community.

The effect of technical education on the economic community is not easy to gauge, but it is clear that in the first place it depended on the level of economic development of the region. It served less to pioneer or to initiate than to diffuse existing technological expertise and to permit backward areas to catch up with those that had forged ahead. Education was of course only one answer to backwardness. The agrarian interest, for example, believed as a rule that the only cure for economic depression, precipitated by a world overproduction of cereals, was protection. Another course was agricultural improvement, shifting away from low-price grain to sugar-beet and dairy farming. But these adaptations were in response to market forces; agricultural education followed and refined these changes, but did not initiate them. There were cases where technical education arrived on the stage too late, already behind by one industrial revolution, still concerned with textiles and metallurgy when the economy had moved on to chemicals, electricity, and petroleum. It is true that professional education responded well to the emergence of a skilled working class and of a managerial stratum; indeed, one of its problems earlier in the century had been to over-educate the cadres. But there was no room for 'business schools' for the *patronat* and the mass of labourers, both industrial and agricultural, remained largely beyond the pale of technical education. Co-operation was the key to improvement in the rural economy, but co-operation was very difficult to achieve.

In the end, the impact of technical education was limited because of its goal. At the end of the nineteenth century, its main objective was not to keep up with the economic dynamism of Germany and the United States but to preserve a traditional social fabric that was being torn apart by violent economic change. Agricultural education was intended above all to dam the tide of rural depopulation, while the Beaux-Arts and *écoles pratiques* were to maintain the traditional standards of the *artisanat* that were being eroded by mechanization and large-scale production, with consequent embarrassment for the small workshop, modest *patron*, and his few employees. And yet technical education, school-based and theoretical as it was bound to be, aiming to turn out 'rounded workers', inevitably produced apprentices who were not immediately employable. Abstract education also tended to draw graduates towards the status of engineers, managers, and teachers, who to some extent could be absorbed in the expanding sector of company employees, but might be tempted by ambitions that had nothing to do with the productive side of the economy.

Conclusion

What then do we learn from the provincial perspective? As far as the politics of education are concerned, it is clear that national legislation cannot be considered as taking immediate effect, neither can resistance to it adequately be explained in terms of 'apathy', 'inertia', or 'problems of communication'. Legislation was enforced not unilaterally, but as the result of struggle and compromise, local initiatives being as significant as initiatives in Paris. Developments in education must be woven back into the fabric of provincial politics from which they were inseparable.

Any given article of legislation was confronted not by a blank sheet of paper—provincial France—but by a geography of attitudes that had been forged by historical conflict. The weight of the past was by far the most important determining factor. Economic developments of course there were during the nineteenth century, but the study of industrial centres as far apart as Flanders and the basin of Alès demonstrates that industrialization did not undermine strongholds of Catholicism, while rural populations were in no sense uniformly pious. What mattered were the ideological contours that had been moulded by the Revolution, by Reformation and Counter-Reformation, and by the annexation of foreign territories.

This is not to say that legislation changed nothing in the nineteenth century. But several observations about the changes that did take place should be made. Firstly, the public system of education, at both secondary and primary levels, the foundations of which were laid at the beginning of the century, was not necessarily antithetical to the Churches, Catholic or Protestant. The University and the écoles communales were empty vessels which could be and were appropriated in the period 1800–60 by the Catholic and Protestant Churches for their own purposes. Only after 1860 did public education become increasingly identified with an anticlerical position.

Secondly, a distinction must be made between public and private education, for whereas by the 1850s the University and the *écoles communales* had been substantially clerical-ized, in the next half-century both were purged of clerical influence. But this clerical influence was only displaced, and entrenched itself in the private schools and colleges. Lastly, though the overall population of Catholic schools, private as well as public, was plainly less in 1914 than in 1860, the geo-graphical distribution of lay and clerical education remained virtually constant. Where Catholic education was forced back, it was out of areas where its grip was tenuous into its citadels of strength, Flanders, the Cévennes, the *arrondisse-ment* of Vitré, and the old diocese of Saint-Malo. Ideological contours became if anything more strongly defined. More-over the Catholics grew not less combative but more so, and by the turn of the century had adopted regionalism, if not separatism, as an additional rampart to defend their faith.

The provincial dimension throws considerable light on the links between social class and education. The class hierarchy did not fit neatly into the system of elementary and second-ary education, that of the masses and that of the élite, because there were effectively not two classes in French society but three. Most significant were the *classes moyennes*, independent producers and salaried functionaries situated between the bourgeoisie and the wage-workers. Early in the nineteenth century, given the nature of elementary education, they formed the clientele of the small-town colleges that taught more French than Latin and of the more modest Catholic schools and *petits séminaires*. The 'aristocratic reaction' in the University of the period 1840–50, which sought to limit the 'excess of educated men' and rival the élite Catholic colleges, forced them into private *pensionnats* and 'intermediary schools' developed by the Catholics until the University introduced special education in a bid to recapture them. But as the liberal professions grew over-crowded, as the service sector of the economy expanded, and as the depression made a cheap, rapid, and practical education attractive, the middling sorts abandoned the

University for the E.P.S., the crowning glory of the primary-school system.

There was a marked distinction between the role of education as interpreted by notables and officials and the role of education as envisaged by the clientele. For the first, secondary education inculcated the rising generation over a long period of time with the classical humanism that defined the *honnête homme* and governing élite, while primary education had a civilizing mission that would integrate the masses into productive, ordered, and healthy society. But the point of view of the families and pupils who used the schools was quite different. What mattered for them was a career or employment, useful knowledge not homilies, rapid returns not protracted expenditure, individual success not social harmony. At the upper level, there was pressure to enter the liberal professions and state service through the faculties and *grandes écoles*; at the lower level, there was pressure to find petty office, return to the farm, or take up an apprenticeship, if the demands of child labour had not already determined that schooling would be rudimentary or non-existent. As a result, short cuts were taken and early leaving became endemic. Those destined for government schools or the *baccalauréat* took to private crammers in order to qualify as soon as possible; pupils in special education left after one or two years in order to take up employment in a shop or office and found the E.P.S. much more to their taste; at the elementary level, early leaving meant that many adolescents had experienced only the briefest brush with enlightenment, and could be civilized only by means of a battery of post-school *œuvres*. Even then, what interested audiences was literacy and the knowledge necessary to improve their position, not the sour-faced preaching of republican morality.

Lastly, factors beyond the school determined the extent to which these ambitions would be realized. The geographical analysis is instructive here, for it throws doubt on the notion that literacy and social mobility were higher in prosperous, industrialized areas than in backward, agricultural ones. Rates of school attendance tended to be low in inner-city areas, even when the schooling they provided before the legislation of 1881 was largely free, while it is clear that early industrialization depressed rates of literacy. In the Nord, the peasantry were second only to the tradesmen in literacy, while textile workers and miners were at the bottom of the pile; in Ille-et-Vilaine, the rural population was far and away the least literate, while textile workers were almost as literate as shoemakers. The analysis of social mobility demonstrates that in the north the prosperous *classes moyennes* tended to return to the workshop or farm, while class-differentiation made escape for the wage-earner extremely difficult. Lack of opportunity in the west produced a combination of *immobilisme* and pressure for secondary education, while rates of *déclassement* were always at their highest in the Midi.

The dichotomy between backward and developed France was also of significance for technical or professional education. Just as orthodox education seemed of more benefit to poorer regions in that it offered the possibility of employment in public service, the professions, or the Church, so technical education was better appreciated in areas where agricultural methods were archaic and industry little developed, since it provided some artificial means of catching up lost ground.

Technical education was nevertheless applied only to limited sectors of the economy——to industry rather than to agriculture, to crafts and skilled trades rather than to large-scale industry. The foreign market for luxury goods was of the first importance, the tradition of mercantilism still strong, the notion of *beaux-arts* and *arts mécaniques* still pervasive. This meant also that technical education was the province of a narrow stratum of producers. It was scorned by the entrepreneurs, the self-made men, as of no importance to themselves. The mass of the work-force was also beyond the pale of technical education: agriculture was run according to traditional practices, the division of labour and mechanization made apprenticeship less and less relevant for large-scale industry, and no means were devised to reach the big battalions of labour before the end of the century. Technical education was essentially for an intermediary class, between owner and work-hand, the draughtsman and

designer, foreman, manager, and technician, and the substantial peasant-farmer.

In many cases, technical education was firmly based in
local and municipal enterprise: the *écoles académiques* that
predated the Revolution, the agricultural schools before
1848, the *écoles manuelles d'apprentissage* of the early
Third Republic. As a rule, it may be said that the local
schools, founded with specific needs in mind, were more
successful than those founded according to a national blueprint. Indeed, there were instances of the Catholic *patronat*
financing Catholic technical schools in order to fight the
government's economic imperialism. Only too often technical
schools had little relevance to the local economy and were
quite marginal to it. The school of master-miners was set
up at Alès instead of Douai, the third École d'Arts et
Métiers at Aix-en-Provence instead of Lille. The Institut
industriel of Lille was better adapted to the first industrial
revolution of coal and textiles than the second industrial
revolution based on engineering and chemicals. *Écoles
pratiques* and *écoles des Beaux-Arts* were concerned to
train the *ouvrier complet*, the idealized artisan rather than
the apprentice who was immediately employable at the lathe,
with the result that many of their graduates remained
unemployed.

Lastly, the purpose of technical education coincided
with that of education in general in that its task was less
to initiate change than to maintain the integrity of the
old order against forces that were bringing about its disintegration. Rural depopulation, the crisis of apprenticeship,
and *déclassement* towards unproductive employment were all
economic changes that posed threats to the stability of
society. Technical education would maintain the prosperity
of agriculture, the vitality of small industry, the supremacy
of the old élites. It could be argued that it was the conservative rather than the progressive aspects of professional
education that were more important. But this is not what
was required by the students of technical education. They
wanted nothing but knowledge and qualifications to fulfil
their ambitions, with the result that ironically technical
schools produced as many cadres, white-collar workers,

and teachers as they did skilled workers and farmers. It stimulated the *déclassement* that it was specifically designed to prevent.

Bibliography

A. MANUSCRIPT SOURCES

A list of all the dossiers, *cartons*, and *liasses* consulted for this work would be both interminable and only a superficial guide to the documentation. Instead, a short appreciation of the nature of the archives is offered here; readers seeking more specific references should consult the footnotes.

1. Public Archives

(a) *Archives Nationales, Paris*

A collection of the first importance, but consulted in each case after using the respective departmental archives.

Series C, the archives of the parliamentary assemblies, includes inquiries such as the Enquête sur le travail agricole et industriel of 1848 (C 953, 954, 960) and petitions concerning the politics of education, such as those of 1872 (C 4131, 4136).

The archives of the various ministries are catalogued under F.

F10 (Ministry of Agriculture) includes material on the agricultural schools of Trois-Croix (2589), Coëtlogon (2632), and the École nationale d'agriculture at Rennes (2483), and the *écoles ménagères ambulantes* (2546).

F12 (Ministry of Commerce) includes reports on child labour under the laws of 1841 (4717) and 1874 (4765).

F17 (Ministry of Public Instruction) is the main source.

F17* includes Guizot's Enquête sur la situation des écoles primaires, although that for the Gard (105) is the only one extant for our departments.

Among other cartons of interest are:

Secondary education

2649–50, confidential reports of Rectors during the Second Empire.

6845, 6846, 6848, inquiry into secondary education, 1864.

6850, 6874, syllabus reforms of 1872 and 1880.

6890–1, special courses annexed to colleges, 1832–43, 1856–7.

7532, reform of Victor Cousin, 1840.

Reports on *lycées* and municipal colleges are classified by the establishment concerned, and are very full.

8701–2 and 8707–8, special education.

8753–5, 8763–4, 8711, 8771–8, secondary education of girls.

8819, 8821, 8823, 8827, *petits séminaires* and private Catholic schools in the first half of the nineteenth century.

8957, institutions and *pensions* in the nineteenth century.

9099, bifurcation.

Primary education

9125^2, hostility against the *école laïque*.

9125^{5,6}, campaign against textbooks used in *écoles communales*.

9195-6, options of municipalities between lay teachers and congregations, 1861-4, 1870s.

9306-19, reports of primary-school-inspectors, 1832-55.

9325-6, 9330, 9338, 9343, 9348, reports of primary-school-inspectors, 1855-60. There is plentiful documentation of *écoles normales d'instituteurs*, less on *écoles normales d'institutrices*, a good deal on E.P.S.

10366, situation of public primary education, 1913.

10776, 10781, 10792, memoranda submitted by *instituteurs* at the invitation of the Minister of Public Instruction, 1860-1.

10804, 10807, 10842, 10845-9, 10864-9, *salles d'asile* and écoles maternelles.

11681-90, statistics of *cours complémentaires*, 1894-8.

11706-8, *écoles professionnelles*.

11765, 11772, 11778, mutual education.

11824-5, 11857-8, 11864, evening classes and popular lectures; reorganized 1898, 11919-21, 11926.

Material on private education is weaker, but one may note:

12434$^{1-4}$, *pensionnats* run by religious congregations, 1834-81.

12442, private ecclesiastical schools, 1874-85.

12474, Frères de Ploërmel.

12508, Protestant primary schools, Academy of Nîmes, 1828-34.

F19 (Ministry of *Cultes*) clearly includes relevant material:

3969-70, 3972, *Liberté de l'enseignement*.

4064, Petit Séminaire of Beaucaire.

4091, Catholic universities at time of law of 1875.

'Police des cultes' has material on the political implications of religious practice, including 5610, reports and notes on the attitude of the clergy and particularly the episcopate, 1872-1906.

6072-4, decrees of 1880.

6072, 6080, 6083, decrees of 1902.

7947, 7955-6, 7976-8, congregations suppressed 1901-4.

(b) *Departmental Archives*

The starting-point and centre of gravity of this work has been the departmental archives of Ille-et-Vilaine, the Gard, and the Nord.

Series M, a broad spectrum including 'police politique': conflicts over politics, religion, and education.

I-et-V 1M, the state and the Catholic Church, 1878-1912.

Gard 6M 341-2, 345-6, 349. The Fonds du Cabinet du préfet, 80-2, includes material about political conflict over education in the Gard, 1884-1909.

Child labour and the law of 1841, Nord M611, 613.

sociétés de patronage and associations d'anciens élèves, Nord, M222.

agricultural education in the Gard (13M) and Ille-et-Vilaine (26M1-3, 5-7), with I-et-V 53 Mc dealing with the École pratique d'industrie of Rennes.

Series 1R, registers of the *conseils de révision*, the source used to study the literacy of conscripts.

Series T, the main source for education, divided into the archives of the *préfecture* (1T in Nord and Ille-et-Vilaine) and the *Rectorat* (2T in Nord); in the Gard, the distinction is between primary (1T) and secondary (2T) education.

The main areas of concern are:

the politics of the *écoles centrales, lycées, collèges royaux*, municipal colleges, *petits séminaires*, and private institutions, especially well documented in the first half of the century.

mutual education.

administration by *comités cantonaux, comités d'arrondissement* (including the proceedings of that of Le Vigan, 1836-50 in Gard 1T 20-1).

conseils académiques (classified in 10T in the Gard) and *délegations cantonales*.

écoles normales, and *conférences pédagogiques*. The Archives du Château de Clausonne Meynes 90, 93 in A.D. Gard has interesting material on the recruitment and training of Protestant *institutrices*, 1841-50.

écoles primaires supérieures.

salles d'asile and *écoles maternelles.*

post-school *œuvres* (*cours d'adultes, bibliothèques scolaires*, etc.)

The reports of school inspectors—thin in Ille-et-Vilaine, but important in the Gard (1T 718, 720, 724, 811, 814, 827) and the Nord (1T 107, 2T 404-5, 1097).

laicization of schools of congregations, notably Gard 1T 510, 513, 625, Nord 1T 68/1 and 88/7.

closure of schools of congregations under legislation of 1901-4, Gard 1T 882-4, Nord 1T 123/7.

textbook controversy of 1909, notably Gard 1T 68/11, 1038.

technical education, in the Gard, the École pratique of Nîmes (1T 677-8), the *écoles académiques* and Beaux-Arts (4T 44-5), and agricultural education (1T 929), and in the Nord, 1T 135-92.

enrolment registers are extant in 10T for the Lycée of Rennes (1834-90) and, in A.D. Seine-Inférieure, 10T p 84-6, for the Lycée of Rouen (1905-17).

Series V, religious matters, includes relevant material in 'police des cultes', Gard V4-5 (1837-95), V321 (An XIII-1900).

Nord 2V56 (role of clergy in elections 1888–1902) and 2V76 (use of
Flemish in catechism, 1806–1902).
seminaries in Gard (V89) and Nord (3V 58).
teaching congregations, their activities and closure under legislation of
1901 (Gard V 325, Nord 6V 29) and 1904 (Gard V 326, 330, 335–8,
Nord, 6V 38–9, 53).

(c) *Municipal Archives*

These have proved to be of especial importance in the field of tech-
nical education.
Angers 26R2 École des Arts et Métiers.
Nîmes MIV École pratique de la Calade.
    Conseil de perfectionnement de l'école pratique de commerce
    et d'industrie, Nîmes, *délibérations*, 1897–1918.
  R 265 public primary education, 1792–1895.
  R 271 École de dessin et des Beaux-Arts, 1820–1929.
  R 272 public courses and education lectures.
Rennes 4F 18–21, agricultural education.
  R 11 primary education.
  R 12 primary schools of congregations.
  R 13 lay municipal schools.
  R 14 primary schools, special courses.
  R 16 school and post-school *œuvres*.
  R 27 *salles d'asile*.
  R 28 *écoles primaires supérieurs*.
  R 30, 33 École pratique d'industrie.
  R 54, École régionale des Beaux-Arts.
  R 63 École régionale d'architecture.
  R 69, 72 Lycée de jeunes filles.
  R 93 public educational courses, 1825.
  R 94 technical education.

(d) *Municipal Libraries*

Apart from the indispensable fund of printed material providing guidance
for the local historian, these libraries contain some manuscript col-
lections.
B.M. Nîmes, M.S.S. 524, Germer-Durand papers.
B.M. Lille, Fonds Humbert 37–9.
    Fonds Lemaire II 53, 57, 73.
    Fonds Mahieu A21, B18.

2. Ecclesiastical Archives

(a) *Archives de l'Archevêché de Lille (transferred from Cambrai).*
  Petit Séminaire of Cambrai, règlements, 1807.
  registre d'inscription, 1828–51

(b) *Archives de l'Archevêché de Rennes.*

series C (education):
1. general matters.
2. seminaries.
3. Lycée and École normale of Rennes.
4. Colleges of Saint-Malo and Vitré.
5. primary education.
6. miscellaneous, including teaching curates.

(c) *Archives de l'Évêché de Nîmes.*

register of ordinations, 1817–1913.
Petit Séminaire of Beaucaire, 1822–1925.
speech of Mgr Plantier at Collège Saint-Stanislas, 1864.

(d) *Archives consistoriales, Nîmes.*

K 13 *écoles communales*, 1815–20.
K 18 grievances of Consistory over ordinance of 27 Feb. 1821.
K 19 Protestant pastor at Lycée.
K 28 Sunday schools.
K 29 Protestant primary schools.
K 34 *Pensionnat normal* for girls.

(e) *Archives des Eudistes.*

Livre des visites ecclésiastiques, Collège Saint-Sauveur, Redon.
MF 24 Collège Saint-Sauveur, Redon.
MF 28 Institution Saint-Martin, Rennes.

(f) *Archives des Augustins de l'Assomption.*

CK2 Règlement du Collège de Nîmes, n.d.
DN2 R. P. Touveneraud, 'Évolution de la population scolaire de l'Assomption, 1843–80'.
DN3 R. P. Touveneraud, 'Notes relatives à la maison de l'Assomption de Nîmes'.
répertoire des élèves, 1845–52.

(g) *Archives des Frères des Écoles Chrétiennes.*

Dossiers by districts, NC Avignon, 287 Cambrai, 296 Nantes, 304 Saint-Omer.
Dossiers by establishments, NC 377 Alès, 458 Cambrai, 504 Douai, 617 La Grand 'Combe, 654 Lille, 673 Martigné-Ferchaud, 716 Nîmes, 777 Rennes, 788 Roubaix, 818 Saint-Malo, 871 Tourcoing, 888 Valenciennes.
NG 326 Estaimpuis, 237 Froyennes.

(h) *Archives des Frères Maristes.*

'F.F.J.' 'Historique de Beaucamps et de la Province du Nord, 1842–1932' (typed M.S. 1932).

Ch. Fr. Norbert, 'Historique de la Province de Beaucamps, 1838–1944' (typed M.S., 1944).
General, 211.8 Annales d'Aubenas. Maison de Bessèges-Forge.
    211.9 Annales d'Aubenas. Maison de Bessèges-Mine.
By provinces:
AUB 660 Anduze, Barjac, Castillon-de-Gagnières.
Chamborigaud, Les Mages, Molières-sur-Cèze, N.D. de la Rouvière.
Robiac, Rochessadoule-Alès, Saint-Ambroix, Saint-Florent.
BEA 660 Croix, Halluin, Haubourdin, Lille, Quesnoy-sur-Deûle, Seclin, Vieux-Condé.
BEL 660 Péruwelz, Pommereul.

(i) *Archives des Frères de Ploërmel.*

Noviciat de Ploërmel, registres d'entrée, 1823–56.
  80 Origines de l'Institut.
  100 Idées pédagogiques de J.-M. de la Mennais.
  101 Législation scolaire.
  102–112 Les Écoles, 1820–60.
By communes: Dinard, Fougères, La Guerche, Livré, Redon, Rennes, Saint-Briac, Saint-Servan, Vitré.

(j) *Archives des Pères de l'Immaculée Conception de Saint-Méen.*

8A II 00 Abbé Corvaisier, Livre de paroisse, Saint-Méen.
8A II 08 Proceedings of general chapter, 1872, 1885.
558-A-3 E. Feildel, Annales de Saint-Méen.
The archives of female religious congregations is the main lacuna in the source material. The starting point for such a study must be Charles Molette, *Guide des sources de l'histoire des congrégations féminines françaises* (Paris, 1974).

3. Archives of Schools

The principal manuscript source consulted—when extant—were the enrolment registers of the secondary schools, other than those of the Lycées of Rennes and Rouen, kept in the departmental archives.
École nationale supérieure agronomique, Rennes, enrolment registers, 1871–1913.
École nationale des mines, Alès.
  cahiers des notes particulières des élèves, 1851–64.
  procès-verbaux des séances du conseil d'administration, 1843–70, 1870–92.
École normales d'instituteurs.
  Douai, enrolment registers, 1856–1913.
  Nîmes, enrolment registers, 1842–1913.
  Rennes, enrolment registers, 1831–1913.
Institutions libres, Marcq-en-Baroeul, enrolment registers, 1840–52.
            Saint-Martin, Rennes, enrolment registers, 1875–90.

Saint-Vincent, Rennes, bursars' registers, 1875–90.
Sacré-Cœur, Tourcoing, enrolment registers, 1845–51.
correspondence, 1848–50, 1850–3.
'correspondance avec l'archevêché de Cambrai'.

B. PRINTED SOURCES

(a) *Primary*

Archivium provinciae Campaniae Societatis Jesu, *Liste alphabétique des anciens élèves du Collège de Brugelette, 1835–54* (Paris, 1875).

Associations d'anciens élèves, reports of proceedings of those of Colleges of Saint-Martin, Saint-Vincent, and Vitré, and Petit Séminaire of Saint-Méen in B.M. Rennes.

Chambre de Commerce, Lille, library in A.D. Nord CC. Strong on technical education, including CC 170, Congrès national de l'apprentissage. Roubaix, 1911.

Conseils généraux of Gard, Ille-et-Vilaine, and Nord in A.D. series N. The turntable of departmental history, with votes of *conseils-généraux* after 1871 and annual reports of Inspectors of Academies and departmental professors of agriculture.

Flahaut, R., *Notes et documents pour servir à l'histoire des institutions ecclésiastiques de l'enseignement secondaire à Dunkerque*, fasc. 3. (Dunkerque, 1902).

Gachet, Édouard, *Œuvres diverses* (ed. H. Lefebvre, Lille, 1846).

Maison de l'Assomption, Nîmes, reports of d'Alzon, 1846, 1847, 1848, 1849, 1850.

Mandements de Carême of bishops such as Brossays Saint-Marc (A.D. I-et-V IV 27) and Regnier (A.D. Nord 2V 111, 114). Completed by L. Besson, *Œuvres pastorales*, ii (Paris, 1879), and H. Plantier, *Instructions, lettres pastorales et mandements*, ii, iii (Paris, 1867).

Palmarès de distribution des prix, Maison de l'Assomption, Collège Saint-Stanislas, Lycées of Lille, Nîmes, and Rennes.

(b) *Secondary*

Anderson, Robert, 'The Conflict in education: Catholic secondary schools, 1850–70' in Zeldin, Theodore, ed., *Conflicts in French Society* (London, 1970).

— *Education in France, 1848–70* (Oxford, 1975).

— 'Some Developments in French Secondary Education during the Second Empire' (Oxford D.Phil. thesis, 1967).

Appert, Benjamin, *Dix Ans à la Cour du Roi Louis-Philippe et souvenirs du temps de l'Empire et de la Restauration* (3 vols., Berlin–Paris, 1846).

— *Traité d'éducation élémentaire d'enseignement mutuel pour les prisonniers, les orphelins et les adultes des deux sexes* (Paris, 1822).

Ardouin-Dumazet, V.-E., *Voyages en France*, 18, 19 (Paris-Lille, 1899).

Ariès, Philippe, *Centuries of Childhood* (London, 1962).

— *Histoire des populations françaises et de leur attitude devant la vie depuis le XVIIe siècle* (Paris, 1948).

Artz, Frederick, *The Development of Technical Education in France, 1500-1850* (Cambridge, Mass. 1966).

Audiganne, A., 'Du Mouvement intellectuel parmi les populations ouvrières', *Revue des deux mondes*, 10, 1851.

— *Les Populations ouvrières et les industries de la France*, (Vol. ii, Paris, 1860).

Aulard, Alphonse, 'Un discours de l'ex-Constituant Mounier sur l'instruction publique en l'an XII', *La Révolution française*, 55, July–Dec. 1908.

— *Napoléon 1er et le monopole universitaire* (Paris, 1911).

Azaïs, P., *Vie de Mgr Jean-François-Marie Cart, évêque de Nîmes* (Nîmes, 1857).

Baker, Donald, and Harrigan, Patrick, *The Making of Frenchmen, Current Directions in the History of Education in France, 1679-1979, Historical Reflexions*, vol. 7, nos. 2 and 3 (Waterloo, Ontario, 1980).

Balmont, Joseph, *L'École nationale des arts industriels de Roubaix* (Lille, 1909).

Barre, Catherine, 'La Chambre de lecture de Rennes, 1775-1875' (*mémoire de maîtrise*, Rennes, 1972).

Bascoul, Louis, *Vie de Mgr Besson, évêque de Nîmes, Uzès et Alais* (2 vols., Arras–Paris, 1902).

Baudrillart, Henri, *Les Populations agricoles de la France*, iii (Paris, 1893).

Baudu, F., *Les Origines de la Congrégation des Sœurs de l'Instruction Chrétienne de Saint-Gildas des Bois. La fondation et les fondateurs, 1807-42* (Vannes, 1948).

Baunard, L., *Les Deux Frères: cinquante années de l'Association catholique à Lille, Philibert Vrau. Camille Féron-Vrau* (2 vols., Paris, 1911).

Bécus, Édouard, *Mathieu de Dombasle, sa vie et ses œuvres* (Paris, 1874).

Bellet, Roger, 'Une Bataille culturelle, provinciale et nationale, à propos des bons auteurs pour les bibliothèques populaires', *Revue des sciences humaines*, fasc. 135, 1969.

Benaerts, Louis, *Le Régime consulaire en Bretagne: le département d'Ille-et-Vilaine durant le Consulat, 1799-1804* (Paris, 1914).

Berger, Suzanne, *Peasants against Politics: Rural Organization in Brittany, 1911-67* (Cambridge, Mass, 1972).

Bernoville, Gaéton, *Les Religieuses de Saint-Thomas de Villeneuve* (Paris, 1953). *Terre de Bretagne: les Sœurs de Rillé* (Paris, 1957).

Beylard, Hugues, 'Les Jésuites à Lille au XIXe siècle', *Revue du Nord*, liii, no. 208, 1971.

Blanchard, Raoul, *La Densité de la population du département du Nord*, (Lille, 1906). *La Flandre, étude géographique de la plaine flamande en France, Belgique et Hollande* (Paris, 1906).

Bois, Paul, *Paysans de l'Ouest: des structures économiques et sociales aux options politiques depuis l'époque révolutionnaire dans la Sarthe* (Le Mans, 1960).

Bollême, Geneviève, *La Bibliothèque bleue. La littérature populaire en France du XVIe au XIXe siècle* (Paris, 1971).

Bonnafous, A., 'Les Royalistes dans le Nord et le Ralliement', *Revue du Nord*, xlvii, no. 184, 1965.

Boudon, Raymond, *L'Inégalité des chances. La mobilité sociale dans les sociétés industrielles* (Paris, 1973).

Boulard, Fernand. *Premiers Itinéraires en sociologie religeuse* (Paris, 1954).

—— and Pérry, J., *Pratique religieuse urbaine et régions culturelles* (Paris, 1968).

Bouquet, Daniel, and Vitumi, Henri, 'Pluralisme religieux et instruction primaires: la loi Guizot et son application dans le Gard, 1833-49' (D.E.S., Montpellier, 1970).

Bourdieu, Pierre, 'L'École conservatrice: les inégalités devant l'école et devant la culture', *Revue française de sociologie*, vii, 1966.

—— and Passeron, Jean-Claude, *Les Héritiers: les étudiants et la culture* (Paris, 1964).

Boussemart, A., *Histoire du Petit Séminaire de Cambrai* (Cambrai, 1902).

Bréal, Michel, *Quelques Mots sur l'instruction publique en France* (Paris, 1872).

Bricaud, Jean, *L'Administration du département d'Ille-et-Vilaine au début de la Révolution, 1790-1* (Rennes, 1965).

Brochon, Pierre, *Le Livre de colportage en France depuis le XVIe siècle* (Paris, 1954).

Bruhat, Jean, 'Anticléricalisme et mouvement ouvrier en France avant 1914', *Mouvement social*, 57, 1966.

Brun, H., 'Les Ordres religieux du diocèse de Nîmes', *Bulletin du Comité de l'Art Chrétien de Nîmes*, ix, no. 57, 1907.

Bruyère, Marcel, *Alès, capitale des Cévennes* (Nîmes, 1948).

—— 'Le Collège royal de Nîmes sous la Réstauration', *Mémoires de l'Académie de Nîmes*, li, 1936-8.

Capelle, F., *Éloge historique de M. de Forest de Lewarde* (Douai, 1852). *Vie du Cardinal P. Giraud* (Lille, 1852).

Caron, François, *Histoire de l'exploitation d'un grand réseau: la compagnie du chemin de fer du Nord, 1846-1937* (Paris-The Hague, 1973).

Catta, Tony, *Le Père Dujarié, 1767-1838* (Paris-Montréal, 1958).

Centre d'histoire contemporaine du Languedoc mediterranéen et du Roussillon,

—— *Droite et gauche de 1789 à nos jours* (Montpellier, 1975).

—— *Économie et société en Languedoc-Roussillon de 1789 à nos jours* (Montpellier, 1978).

Certeau, Michel de, Julia, Dominique, and Revel, Jacques, *Une politique de la langue: la Révolution française et le patois* (Paris, 1975).

Chalmel, Théodore, *Saint-Père-Marc-en-Poulet* (Rennes, 1931).

Chamson, André, *Le Chiffre de nos jours* (fifteenth edn., Paris, 1954).

Charton, Édouard, *Dictionnaire des professions* (third edn., Paris, 1880).

Choleau, Jean, *Condition actuelle des serviteurs ruraux bretons* (Vannes, 1907).

— *Questions bretons du temps présent* (2 vols., Vitré, 1942).

Cholvy, Gérard, 'Enseignement religieux et langues maternelles en France au XIXe siècle', *Revue des langues romanes*, lxxxii, 1976.

— 'La Question scolaire au XIXe siècle: le conflit pour le contrôle de l'école dans l'Hérault, 1866-90', *Annales du Midi*, 87, no. 124, 1975.

— 'Une école des pauvres au début du XIXe siècle: "pieuses filles", béates ou sœurs des campagnes', in D. Baker and P. Harrigan, *The Making of Frenchmen* (1980).

— 'Religion et politique en Languedoc méditerranéen et Roussillon à l'époque contemporaine' in *Droite et gauche de 1789 à nos jours* (Montpellier, 1975).

— 'Religion et Révolution: la déchristianisation de l'An II', *Annales historiques de la Révolution française*, 50, no. 233, 1978.

Clastron, J., *Vie de Sa Grandeur Mgr Plantier, évêque de Nîmes* (2 vols., Nîmes, 1882).

Cobban, Alfred, 'The influence of the clergy and the *instituteurs primaires* on the elections to the French Constituent Assembly, April 1848', *English Historical Review*, lvii, 1942.

Colette, Aubain, 'L'idéal pédagogique du Père d'Alzon' in *Mélanges Emmanuel d'Alzon* (Saint-Gérard, 1950).

Coornaert, E., 'Flamand et français dans l'enseignement en Flandre française des annexations au XXe siècle', *Revue du Nord*, liii, no. 209, 1971.

Corbin, Alain, 'Pour une étude sociologique de la croissance de l'alphabétisation au XIXe siècle. L'Instruction des conscrits de l'Eure-et-Loir, 1833-1883', *Revue d'histoire économique et sociale*, 53, 1975.

Cosson, A., 'Industrie de soie et population ouvrière à Nîmes de 1815 à 1848' in *Économie et société en Languedoc-Roussillon de 1789 à nos jours* (Montpellier, 1978).

Couderc de Latour-Lisside, Félix-Adrien, *Vie de Mgr de Chaffoy, ancien évêque de Nîmes* (2 vols., Nîmes, 1856-7).

Courson, Aurélien de, *La Division de Vitré en 1832* (Vannes, 1899).

Crubellier, Maurice, *L'Enfance et la jeunesse dans la société française, 1800-1950* (Paris, 1979).

Dainville, François de, 'Effectifs des collèges et scolarité au XVIIe et XVIIIe siècle dans le nord et le nord-est de la France', *Population*, 10, 1955.

Dalimier, M.-J.-M., *La Pédagogie des écoles rurales, principalement à l'usage des institutrices* (Rennes, 1843).

Dargis, André, 'La Congrégation de Saint-Pierre' (thesis, Louvain, 1971).

Darmon, J.-J., *Le Colportage de librairie en France sous le Second Empire* (Paris, 1972).

Daudet, Alphonse, *Le Petit Chose* (first edn., 1868; Paris, 1947).

Dauphin, Joseph, *Les Eudistes dans le diocèse de Rennes* (Rennes–Paris, 1910). *Le R. P. Louïs de la Morinière* (Rennes–Paris, 1899).

David, Jean-Philippe, *L'Etablissement de l'enseignement primaire au XIXe siècle dans le département de Maine-et-Loire, 1816–79* (Angers, 1969).

David, Paul, 'Les Origines du syndicalisme agricole en Ille-et-Vilaine, 1885–1910' (*mémoire de maîtrise*, Rennes, 1972).

Day, C. R., 'Technical and Professional education in France: the rise and fall of "enseignement secondaire spécial"', *Journal of Social History*, 6, 1872–3.

—— 'The Third Republic and the development of intermediate education in France, 1870–1914', *Proceedings of the Fourth Annual Meeting of the Western Society for French History*, 1976 (Santa Barbara, 1977).

Delattre, P., *Les Établissements des Jésuites en France depuis quatre siècles* (vols. 1 and 2, Enghien, 1948, 1953).

Delcourt, Raymond, *La Condition des ouvriers dans les mines du Nord et du Pas-de-Calais* (Paris, 1906).

Delourme, Paul (pseudonym of abbé Trochu), *Trente-cinq Années de politique religieuse, ou l'histoire de 'l'Ouest-Eclair'* (Paris, 1936).

Delumeau, Jean, ed., *Histoire de la Bretagne* (Toulouse, 1969).

Demangeon, Albert, *La Picardie et les régions voisines. Artois, Cambrésis, Beauvaisis* (Paris, 1905).

Denis, Michel, 'Rennes au XIXe siècle: ville parasitaire?', *Annales de Bretagne*, lxxxx, 1973.

Denniée, Victor, *De l'enseignement professionnel* (Paris, 1852).

Dervaix, J.-F., *Le Doigt de Dieu. Les Filles de la Sagesse après la mort des fondateurs*, vol. ii, 1800–1900 (Cholet, 1955).

Deschuytter, Joseph, *L'Esprit public et son évolution dans le Nord de 1791 au lendemain de Thermidor An II* (vol. i, Gap, 1959).

Desert, Gabriel, 'Alphabétisation et scolarisation dans le Grand-Ouest au XIXe siècle', in D. Baker and P. Harrigan, *The Making of Frenchmen* (1980).

Dessal, M., *Un Révolutionnaire jacobin: Charles Delescluze* (Paris, 1952).

Destombes, C.-J., *Vie de Son Eminence le Cardinal Regnier, archévêque de Cambrai* (2 vols., Lille–Paris, 1885).

Destutt de Tracy, *Observations sur le système actuel de l'instruction publique* (Paris, An IX).

Detrez, L., *Mère Natalie, fondatrice de la Congrégation des Filles de l'Enfant-Jésus* (Paris–Lille, 1930).

Dubois, Paul, *Un patriarche: vie de M. Dubois-Fournier, 1768–1844* (Lille, 1899).

Dugrand, Raymond, *Villes et campagnes en Bas-Languedoc* (Paris, 1963).

Dupré, Guy, *Formation et rayonnement d'une personnalité catholique au XIXe siècle: le père Emmanuel d'Alzon, 1810-80* (Lille, 1975).

Dupuy, Roger, *La Garde nationale et les débuts de la Révolution en Ille-et-Vilaine, 1789-mars. 1793* (Paris, 1972).

Durand, Albert, *Histoire religieuse du département du Gard pendant la Révolution française* (Nîmes, 1918).

Durieux, A., *Le Collège de Cambrai, 1270-1882* (Cambrai, 1882).

Duroselle, Jean-Baptiste, *Les Débuts du Catholicisme social en France, 1822-70* (Paris, 1951).

Durtelle de Saint-Sauveur, Geneviève, 'Le Collège de Rennes depuis sa fondation jusqu'au départ des Jésuites, 1536-1762', *Bulletin de la Société archéologique d'Ille-et-Vilaine*, xlvi, 1918.

École normale supérieure, Paris, *Niveaux de culture et groupes sociaux*, in Actes du Colloque, 7-9 May 1966 (Paris–The Hague, 1967).

Emerit, Marcel, 'Du Saint-Simonisme au Catholicisme: Ignace Plichon, député du Nord', *Revue du Nord*, lvi, no. 220, 1974.

Engrand, Monique and Charles, 'Épidémie et paupérisme: le choléra à Lille en 1832', in Gillet, Marcel, ed., *L'Homme, la vie et la mort dans le Nord au XIXe siècle* (Lille, 1972).

Falcucci, C., *L'Humanisme dans l'enseignement secondaire* (Paris, 1936).

Faury, Jean, *Cléricalisme et anticléricalisme dans le Tarn, 1848-1900* (Toulouse, 1980).

Féval, Paul, *Bouche de fer* (fourth edn., Paris, 1885).

Fitzpatrick, Brian, 'Catholic royalism in the departments of the Gard, 1814-51', (Warwick Ph.D. thesis, 1977).

Fleury, Michel, and Valmary, Pierre, 'Les Progrès de l'instruction élémentaire de Louis XIV à Napoléon III', *Population*, 12, 1957.

Fohlen, Claude, *L'Industrie textile au temps du Second Empire* (Paris, 1956).

Fontaine de Resbecq, *Histoire de l'enseignement primaire avant 1789 dans les communes qui ont formé le département du Nord* (Lille–Paris, 1878).

Foucault, Michel, *Surveillir et punir: naissance de la prison* (Paris, 1975).

Fouqueray, P. H., *La Mère Saint-Félix, Fondatrice des Sœurs de l'Immaculée Conception de Saint-Méen* (Saint-Méen, 1924).

Frainaud, Lucien, 'Propos sur la marine et le Collège royal naval d'Alès', *Mémoires de l'Académie de Nîmes*, VIIe série, lviii, 1971-3.

Frandon, F., *Le Collège d'Uzès, 1566-1793, 1803-1903* (Toulouse, 1907).

Freyssinet-Dominjon, Jacqueline, *Les Manuels d'histoire de l'École libre, 1882-1959* (Paris, 1969).

Frijoff, Willem, and Julia, Dominique, *École et société dans la France d'ancien régime* (Paris, 1975).

Furet, François, and Ozouf, Jacques, *Lire et écrire. L'alphabétisation des Français de Calvin à Jules Ferry* (2 vols., Paris, 1977).

—— 'Literacy and industrialization: the *Département du Nord* in France', *Journal of European Economic History*, 5, no. 1, 1976.

Gadille, Jacques, 'Monsieur Duquesnay et la République, 1872-84', *Revue du Nord*, xlv, no. 178, 1963.

—— *La Pensée et l'action politique des évêques français au début de la Troisième République, 1870-83* (Paris, 1967).

Gaillard, J.-M., 'La Pénétration du socialisme dans le bassin houiller du Gard' in *Droite et gauche de 1789 à nos jours* (Montpellier, 1975).

Garlan, Yvon, et Nières, Claude, *Les Révoltes bretonnes de 1675: papier timbré et bonnets rouges* (Paris, 1975).

Garnier, Adrien, *Frayssinous, son rôle dans l'Université sous la Restauration, 1822-28* (Paris–Rodez, 1925).

Gautier, Elie, *Un siècle d'indigence. Pourquoi les Bretons s'en vont* (Paris, 1950).

Gerbod, Paul, *La Condition universitaire en France au XIXe siècle* (Paris, 1965). *Paul-François Dubois, universitaire, journaliste et homme politique* (Paris, 1967).

—— *La Vie quotidienne dans les lycées et collèges au XIXe siècle* (Paris, 1968).

Gillet, Marcel, *Les Charbonnages du Nord de la France au XIXe siècle* (Paris–The Hague, 1973).

Girard, Alain, 'L'Origine sociale des élèves des classes de 6e', *Population*, 17, 1962.

—— *La Réussite sociale en France, ses caractères, ses lois, ses effets* (Paris, 1961).

Goblot, Edmond, *La Barrière et le niveau* (Paris, 1925).

Goiffon, Étienne, *L'Instruction publique à Nîmes: le Collège des Arts, les Jésuites, les Doctrinaires* (Nîmes, 1876).

Gontard, Maurice, *Les Écoles primaires de la France bourgeoise, 1833-75* (Toulouse, 1964).

—— *L'Enseignement primaire en France de la Révolution à la loi Guizot 1789-1833* (Paris, 1959).

—— 'Guizot et l'instruction populaire: la loi du 28 juin 1833' in *Actes du colloque François Guizot*, 22-25 Oct. 1974 (Paris, 1976).

Goodwin, Albert, 'Counter-revolution in Brittany: The royalist conspiracy of the marquis de la Rouërie, 1791-3', *Bulletin of the John Rylands Library*, 39, no. 2 (mar. 1957).

Gossez, A.-M., *Le Département du Nord sous la Deuxième République, 1848-52* (Lille, 1904).

Gouallou, Henri, 'Les Élections à l'Assemblée législative en Ille-et-Vilaine, 13 May 1849', *Annales de Bretagne*, lxxx, no. 2, 1973.

—— 'L'Évolution politique de l'Ille-et-Vilaine, 1851-79' (thesis, Rennes, 1971).

—— *Hamon, commissaire du gouvernement, puis préfet d'Ille-et-Vilaine, 1848-9* (Paris, 1973).

Gouallou, Henri, 'Pratique religieuse et options politiques en Ille-et-Vilaine à la fin du XIXe siècle', *Annales de Bretagne*, lxxii, 1965.

Goyet, Martine, 'Les Comices agricoles et les sociétés d'agriculture en Ille-et-Vilaine, 1831–1905' (*mémoire de maîtrise*, Rennes, 1968).

Grandidier, F., *Vie du R. P. Achille Guidée de la Compagnie de Jésus* (Amiens–Paris, 1867).

Grangent, S.-V., *Description abrégée du département du Gard* (Nîmes, An VII).

Grignon, Claude, 'L'Orientation scolaire des élèves d'une école rurale', *Revue française de sociologie*, ix, no. spécial, 1968.

Grimaud, Louis, *Histoire de la liberté de l'enseignement en France, v, vi,* (Paris, 1954).

Guéhenno, Jean, *Changer la vie: mon enfance et ma jeunesse* (Paris, 1961).

Guillou, Adolphe, and Revillon, Armand, *La Vente des biens nationaux dans les districts de Rennes et de Bain* (Rennes, 1911).

Guinot, J.-B., *Formation professionnelle et travailleurs qualifiés depuis 1789* (Paris, 1948).

Haize, Jules, *Histoire du Collège de Saint-Servan* (Saint-Servan, 1980).

Harrigan, Patrick, 'Catholic Secondary Education in France, 1851–82' (University of Michigan Ph.D. thesis, 1970).

— (with Victor Neglia), *Lycéens et collégiens sous le Second Empire. Étude statistique sur les fonctions sociales de l'enseignement secondaire d'après l'enquête de Victor Duruy, 1864–5* (Paris, 1979).

— *Mobility, Élites and Education in French Society of the Second Empire* (Waterloo, Ontario, 1980).

Hélias, Pierre-Jakez, *Le Cheval d'orgueil, mémoires d'un Breton du pays bigouden* (Paris, 1975).

Hemeryck, Richard, 'La Congréganisation des écoles normales du département du Nord au milieu du XIXe siècle, 1845–1883', *Revue du Nord*, lv, no. 217, 1973, lvi, no. 220, 1974.

— 'La Laïcisation des écoles des Frères à Lille en 1868', *Actes du 95e Congrès national des sociétés savantes*, Reims, 1970 (Paris, 1974).

Herpin, E., Hervot, H., Mathurin, J., Saint-Mleux, G., *Histoire du Collège de Saint-Malo* (Ploërmel, 1902).

Hilaire, Yves-Marie, ed., *Atlas électoral, Nord, Pas-de-Calais, 1876–1936* (Lille, 1977).

— *Une chrétienté au XIXe siècle? La vie religieuse des populations du diocèse d'Arras, 1840–1914* (Lille, 1977).

— 'Les ouvriers de la région du Nord devant l'Église catholique, XIXe–XXe siècle', *Mouvement social*, no. 59, 1966.

Holt, Richard, *Sport and Society in Modern France* (London, 1981).

Hood, James, N., 'Patterns of Popular protest in the French Revolution: the conceptual contribution of the Gard', *Journal of Modern History*, 48, 1976.

— 'Protestant–Catholic relations and the roots of the first popular counter-revolutionary movement in France', *Journal of Modern History*, 43, 1971.

— 'Revival and mutation of old rivalries in revolutionary France', *Past and Present*, 82, 1979.

Horvath, Sandra Ann, 'Victor Duruy and the controversy over secondary education for girls', *French Historical Studies*, ix, no. 1, 1975.

Huard, E., *Edouard Lefort, président de la Société de Saint-Joseph de Lille* (Paris, 1893).

Huard, Raymond, *La Bataille pour l'école primaire dans le Gard 1866-72* (Nîmes, 1966).

— 'Montagne rouge et montagne blanche en Languedoc-Roussillon sous la Seconde République, in *Droite et gauche de 1789 à nos jours*, (Montpellier, 1975).

— *Le Mouvement républicain en Bas-Languedoc, 1848-1881* (Paris, 1982).

— 'La Préhistoire des partis: le parti républicain et l'opinion républicaine dans le Gard de 1848 à 1881' (thesis, Paris, 1977).

Huot-Pleuroux, Paul, *Le Recrutement sacerdotal dans le diocèse de Besançon de 1801 à 1960* (Paris, 1966).

Isambert-Jamatie, Vivianne, *Crises de la société, crises de l'enseignement* (Paris, 1970).

— 'Permanence ou variations des objectifs poursuivis par les lycées depuis cent ans', *Revue française de sociologie*, viii, no. spécial, 1967.

— 'Une Réforme des lycées et collèges. Essai d'analyse sociologique de la réforme de 1902', *L'Année sociologique*, 20, 1969.

Jégo, J.-B., *L'Institution Saint-Martin et les Eudistes à Rennes* (Rennes, 1954).

Jégouzo, G., and Brangeon, J.-L., *Les Paysans et l'école* (Paris, 1976).

Jénouvrier, L., *Un collège français et chrétien, Saint-Vincent de Paul de Rennes* (Rennes, 1924).

Johnson, Douglas, *Guizot: Aspects of French History, 1787-1874* (London–Toronto, 1963).

Johnson, Richard, 'Educational policy and social control in early Victorian England', *Past and Present*, no. 49, 1970.

Joutard, Philippe, *La Légende des Camisards: une sensibilité au passé* (Paris, 1977).

Jovenaux, J.-B., 'Les Congrès Catholiques du Nord et du Pas-de-Calais, 1873-98' (D.E.S. Lille, 1967).

Judt, Tony, *Socialism in Provence, 1871-1914* (Cambridge, 1971).

Julia, Dominique, and Pressly, Paul, 'La Population scolaire en 1789. Les extravagances statistiques du ministère Villemain', *Annales E.S.C.*, 30e ann., no. 6., 1975.

Labbé, M., *Rapport sur l'enseignement professionnel* (Lille, 1913).

La Borderie, Arthur de, *Essai sur la géographie féodale de la Bretagne* (Rennes, 1889).

Labuda, A. W., 'La Langue de l'Empereur: la culture littéraire des lycées sous le Second Empire', *Littérature*, no. 22, 1976.

Laget, M., 'Petites écoles en Languedoc au XVIIIe siècle', *Annales E.S.C.*, 26e ann., 1971.

La Gorce, P.-M. de, 'Un Grand évêque et un grand diocèse, le Cardinal Regnier, 1850-81', *Revue des questions historiques*, no. 119, 1934.

Lagrée, Michel, 'Aspects de la vie religieuse en Ille-et-Vilaine, 1814-48' (thesis, Rennes, 1974).

— *Mentalités, religion et histoire en Haute Bretagne au XIXe siècle: le diocèse de Rennes, 1815-48* (Paris, 1977).

— 'La structure pérenne: événement et histoire en Bretagne orientale, XVIe-XXe siècle', *Revue d'histoire moderne et contemporaine*, xxiii, 1976.

Lajusan, Alfred, 'La Carte des opinions françaises', *Annales E.S.C.*, 4, 1949.

Lambert-Dansette, J., *Quelques Familles du patronat textile de Lille-Armentières, 1789-1914* (Lille, 1954).

Lamorisse, René, *Recherches géographiques sur la population de la Cévenne languedocienne* (Montpellier, 1975).

Landes, David, 'French entrepreneurship and economic growth in the nineteenth century', *Journal of Economic History*, 9, 1949.

Langlois, Claude, *Un diocése breton au début du dix-neuvième siècle: Vannes 1800-30* (Paris, 1974).

— *Structures religieuses et célibat féminin au XIXe siècle: les tiers ordres dans le diocèse de Vannes* (Lyon, 1972).

Lanse	zeur, Yves, 'Étude démographique et sanitaire de la ville de Rennes' (medical thesis, Paris, 1902).

Latreille, André, and Siegfried, André, *Les Forces religieuses et la vie politique* (Paris, 1951).

Laurent, Robert, 'Droite et Gauche en Languedoc: mythe ou réalité?' in *Droite et gauche de 1789 à nos jours* (Montpellier, 1975).

— 'Les Quatre ages du vignoble' in *Économie et société en Languedoc-Roussillon de 1789 à nos jours* (Montpellier, 1978).

Laveille, A., *Jean-Marie de la Mennais (1780-1860)* (2 vols., Paris, 1903).

Leblanc, H.-J., *Le Collège communal de Tourcoing pendant les vingt-cinq dernières années du régime universitaire, 1858-1881* (Lille, 1885).

— *Histoire du Collège de Tourcoing, principalement sous l'adminstration de M. l'abbé Lecomte* (Tourcoing, 1870).

Leblond, M., 'La Scolarisation dans le département du Nord au XIXe siècle', *Revue du Nord*, lii, no. 206, 1970.

Leduc, Robert, 'La vie religieuse dans les arrondissements de Lille, Hazebrouck et Dunkerque d'après les chroniques de missions pendant la seconde moitié du XIXe siècle' (*mémoire de maîtrise*, Lille, 1969).

Lefebvre, Georges, *Les Paysans du Nord pendant la Révolution française* (Paris–Lille, 1924).

Leflon, Jean, 'Les Petits séminaires en France au XIXe siècle', *Revue d'histoire de l'Église de France*, lxi, no. 166, 1975.

Le Gall, Hervé, and Bessec, Alain, 'L'École des Trois-Croix: un établissement au service du progrès agricole, 1835–89' (*mémoire de maîtrise*, Rennes, 1972).

Le Glay, André, *Mémoire sur les bibliothèques publiques et les principales bibliothèques particulières du département du Nord* (Lille, 1841).

Le Goff, T. J. A., and Sutherland, D. M. G., 'The Revolution and the rural community in eighteenth-century Brittany', *Past and Present*, no. 62, 1974.

Legrand, Louis, *L'Influence du positivisme dans l'œuvre scolaire de Jules Ferry* (Paris, 1961).

Lemire, J., *L'Abbé Dehaene et la Flandre* (Lille, 1891).

Lemoine, René, *La Loi Guizot, son application dans le département de la Somme* (Abbéville, 1933).

Lennel, F., *L'Instruction primaire dans le département du Nord pendant la Révolution* (Paris, 1912).

Lentaker, F., 'Les ouvriers belges dans le département du Nord au milieu du XIXe siècle', *Revue du Nord*, xxxviii, 1956.

Léonard, Émile, *Le Protestant français* (Paris, 1953).

Léonard, Jacques, 'L'Exemple d'une catégorie socio-professionnelle au XIXe siècle: les médicins français', in Daniel Roche, ed., *Ordres et classes*, Colloque d'histoire sociale, Saint-Cloud, 24–5 May 1967 (Paris–The Hague, 1973).

—— *La Médecine entre les pouvoirs et les savoirs* (Paris, 1981).

Le Priellec, J.-Y., 'Les Élections législatives en Ille-et-Vilaine, 1881–97' (*mémoire de maîtrise*, Rennes, 1974).

Lequin, Yves, 'Labour in the French Economy since the Revolution', in *Cambridge Economic History of Europe*, vii (1), 1978.

Leroy, A., *Fleur de Bretagne: Mère Marie-Amélie Fristel, 1798–1866* (Paris, 1960).

Le Roy Ladurie, Emmanuel, *Paysans de Languedoc* (Paris, 1966).

Le Taillandier, J.-B., *Vie des fondateurs et annales de la Congrégation des Religieuses Adoratrices de la Justice de Dieu* (Rennes, (1899).

Leuridan, T., *Histoire de l'Institution Notre-Dame des Victoires à Roubaix* (Roubaix, 1891).

Lévy-Leboyer, Claude, *L'Ambition professionnelle et la mobilité sociale* (Paris, 1971).

Lewis, Gwynne, *The Second Vendée. The Continuity of Counter-Revolution in the Department of the Gard 1789–1815* (Oxford, 1978).

—— 'The White Terror in the department of the Gard, 1789–1820' (Oxford D.Phil. thesis, 1966).

Ligou, Daniel, 'L'Église réformée du Désert, fait économique et sociale', *Revue d'histoire économique et sociale*, xxxii, 1954.
— *Frédéric Desmons et la franc-maçonnerie sous la Troisième République* (Paris, 1966).
Lorain, Paul, *Tableau de l'instruction primaire en France* (Paris, 1837).
Loubère, Léo-A, 'The Emergence of the Extreme Left in Lower Languedoc, 1848-51, social and economic factors in politics', *American Historical Review*, 73 (2),
— *Radicalism in Mediterranean France, 1848-1914* (Albany, 1974).
— *The Red and the White, A History of Wine in France and Italy in the Nineteenth Century* (Albany, 1978).
Louvière, Luc, *Tel l'ajonc sous la neige* (Chateaulin, 1963).
Maillard, J., *L'Oratoire à Angers au XVIIe et XVIIIe siècle* (Paris, 1975).
Mahieu, Léon, *Mgr Louis Belmas, ancien évêque constitutionnel de l'Aude, évêque de Cambrai, 1757-1841* (2 vols., Paris, 1934).
— 'Mouvement des ordinations et tableau de l'augmentation du clergé dans le diocèse de Cambrai', *Bulletin de la Société d'études de la province de Cambrai*, xliii, 1956-7.
Marchand, Philippe, 'Le Réseau des collèges dans le Nord de la France en 1789', *Revue du Nord*, lviii, no. 229, 1976.
Marchant, Lucien, *L'Institution libre de Marcq-en-Barœul* (Lille, 1948).
Masure, E., *Le Clergé du diocèse de Cambrai, 1802-1913* (Roubaix, 1913).
Maubon, A., *L'École des Mines d'Alès et son histoire* (Alès, 1950).
Maurain, Jean, *La Politique ecclésiastique du Second Empire de 1852 à 1869* (Paris, 1930).
Maurin, Georges, 'L'Instruction publique sous le Premier Empire', *Revue du Midi*, 39, 1906, 40, 1907.
Mayeur, Françoise, *L'Enseignement des filles en France au XIXe siècle* (Paris, 1977).
— *L'Enseignement secondaire des jeunes filles sous la Troisième République* (Paris, 1977).
— 'Les Évêques français et Victor Duruy: les cours secondaires de jeunes filles', *Revue d'histoire de l'Église de France*, lvii, no. 159, 1971.
Mayeur, Jean-Marie, *Un prêtre démocrate: l'abbé Lemire, 1853-1928* (Paris, 1968).
Mazoyer, Louis, 'La Jeunesse villageoise du Bas-Languedoc et des Cévennes en 1830', *Annales H.E.S.*, x, 1938.
— 'La Rénovation intellectuelle et problèmes sociaux: la bourgeoisie du Gard et l'instruction au début de la Monarchie de Juillet', *Annales H.E.S.*, vi, 1934.
Ménager, Bernard, *La Laïcisation des écoles communales dans le département du Nord, 1879-99* (Lille, 1971).
— 'Prélude à la bataille scolaire: les pétitions de 1872 concernant l'enseignement primaire dans le département du Nord', *Revue du Nord*, lv, no. 217, 1973.

Mettrier, M., *L'École des maîtres-mineurs de Douai* in *Lille et la région du Nord* (Lille, 1909).

Meril, Joseph, 'Les Élections législatives en Ille-et-Vilaine, 1898–1914' (*mémoire de maîtrise*, Rennes, 1974).

Michel, Henry, *La Loi Falloux* (Paris, 1906).

Mora, C., 'La diffusion de la culture dans la jeunesse des classes populaires depuis un siècle: l'action de la Ligue de l'Enseignement', in *Niveaux de culture et groupes sociaux* (Paris–The Hague, 1967).

Morange, Jean, and Chassaing, Jean-François, *Le Mouvement de réforme de l'enseignement en France, 1760–1798* (Paris, 1974).

Néré, Jacques, 'Les Élections de Boulanger dans le département du Nord' (thesis, Paris, 1959).

Nora, Pierre, 'Ernest Lavisse, son rôle dans la formation du sentiment nationale', *Revue historique*, 228, 1962.

Oberlé, Raymond, *L'Enseignement à Mulhouse de 1798 a` 1870* (Paris, 1961).

O'Boyle, Lenore, 'The Problem of an excess of educated men in Western Europe, 1800–50', *Journal of Modern History*, 4, 1970.

Ong, W. S., 'Latin language study as a Renaissance puberty rite', *Studies in Philology*, lvi, no. 2, 1959.

Orhand, le R. P., *Le R. P. Pillon et les Collèges de Brugelette, Vannes, Sainte-Geneviève, Amiens et Lille* (Lille, 1888).

Ozouf, Jacques, *Nous les maîtres d'école* (Paris, 1967).

Padberg, John W., *Colleges in Controversy: the Jesuit Schools in France from Revival to Suppression, 1815–80* (Cambridge, Mass., 1969).

Palmer, Michael, 'Some Aspects of the French press during the rise of the popular daily' (Oxford D.Phil. thesis, 1973).

Paquier, J.-B., *L'Enseignement professionnelle en France* (Paris, 1908).

Partin, M. O., *Waldeck-Rousseau, Combes and the Church: the Politics of anticlericalism, 1899–1905* (Duke University, 1969).

Pelen, Jean-Noël, 'La vallée longue en Cévenne: vie, traditions et proverbes du temps passé', *Causses et Cévennes*, no. spécial, 1975.

Perrée, T., *Le Tiers Ordre de Notre-Dame du Mont-Carmel d'Avranches* (Coutances, 1965).

Perrot, Michelle, *Les Ouvriers en grève, 1871–90* (2 vols., Paris–The Hague, 1974).

Peter, Joseph, *L'Enseignement secondaire dans le département du Nord pendant la Révolution, 1789–1802* (Lille, 1912).

Peter, Joseph, and Poulet, C., *Histoire religieuse du département du Nord pendant la Révolution, 1789–1902* (2 vols., Lille, 1930–3).

Piacentini, R., *Les Filles de Jésus* (Bar-le-Duc, 1952).

Pichon, L., *M. le Chanoine Henry Ceillier* (Rennes, 1912).

Pierrard, Pierre, 'L'Enseignement primaire à Lille sous la Monarchie du Juillet', *Revue du Nord*, lvi, no. 220, 1974.

—— 'L'Enseignement primaire à Lille sous la Restauration', *Revue du Nord*, lv, no. 217, 1973.

Pierrard, Pierre, 'L'Établissement des Frères des Écoles Chrétiennes à Valenciennes, 1824–48' (*mémoire d'études supérieures*, Lille, 1949).

— 'Un grand bourgeois à Lille: Charles Kolb-Bernard, 1798–1888', *Revue du Nord*, xlviii, no. 190, 1966.

— *La Vie ouvrière à Lille sous le Second Empire* (Paris, 1965).

Plongeron, Bernard, 'Du Modele jésuite au modèle oratorien dans les collèges français à la fin du XVIIIe siècle', in *Église et enseignement*, Actes du Colloque du Xe anniversaire de l'Institut d'histoire du Christianisme de l'Université de Bruxelles, ed. Jean Préaux (Brussels, 1977).

— *Théologie et politique au siècle des lumières 1770–1820* (Paris, 1973).

Pocquet du Haut-Jussé, B.-A., 'Correspondance politique du Colonel Carron, 1871–85', *Mémoires de la Société historique et archéologique de Bretagne*, xl, 1960.

Poirier, J., 'L'Universitaire provisoire, 1814–21', *Revue d'histoire moderne*, i, 1926, ii, 1927.

Poitrineau, F., *Les Écoles de hameau* (Paris, 1889).

Pouthas, Charles, *Une famille bourgeoise française de Louis XIV à Napoléon* (Paris, 1934).

Pressly, Paul, 'The Personnel of French public education, 1809–30, a study of Angers and Paris during the Empire and Restoration' (Oxford, D.Phil. thesis, 1971).

Prévot, André, *L'Enseignement technique chez les Frères des Écoles Chrétiennes au XVIIIe et XIXe siècle* (Paris, 1964).

Prost, Antoine, *Histoire de l'enseignement en France, 1800–1967* (Paris, 1968).

Quinion-Hubert, L., *Écoles académiques et professionnelles de la ville de Douai* (Douai, 1897).

Raphaël, Paul, and Gontard, Maurice, *Un Ministre de l'Instruction publique sons l'Empire autoritaire: Hippolyte Fortoul, 1851–6* (Paris, 1975).

Rayez, André, 'Clorivière et ses fondations, 1790–2', *Revue d'histoire de l'Eglise de France*, liv, 1968.

— *Formes modernes de la vie consacrée* (Paris, 1966).

Rebillon, Armand, *La Situation économique du clergé à la veille de la Révolution dans les districts de Rennes, Fougères et Vitré* (Rennes, 1913).

— 'L'Université et l'Église à Rennes au temps de Louis-Philippe', *Annales de Bretagne*, lii, 1945.

Reinhard, M., 'Élite et noblesse dans la seconde moitié du XVIIIe siècle', *Revue d'histoire moderne et contemporaine*, iii, 1956.

Renan, Ernest, *Souvenirs d'enfance et de jeunesse* (Paris, 1883).

Resnick, Daniel, P., *The White Terror and the Political Reaction after Waterloo* (Cambridge, Mass., 1966).

Ribot, Alexandre, *La Réforme de l'enseignement secondaire* (Paris, 1900).

Richter, Noë, *Les Bibliothèques populaires* (Paris, 1978).

Rigault, Georges, *Histoire générale de l'Institut des Frères des Écoles Chrétiennes*, v, vii (Paris, 1945, 1949).

Rivoire, Hector, *Statistique du département du Gard* (2 vols., Nîmes, 1842).

Robert, Daniel, *Les Églises réformées de France, 1800-30* (Paris, 1961).

Rocque, J.-D., 'Positions et tendances des Protestants nîmois au XIXe siècle', in *Droite et gauche de 1789 à nos jours* (Montpellier, 1975).

Rohr, J., *Victor Duruy, ministre de Napoléon III* (Paris, 1967).

Rouvière, François, *L'Aliénation des biens nationaux dans le département du Gard* (Nîmes, 1900).

—— *Histoire de la Révolution française dans le département du Gard* (4 vols., Nîmes, 1887-9).

Roux, Henri, *La Loi Guizot et son application dans un coin de Languedoc* (Nîmes, 1912).

Rovolt, J.-B., *Vie du T.R.P. Ange Le Doré* (2 vols., Besançon, 1925).

Roy, J.-A., and Lambert-Dansette, J., 'Origine et évolution d'une bourgeoisie: le patronat textile du bassin lillois', *Revue du Nord*, xxxviii (1955), xxxix (1957), xl (1958), xli (1959).

Ruffet, Jean, 'La Liquidation des Instituteurs-Artisans', *Les Révoltes logiques*, No. 3, Autumn 1976.

Rulon, H.-C., and Friot, P., *Un siècle de pédagogie dans les écoles primaires 1820-1940* (Paris, 1962).

Sagnac, Philippe, 'Le Serment à la Constitution civile du clergé en 1791 dans le région du Nord', *Annales de l'Est et du Nord*, 3, 1907.

Sancier, Raymond, 'L'Enseignement primaire en Bretagne de 1815 à 1850', *Bulletin de la Société d'histoire et d'archéologie de Bretagne*, xxxii, 1952.

Sanderson, Michael, 'Literacy and social mobility in the Industrial Revolution in England', *Past and Present*, no. 56, 1972.

Schmidt, Charles, *La Réforme de l'Université impériale en 1811* (Paris, 1905).

Schnapper, Bernard, *La Remplacement militaire en France. Quelques aspects politiques, économiques et sociaux du recrutement au XIXe siècle* (Paris, 1968).

Schram, Stuart, *Protestantism and Politics in France* (Alençon, 1954).

Secondy, Louis, 'Les Établissements secondaires libres et les petits séminaires de l'Académie de Montpellier de 1854 à 1924' (thesis, Montpellier, 1974).

Serman, William, *Les Origines des officiers français, 1848-70* (Paris, 1979).

Siegfried, André, 'Le Groupe protestant cévenol' in Boegner, Marc, and Siegfried, André, *Le Protestantisme français* (Paris, 1945).

Simon, Jules, *Premières Années* (Paris, 1901).

Sorlin, Pierre, *"La Croix" et les juifs, 1880-99* (Paris, 1967).

—— *Waldeck-Rousseau* (Paris, 1966).

Spivak, M., 'Le développement de l'éducation physique et du sport de 1852 à 1914' *Revue d'histoire moderne et contemporaine*, xxiv, 1977.

Sussman, George D., 'The Glut of doctors in mid-nineteenth-century France', *Comparative Studies in Society and History*, vol. 19, 1977.

Taine, Hippolyte, *Carnets de voyage: notes sur la province, 1863-5* (Paris, 1895).

Talmy, Robert, 'Les Tendances anticléricales des socialistes dans le département du Nord, 1860-1900' (*mémoire de licence*, Université Catholique de Lille, 1952).

Tiercelin, Louis, *Le Cardinal Godefroy Saint-Marc, archévêque de Rennes* (Rennes, 1875).

Tortelier, Henry, *Le Collège Saint-Augustin de Vitré* (Vitré, 1908).

Trénard, Louis, *Histoire des Pays-Bas français* (Toulouse, 1972).

—— *Histoire d'un métropole: Lille, Roubaix, Tourcoing* (Toulouse, 1977).

Trévet, J., 'L'Instruction primaire dans l'arrondissement de Fougères sous le régime de la loi du 28 juin 1833', *Annales de Bretagne*, xxix, 1913-14.

Tronchot, Raymond, *L'Enseignement mutuel en France de 1815 à 1833* (Lille 1973).

Tudesq, André-Jean, *Les Grands Notables en France, 1840-9* (Paris, 1964).

—— 'L'Opposition légitimist en Languedoc en 1840', *Annales du Midi*, lxviii, 1956.

Vailhé, Siméon, *Vie du P. Emmanuel d'Alzon* (2 vols., Paris, 1926, 1934).

Vaillant, Annette, 'Le Clergé séculier en Ille-et-Vilaine au XIXe siècle' (*mémoire de maîtrise*, Rennes, 1972).

Vandenbussche, Robert, 'Aspects de l'histoire du Radicalisme dans le département du Nord, 1870-1905', *Revue du Nord*, xlvii, no. 185, 1965.

Van Elslande, A.-M., 'Les Congrès catholiques du Nord et du Pas-de-Calais, 1897-1912' (D.E.S., Lille, 1967).

Varillon, P., 'Les Conférences de Saint-Vincent de Paul dans le département du Gard au XIXe siècle, 1834-1914' (*mémoire de maîtrise*, Montpellier, 1972). .

Vial, Francisque, *Trois Siècles d'histoire de l'enseignement secondaire* (Paris, 1936).

Vidalenc, Jean, *La Société française de 1815 à 1848: le peuple des campagnes* (Paris, 1970).

Villeneuve-Bargemon, Alban de, *Économie politique chrétienne, ou recherches sur la nature et les causes du paupérisme en Europe et les moyens de le soulager et de le prévenir* (3 vols., Paris, 1834).

Villermé, L.-R., *Tableau de l'état physique et moral des ouvriers employés dans les manufactures de coton, de laine et de soie* (2 vols., Paris, 1840).

Vovelle, Michel, *Religion et révolution: la déchristianisation de l'An II* (Paris, 1976).

Weber, Eugène, 'Gymnastics and Sports in *fin de siècle* France', *American Historical Review*, 76, 1971.

Weber, Eugène, *Peasants into Frenchmen. The Modernization of Rural France, 1870-1914* (London, 1977).

Weiss, G. H., 'Origins of a Technological Élite: engineers, education and social structure in nineteenth-century France' (Harvard D.Phil. thesis, 1977).

Weisz, G., 'The Politics of medical professionalization in France, 1845-48', *Journal of Social History*, 12, 1978.

Winter, Z. de, *Pharaon de Winter, sa vie, son enseignement, son œuvre 1849-1924* (Lille, 1926).

Wolff, Philippe, ed., *Histoire du Languedoc* (Toulouse, 1967).

Young, Arthur, *Travels in France* (2 vols., Dublin, 1793).

Zeldin, Theodore, ed., *Conflicts in French Society* (London, 1970). *France, 1848-1945* (2 vols., Oxford, 1973, 1977).

Zind, Pierre, *Les Nouvelles Congrégations enseignantes en France, 1800-30* (Lyon, 1969).

Index